Pricing Models of Volatility Products and Exotic Variance Derivatives

Chapman & Hall/CRC Financial Mathematics Series

Aims and scope:

The field of financial mathematics forms an ever-expanding slice of the financial sector. This series aims to capture new developments and summarize what is known over the whole spectrum of this field. It will include a broad range of textbooks, reference works and handbooks that are meant to appeal to both academics and practitioners. The inclusion of numerical code and concrete real-world examples is highly encouraged.

Introductory Mathematical Analysis for Quantitative Finance
Daniele Ritelli, Giulia Spaletta

Handbook of Financial Risk Management
Thierry Roncalli

Optional Processes: Stochastic Calculus and Applications
Mohamed Abdelghani, Alexander Melnikov

Machine Learning for Factor Investing: R Version
Guillaume Coqueret, Tony Guida

Malliavin Calculus in Finance: Theory and Practice
Elisa Alos, David Garcia Lorite

Risk Measures and Insurance Solvency Benchmarks: Fixed-Probability Levels in Renewal Risk Models
Vsevolod K. Malinovskii

Financial Mathematics: A Comprehensive Treatment in Discrete Time, Second Edition
Giuseppe Campolieti, Roman N. Makarov

Pricing Models of Volatility Products and Exotic Variance Derivatives
Yue Kuen Kwok, Wendong Zheng

For more information about this series please visit: https://www.crcpress.com/Chapman-and-HallCRC-Financial-Mathematics-Series/book series/CHFINANCMTH

Pricing Models of Volatility Products and Exotic Variance Derivatives

Yue Kuen Kwok
Hong Kong University of Science and Technology,
Hong Kong

Wendong Zheng
Credit Suisse, Hong Kong

CRC Press is an imprint of the
Taylor & Francis Group, an **informa** business
A CHAPMAN & HALL BOOK

First edition published 2022
by CRC Press
6000 Broken Sound Parkway NW, Suite 300, Boca Raton, FL 33487-2742

and by CRC Press
4 Park Square, Milton Park, Abingdon, Oxon, OX14 4RN

CRC Press is an imprint of Taylor & Francis Group, LLC

ISBN: 978-1-032-19902-3 (hbk)
ISBN: 978-1-032-20432-1 (pbk)
ISBN: 978-1-003-26352-4 (ebk)

DOI: 10.1201/9781003263524

Typeset in CMR10 font
by KnowledgeWorks Global Ltd.

Publisher's note: This book has been prepared from camera-ready copy provided by the authors.

Dedication from Yue Kuen Kwok:
To Oi Chun, Grace, Joyce, Yinan and Emma, for their love and support.

Dedication from Wendong Zheng:
To my wife Sherry and daughter Chloe, who have brought enormous love and happiness into my life.

Contents

Preface

Scope and Audience

Variance derivatives are financial securities whose payoff depends on the realized variance of an underlying asset or index return. With the introduction of variance derivatives in the financial markets since the mid-1990s, variance becomes in effect tradable asset and can be conceptualized as an asset class. The variance swaps on stock indexes began to become popular in 1998 and later extended to products on individual stocks. Starting in 2005, the third generation variance derivatives were structured, including the gamma and corridor swaps. The timer options were introduced in 2007. The VIX is a volatility index launched by the Chicago Board Options Exchange, which informs investors about the expected market volatility of the S&P500 index in the next 30 calendar days. The current VIX, launched in 2003, is based on the model-free volatility measure.

With the growth of trading volumes of variance and VIX derivatives, numerous pricing models have been developed in the literature in the last two decades that address pricing and hedging issues of various variance and VIX products. The earlier pricing models consider payoffs that are based on continuous realized variance, under which pricing procedures can be handled easier. However, actual contractual specifications of variance products are based on discrete realized variance measured on discrete monitoring dates. It has been observed that the discretization effects on the fair prices can be significant. The later pricing models focus on products with discrete realized variance payoffs. Most of these pricing models assume joint stochastic process of asset price and variance, or time changed Lévy process. Since discrete realized variance payoffs are highly nonlinear, these pose technical challenges in solving these pricing models.

In this book, we summarize most of the recent research results in pricing models of derivatives on discrete realized variance and VIX. We start with the presentation of volatility trading and uses of variance derivatives. We discuss the robust replication strategy of continuously monitored variance swaps using portfolio of options, which is one of the major milestones in pricing theory of variance derivatives. The replication procedure provides the theoretical foundation of the construction of VIX. We also discuss the approximate replication strategies for discretely monitored variance swaps and their exotic extensions. This book is made self-contained by adding one chapter on the characterization of Lévy processes and stochastic volatility models, the building block in the modeling of asset price dynamics in pricing models of variance and VIX derivatives. In the chapter on pricing of VIX derivatives, most of the pricing models in the literature include both the consistent approach and direct approach. In two separate chapters, we provide detailed exposition on deriving analytic pricing formulas of discrete variance swaps using various analytic methods and developing analytic approximation formulas for pricing options on discrete realized variance. The last chapter discusses various pricing approaches of the timer options, including the perturbation method and Fourier transform algorithms. We follow the rigorous style of showing detailed derivation of pricing formulas, providing sound formulation of pricing models and formal proofs of most technical results. Some illustrative numerical examples are included to show accuracy and effectiveness of analytic and approximation methods.

This book will be most valuable to researchers interested in pricing and hedging issues of variance derivatives and VIX products. It offers a single source for information on the research results on various technical issues. Practitioners and quants in the financial industry can find this book useful in helping them to make their choices of pricing models of variance derivatives and assessing their relevance. It is also well suited as a textbook in a topic course on pricing variance derivatives in universities.

Guide to Chapters

This book consists of six chapters. In the first chapter, we present the notion of volatility trading and analyze volatility exposure generated by delta hedging options. We introduce different types of variance and volatility derivatives, including the third generation products, like the corridor variance swaps and timer options. The ability of extracting the expected value of the future variance from traded option prices provides the theoretical basis of VIX on the S&P500 index. We discuss the construction of the VIX formula and examine the sources of errors associated with its practical implementation. We also consider the approximate replication of exotic swaps on discrete realized variance using traded options.

In the second chapter, we start with review of the compound Poisson processes, jump-diffusion models and Lévy processes. We illustrate the relation between the infinitely divisible distribution and the Lévy process. The highlights are the Lévy-Khintchine representation and Lévy-Itô decomposition theorem. The notion of time change in infinite activity Lévy process is introduced. It is then used to obtain several prototype Lévy models via subordination of a Brownian motion, like the Variance Gamma model and Normal Inverse Gaussian model. We also illustrate the versatility of applying time change via stochastic activity rate to generate stochastic volatility in exponential Lévy models. Next, we present an overview of various formulations and analytic properties of stochastic volatility models with jumps. We show the technical procedures of deriving analytic formulas for the joint moment generating functions for the affine stochastic volatility models and $3/2$-model via the solution of the system of Riccati ordinary differential equations and partial Fourier transform method.

Pricing models of VIX futures and options are presented in Chapter 3. We start with the discussion of the relation between the variance swap rate and VIX^2 under jumps. New VIX products, like VVIX and VXX, are introduced. There are two approaches for pricing VIX derivatives. In the consistent model, the dynamics for VIX is derived based on the joint dynamics of the index value process and stochastic volatility. Pricing formulas for VIX futures and options are derived under the affine models, $3/2$-model, Barndorff-Nelsen and Shephard model and GARCH-type models. The other pricing approach directly models the dynamics of VIX, which include the multifactor affine jump-diffusion model and $3/2$ plus model. A summary of the most recent direct models is presented.

In Chapter 4, we focus on the discussion of pricing models of swap products on discrete realized variance and volatility under stochastic volatility models and Lévy models. Three analytic methods for finding the fair strikes of various swap products under the affine stochastic volatility model and $3/2$-model are presented. These include the direct expectation approach, nested expectation via partial integro-differential equation and moment generating function method. We also present pricing models of exotic variance swaps on discrete realized variance under time changed Lévy processes. Though the discrete realized variance converges to its continuous counterpart in probability, this does not guarantee convergence in expectation. We discuss the technical conditions of taking the continuous limit under which the fair strike formulas of variance swap products under continuous realized variance can be deduced from those under discrete realized variance.

In the next chapter, we present various analytic approximation and semi-analytic methods for pricing options on discrete realized variance. One method involves mixing the discrete variance in a lognormal model and the quadratic variance in the Heston model. An adjustment term is derived for the discretization effect via the lognormal distribution. Conditional on the realization of the stochastic variance, the discrete realized variance is asymptotically normal as the number of monitoring instants goes to infinity. Based on this theoretical property, one can derive the conditional Black-Scholes method for pricing options on discrete realized variance based on a simulation path of the stochastic variance process. In addition, the partially exact and bounded approximation (based on the conditioning variable approach) is used to derive approximation formulas for pricing options on discrete realized variance. In the last approximation method, we explore the small time limit of the price difference between prices of options on discrete and continuous realized variance and use the result to derive the adjustment term required due to discrete sampling of realized variance.

In the last chapter, we consider various pricing models of timer options. A timer option resembles its European vanilla counterpart except that earlier expiration occurs if the accumulated discrete variance exceeds a given preset variance budget. This product provides the buyer the combination of directional bet and volatility bet in single product. Under perpetuity assumption, we manage to derive closed-form analytic formulas under some special cases. Otherwise, timer option prices can be represented as conditional expectation on the Black-Scholes type formulas. Under the Heston model, the perpetual timer option prices can be expressed in terms of integral transform formulas. Also, effective perturbation methods can be used to obtain highly accurate analytic approximation for option prices under the Heston model and $3/2$-model. For finite maturity timer options, we derive the Fourier integral formulas for the price functions. In addition, the Fourier space time stepping algorithms are presented for finding numerical values of the timer option prices.

Chapter 1

Volatility Trading and Variance Derivatives

In this chapter, we present the notion of volatility trading and drawbacks in the use of options on trading volatility. Derivatives on realized variance and volatility of the asset returns are better volatility trading instruments since they provide pure volatility exposure without directional risk of the asset price. Volatility of an asset price process is a latent random process, and there are various specifications to quantify its characteristics and measurements. We start with various definitions of discrete realized variance and volatility used in variance and volatility contracts. The relations between implied volatility and local volatility are discussed. We analyze volatility exposure generated by delta hedging options, where the delta hedged option position generates a profit and loss that is related to the product of cash gamma of the portfolio and difference between the realized variance and implied variance. The resulting profit and loss would be dependent on the realization of the asset price path unless the cash gamma position is made to be constant. We then introduce different variance and volatility derivatives, including exotic products like the gamma variance swaps, corridor variance swaps, conditional variance swaps, and timer options. Under the assumption of no jump in the underlying asset price dynamics, continuously sampled realized variance of the asset price can be replicated by a portfolio of options of continuum of all strikes. The ability of extracting the expected value of the future variance from traded option prices provides the theoretical basis of VIX on the S&P 500 index launched by the Chicago Board of Options Exchange. We discuss the construction of the VIX formula and examine the sources of errors associated with the practical implementation of the formula under the limitation of availability of options with finite set of discrete strikes. Lastly, we consider approximate replication of swaps on discrete realized variance and other exotic variance swaps using portfolio of traded options.

Discrete realized variance and volatility

The volatility of an asset price process is defined quantitatively as the annualized standard deviation of the rate of return or logarithm of return of the asset price process. We present two versions of the definition of discrete realized variance over a time period. The square root of discrete realized variance is taken as discrete realized volatility.

Let S_t denote the asset price process, and the monitoring instants of the process are set on discrete time points: $0 = t_0 < t_1 < \cdots < t_N = T$. For daily monitoring, the time points are successive trading days. The discrete realized variance of the asset

DOI: 10.1201/9781003263524-1

return over the time period $[0,T]$ is defined to be

$$I(0,T;N) = \frac{A}{N}\sum_{k=1}^{N} r_k^2, \tag{1.1}$$

where r_k is the rate of asset return over the kth period $[t_{k-1},t_k]$, $k=1,2,\ldots,N$, and A is the constant annualized factor based on the convention: $\frac{A}{N} = \frac{1}{T}$. For daily monitoring, it is common to take $A = 252$ (day count convention of 252 trading days for one year).

This definition of discrete realized variance is not exactly the sample variance of the returns in statistical sense[1]. Since the sample mean of rates of return is typically small for most asset price processes, the square of the sample mean of rates of return is negligible and can be neglected. Therefore, the discrete realized variance defined in (1.1) is a good approximation of the sample variance. Besides, this definition of the discrete realized variance has the nice additivity property: 5-year realized variance from now is equal to 2-year realized variance from now plus 3-year realized variance 2-year forward. As a result, the discrete realized variance defined in terms of sum of squared returns is more mathematically tractable.

There are two choices for the specification of the asset rate of return r_k, namely, simple rate of return: $r_k = \dfrac{S_{t_k}}{S_{t_{k-1}}} - 1$; or logarithm of return: $r_k = \ln\dfrac{S_{t_k}}{S_{t_{k-1}}}$. Mathematically, since

$$\ln(1+x) = x - \frac{x^2}{2} + \cdots$$

for small x, so when $\frac{S_{t_k}}{S_{t_{k-1}}} - 1$ is sufficiently small, the simple rate of return and logarithm of return are very close in value. The difference in $I(0,T;N)$ based on the two definitions is in cubic order (Carr and Lee, 2009), where

$$\sum_{k=1}^{N}\left(\ln\frac{S_{t_k}}{S_{t_{k-1}}}\right)^2 - \sum_{k=1}^{N}\left(\frac{S_{t_k}}{S_{t_{k-1}}} - 1\right)^2 = -\sum_{k=1}^{N}\left(\frac{S_{t_k}}{S_{t_{k-1}}} - 1\right)^3 + O\left(\frac{S_{t_k}}{S_{t_{k-1}}} - 1\right)^4. \tag{1.2}$$

For the seller of variance swap (payer of discrete realized variance), he may prefer to adopt the logarithm of return instead of the simple rate of return due to negative difference of these two specifications (though in cubic order). Also, the more common use of logarithm of return may be partially due to the frequent use of geometric Brownian motion or its variation as the underlying asset price model. The small difference between the two return specifications vanishes when the monitoring intervals become vanishingly small.

---

[1]The historical variance defined in statistical sense is given by

$$\sigma_{stat}^2 = \frac{252}{N-1}\sum_{k=1}^{N}(r_k - \overline{r})^2,$$

where $\overline{r}$ is the mean rate of return and N is the number of trading days.

1.1 Implied volatility and local volatility

The difficulties of setting volatility value in option price formulas lie in the fact that the input volatility should be the forecast volatility value over the remaining life of the option rather than an estimated volatility value from the past history of the asset price process *(historical volatility)*.

The Black-Scholes model assumes a lognormal probability distribution of the asset price at all future times. Since the volatility parameter σ is the only unobservable parameter in the Black-Scholes model, the model gives the option price as a function of volatility. Let $V_{market}(S,t;K,T)$ denote the market observable option price at time t with strike K and maturity T, and $V_{BS}(S,t;K,T,\sigma)$ is the Black-Scholes option price formula. The Black-Scholes *implied volatility* $\sigma_{imp}(K,T)$ is the unique solution to the algebraic equation:

$$V_{market}(S,t;K,T) = V_{BS}(S,t;K,T,\sigma_{imp}(K,T)). \tag{1.3}$$

The above equation is an answer to: What volatility is implied from the observed option price, if the Black-Scholes model is a valid description of the market conditions of option trading?

Though practitioners are well aware that the Black-Scholes model of constant volatility does not give the correct description of the dynamics of S_t, it remains to be a common practice for traders to quote an option's market price in terms of implied volatility σ_{imp}. The Black-Scholes implied volatility computed from the market option price by inverting the Black-Scholes price formula varies with strike K and maturity T.

The old VIX (volatility index on the S&P 500 index) of Chicago Board of Options Exchange is the model-dependent 22-trading-day volatility measure based on the Black-Scholes implied volatilities on eight near-the-money options at the two nearest maturities. At each of these two maturities, we choose two call options and two put options at the two strike prices that are nearest to and bracket the spot level. At each strike price, the two implied volatilities calculated from the traded prices of the call option and put option are averaged. The estimated at-the-money spot implied volatility is obtained by interpolating linearly between the two average implied volatilities at the two strike prices. Lastly, the 22-trading-day volatility is computed by further interpolation of the at-the-money implied volatilities along the maturity dimension. The method of extracting volatility from traded option prices using the *model-dependent* Black-Scholes implied volatilities has been replaced by a new *model-independent* volatility measure based on the replication procedure since 2004, the details of which are discussed in Sec. 1.5.2.

Volatility skew

The plots of the implied volatility σ_{imp} against moneyness S/K for traded equity call and put options with the same maturity date typically show the skew shape as shown in Fig. 1.1.

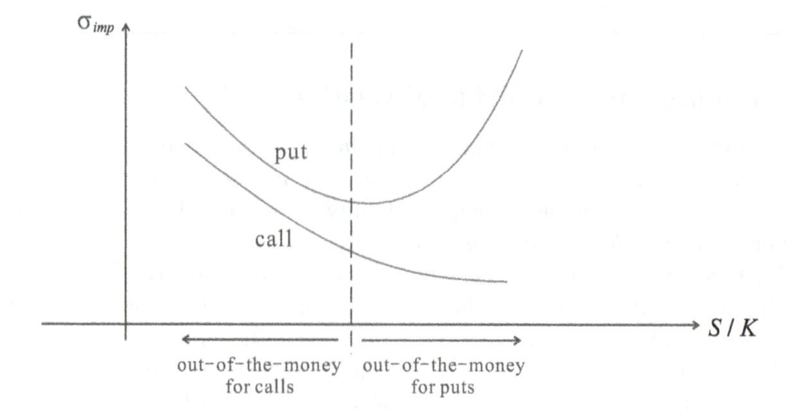

FIGURE 1.1: The volatility skew exists because of increased demand for hedging extreme market move. Empirical studies show that σ_{imp} of puts are normally higher than σ_{imp} of calls. Out-of-the-money equity puts ($S/K > 1$) exhibit higher implied volatility levels since option buyers are willing to pay higher equity put prices to hedge against perceived downside risk.

The volatility skew across moneyness occurs since the equity option prices for deep out-of-the-money options are bid up higher than around-the-money counterparts. Since the stock markets tend to crash downward faster than they move upward, option buyers are willing to bid at higher prices for puts than calls, so σ_{imp} of puts are typically higher than σ_{imp} of calls. Investors have stronger motive to use deep out-of-the-money puts to hedge against drastic market decline, so out-of-the-money puts are trading more expensively than out-of-the-money calls for the same level of out-of-the-moneyness (say, $S/K = 0.8$ for a call and $S/K = 1.25$ for a put). The skew feature has increased significantly since the 1987 crash. For foreign exchange options, the implied volatilities generally display the U-shaped "smile" instead of the "skew" shape. This may be attributed to the occurrences of symmetric jumps in exchange rates rather than most often downward jump in equity prices.

Besides exhibiting variations with respect to moneyness S/K, the implied volatility σ_{imp} shows variations with respect to time to expiration $T-t$, termed as the volatility term structure. The three-dimensional plot of the implied volatilities against moneyness and time to expiration generates the *implied volatility surface*, as characterized by $\sigma_{imp}(S/K, T-t)$. Note that the implied volatility surface for fixed value of maturity T changes as the calendar time t proceeds. Usually, the skew/smile feature flattens out as maturity increases. Also, implied volatility patterns vary less in time when expressed in terms of moneyness S/K than when expressed in terms of K.

Local volatility function

Assuming that the asset price dynamics under a risk neutral measure Q is governed by

$$\frac{dS_t}{S_t} = (r-q)dt + \sigma(S_t,t)dW_t, \tag{1.4}$$

where r and q are constant interest rate and dividend yield, respectively, and W_t is the standard Brownian motion under Q. Instead of constant volatility as assumed in the Black-Scholes model, the local volatility function $\sigma(S_t,t)$ admits both state and time dependence. Given $\sigma(S_t,t)$, one can obtain the price of a European vanilla option $V(S,t)$ by solving the Black-Scholes-type option pricing equation:

$$\frac{\partial V}{\partial t} + (r-q)S\frac{\partial V}{\partial S} + \frac{\sigma(S,t)^2}{2}S^2\frac{\partial^2 V}{\partial S^2} - rV = 0, \tag{1.5}$$

with $V(S,T)$ set equal to the preset terminal option payoff. Conversely, given a spectrum of option prices with varying strikes K and maturity dates T, we would like to find $\sigma(S,T)$ at some future market level S and time T as the future volatility of the asset price process at that market level and time such that the current observable option prices are recovered.

We use $c(K,T)$ to denote the traded call price with strike K and maturity T. Treating K and T as the state variable and time variable, respectively, Dupire (1994) obtains the following relation between $\sigma(K,T)$ and derivatives of European call option prices $c(K,T)$ with respect to strike K and maturity T, where

$$\sigma^2(K,T) = 2\frac{\frac{\partial c}{\partial T} + qc + (r-q)K\frac{\partial c}{\partial K}}{K^2\frac{\partial^2 c}{\partial K^2}}. \tag{1.6}$$

Here, we assume that traded call prices exist for continuum of strikes K and maturity dates T so that the derivatives $\frac{\partial c}{\partial T}$ and $\frac{\partial c}{\partial K}$ are defined. The proof of the Dupire formula (1.6) is presented in Appendix.

On the other hand, the market prices for European options are quoted in terms of their implied volatilities. Note that $c(K,T)$ and $\sigma_{imp}(K,T)$ are related by the Black-Scholes call price formula c_{BS}, where

$$\begin{aligned} c(K,T) &= c_{BS}(S,T;\sigma_{imp}(K,T)) \\ &= Se^{-qT}N(d_1) - Ke^{-rT}N(d_1 - \sigma_{imp}(K,T)\sqrt{T}). \end{aligned}$$

By performing the tedious calculus exercise of relating the higher order derivatives: $\frac{\partial c}{\partial T}$, $\frac{\partial c}{\partial K}$ and $\frac{\partial^2 c}{\partial K^2}$ with those of $\sigma_{imp}(K,T)$, and equating them to $\sigma^2(K,T)$ using the Dupire formula (1.6), we can derive the following relation between $\sigma(K,T)$ and $\sigma_{imp}(K,T)$ (Andersen and Brotherton-Ratcliffe, 1998):

$$\sigma^2(K,T) = \frac{\sigma_{imp}^2 + 2T\sigma_{imp}\frac{\partial \sigma_{imp}}{\partial T} + 2(r-q)KT\sigma_{imp}\frac{\partial \sigma_{imp}}{\partial K}}{\left(1 + Kd_1\sqrt{T}\frac{\partial \sigma_{imp}}{\partial K}\right)^2 + K^2T\sigma_{imp}\left[\frac{\partial^2 \sigma_{imp}}{\partial K^2} - d_1\sqrt{T}\left(\frac{\partial \sigma_{imp}}{\partial K}\right)^2\right]}, \tag{1.7}$$

where

$$d_1 = \frac{\ln \frac{S}{X} + \left[r - q + \frac{\sigma_{imp}^2(K,T)}{2} \right] T}{\sigma_{imp}(K,T)\sqrt{T}}.$$

The determination of σ_{imp} from market option prices poses no numerical difficulties since it only involves the root finding procedure of solving the algebraic equation (1.3). However, the extraction of $\sigma^2(K,T)$ using (1.7) results in ill-posed numerical problems. This is because $\frac{\partial \sigma_{imp}}{\partial T}$, $\frac{\partial \sigma_{imp}}{\partial K}$ and $\frac{\partial^2 \sigma_{imp}}{\partial K^2}$ are difficult to be estimated with good accuracy due to sparse data of σ_{imp} at discrete strikes and maturities.

In the above discussion, deterministic volatility has been assumed. Readers may be interested to read the discussion on the relation between implied volatility and local volatility under stochastic volatility in Lee (2001).

1.2 Volatility trading using options

Suppose an investor expects that significant move in the asset price is going to happen but has no idea on which direction it will go, taking long position in volatility would be the best strategy to benefit from the correct realization of this view. Volatility trading bets on the significant fluctuation of the price moves, which is different from traditional trading strategies that bet on the future directional movement of the underlying asset price. The use of options as the useful tools for volatility trading arising since option prices are dependent on the volatility of the underlying asset.

This section only provides a brief discussion of volatility trading using portfolios of options and volatility exposure generated by delta hedging options. There have been numerous books and industrial research reports that discuss volatility trading using options in good details. Interested readers may read the book chapter by Fitzgerald (1996), the books by Connolly (1997) and Bouzoubaa and Osseiran (2010).

1.2.1 Taking volatility position using straddles and strangles

To benefit from an increasing volatility of the underlying asset price process in the future, one can long one at-the-money call and one at-the-money put with the same strike to form a straddle (see Fig. 1.2a). However, to establish the straddle portfolio, one has to pay the upfront premium of two expensive at-the-money options. To establish a less costly portfolio with similar volatility exposure, one may switch to a strangle which consists of one out-of-the-money put and one out-of-the-money call (see Fig. 1.2b). In order to achieve a positive terminal payoff in a strangle, the asset price has to go beyond an interval with bounds specified by the strikes of the out-of-the-money call and put. The cost required to establish the strangle is lower when the interval becomes wider, but there is a higher chance of getting zero terminal payoff under more widened interval.

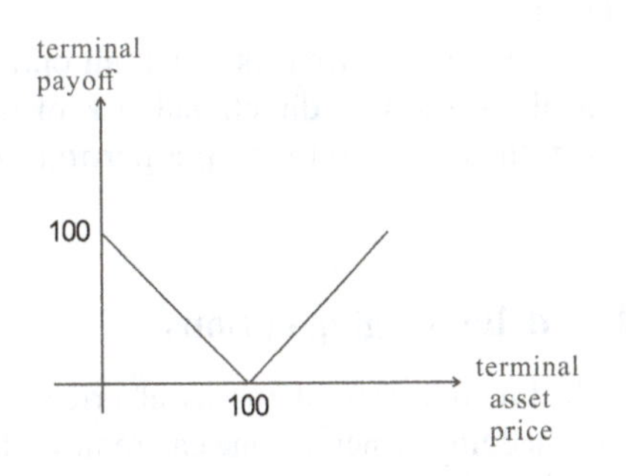

FIGURE 1.2a Terminal payoff of a straddle, which consists of one at-the-money call and one at-the-money put.

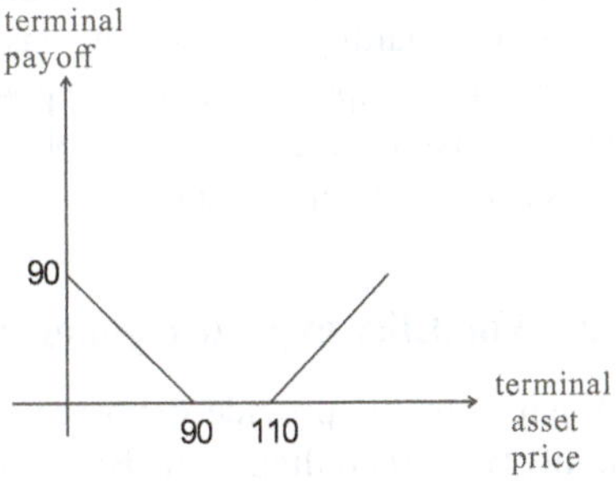

FIGURE 1.2b Terminal payoff of a strangle, which consists of one out-of-the-money call and one out-of-the-money put.

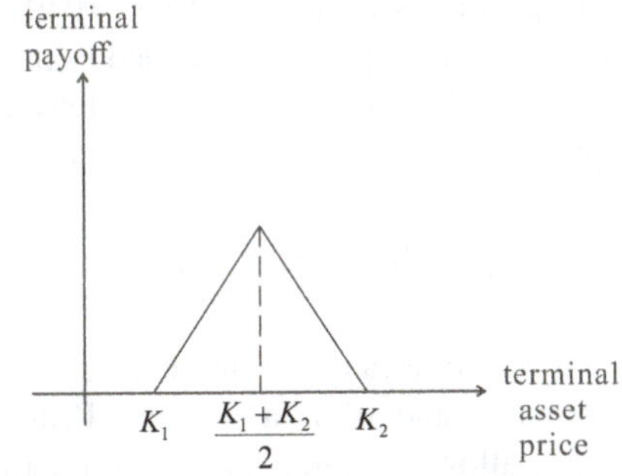

FIGURE 1.2c Terminal payoff of a butterfly, which consists of long position of one call with lower strike K_1, one call with higher strike K_2 and short position of two calls with strike $(K_1+K_2)/2$.

Investors may want to short sell volatility when they expect that the market volatility reverts back to its normal level. Shorting options leads to such an exposure. Yet, shorting options can be exposed to unlimited risk, so one may prefer other less risky alternatives. A butterfly strategy can serve this purpose. A butterfly portfolio is an option portfolio that consists of long position in two call options with lower strike K_1 and higher strike K_2, and short position in two units of call option with strike $(K_1+K_2)/2$, all of them have the same maturity. The butterfly position has positive payoff if the underlying asset price remains within the interval (K_1, K_2) at maturity (see Fig. 1.2c).

The holding of an option generates net profit from changes in implied volatility. More precisely, the net profit is captured by

$$\text{vega} \times (\text{current implied volatility} - \text{original implied volatility}),$$

where vega is the sensitivity of option price to implied volatility. This net profit may arise from the volatility term structure effect (change of implied volatility over time) or due to changing market condition on volatility.

Trading volatility with options has several obvious drawbacks. First, an option position offers impure volatility exposure in the sense that directional risk of the underlying asset price is unavoidable. Second, the cost of setting up a portfolio of options is usually quite expensive.

1.2.2 Volatility exposure generated by delta hedging options

Equity options provide exposure to both direction risk of the asset price and volatility risk. According to the Black-Scholes hedging principle, one can remove the exposure to the asset price risk using delta hedging. However, delta hedging can never be perfect since the real world financial markets violate many of the Black-Scholes model assumptions, like volatility cannot be accurately estimated, asset price may move discontinuously, liquidity may not be available, frequent trading would incur high transaction costs, etc.

We would like to show that delta hedging an option based on some chosen time-dependent hedge volatility generates a profit and loss (P&L) that is related to the realized variance and the cash gamma position (defined as product of option gamma and square of asset price). Consider the underlying asset price which follows a continuous semimartingale process under a risk neutral measure Q as specified by

$$\frac{\mathrm{d}S_t}{S_t} = (r-q)\mathrm{d}t + \sigma_t\,\mathrm{d}W_t, \tag{1.8}$$

where r and q are constant risk-free rate and dividend yield, respectively, σ_t is the instantaneous volatility process, and W_t is a standard Brownian motion under Q. The assumed dynamics of S_t allows no jump. Let σ_t^i be the time-dependent implied volatility derived from traded option prices at varying times. We write the time-t option price as $V_t^i = V(S_t,t;\sigma_t^i)$ with reference to implied volatility σ_t^i. Suppose an option trader sells an option at time zero priced at the current market implied volatility σ_0^i, the option price is given by $V_0^i = V(S_0,0;\sigma_0^i)$. The seller's short position in the option is delta hedged at some chosen time-dependent hedge volatility σ_t^h for the remaining life of the option. In summary, for $t \in [0,T]$, there are 3 volatilities: (i) σ_t is the actual instantaneous volatility of S_t, (ii) σ_t^i is the implied volatility derived from market option price V_t^i at varying time t, (iii) σ_t^h is the time varying hedging volatility adopted by the hedger.

After selling one unit of the option, the option writer replicates the option by long holding a dynamic replicating portfolio consisting of certain units of the underlying asset and money market account. At the initial time t_0, the dollar amount V_0^i received from selling one unit of the option is used to buy Δ_0^h units of the underlying asset and the remaining amount $V_0^i - \Delta_0^h S_0$ is put into the money market account. With the choice of σ_t^h at time t by the hedger, the hedge ratio is set to be $\Delta_t^h = \frac{\partial}{\partial S}V(S_t,t;\sigma_t^h)$, where $V(S_t,t;\sigma_t^h)$ is the Black-Scholes price function evaluated at the volatility value

σ_t^h. Replication requires putting the remaining amount $V_t^i - \Delta_t^h S_t$ into the market money account M_t, where

$$M_t = V_t^i - \Delta_t^h S_t.$$

The delta hedged portfolio consists of short position of one unit of option and long holding Δ_t^h units of the underlying asset plus money market account worth $V_t^i - \Delta_t^h S_t$. The P&L of the delta hedged portfolio over the infinitesimal time interval $[t, t + dt]$ consists of the following three components:

- change of the option value: $-dV_t^i$;
- P&L resulted from the dynamic position of the underlying asset and dividend income: $\Delta_t^h (dS_t + qS_t\, dt)$;
- risk-free interest earned from the money market account: $r(V_t^i - \Delta_t^h S_t)dt$.

Recall the Greek parameters $\Theta_t^h = \frac{\partial}{\partial t} V(S_t, t; \sigma_t^h)$ and $\Gamma_t^h = \frac{\partial^2}{\partial S^2} V(S_t, t; \sigma_t^h)$. Since $V_t^h = V(S_t, t; \sigma_t^h)$ satisfies the Black-Scholes equation with volatility parameter σ_t^h so that

$$\Theta_t^h = -\frac{\Gamma_t^h S_t^2}{2} (\sigma_t^h)^2 + rV_t^h - (r - q)\Delta_t^h S_t. \tag{1.9a}$$

By applying Itô's lemma to the price function $V_t^h = V(S_t, t; \sigma_t^h)$, we have

$$dV_t^h = \Delta_t^h dS_t + \left(\Theta_t^h + \frac{\Gamma_t^h S_t^2}{2} \sigma_t^2 \right) dt. \tag{1.9b}$$

The diffusion term in (1.9b) involves σ_t^2 instead of $(\sigma_t^h)^2$ since S_t follows the actual dynamics as specified in (1.8). Note that the difference in σ_t^2 and $(\sigma_t^h)^2$ weighted by $\frac{\Gamma_t^h S_t^2}{2}$ gives rise to the variance exposure in the procedure of delta hedging of an option using σ_t^h. Substituting Θ_t^h in (1.9a) into (1.9b), we eliminate Θ_t^h and obtain

$$dV_t^h = \Delta_t^h dS_t + \frac{\Gamma_t^h S_t^2}{2} [\sigma_t^2 - (\sigma_t^h)^2] dt + r(V_t^h - \Delta_t^h S_t) dt + q\Delta_t^h S_t\, dt. \tag{1.9c}$$

Let Π_t denote the time-t value of the P&L. Its differential change $d\Pi_t$ over $[t, t + dt]$ is given by the sum of the three components:

$$d\Pi_t = \Delta_t^h (dS_t + qS_t\, dt) + r(V_t^i - \Delta_t^h S_t) dt - dV_t^i.$$

We then substitute (1.9c) into the above equation to eliminate Δ_t^h and obtain

$$\begin{aligned} d\Pi_t &= (dV_t^h - dV_t^i) - r(V_t^h - V_t^i) dt - \frac{\Gamma_t^h S_t^2}{2} [\sigma_t^2 - (\sigma_t^h)^2] dt \\ &= e^{-r(T-t)} \frac{d}{dt} \left[e^{r(T-t)} (V_t^h - V_t^i) \right] dt - \frac{\Gamma_t^h S_t^2}{2} [\sigma_t^2 - (\sigma_t^h)^2] dt. \end{aligned}$$

The total P&L at maturity T is given by the accumulated sum of the forward value at T of the differential P&L over the whole time period $[0,T]$, where

$$\begin{aligned}\Pi_T &= \int_0^T e^{r(T-t)}\,\mathrm{d}\Pi_t \\ &= (V_T^h - V_T^i) - e^{rT}(V_0^h - V_0^i) - \int_0^T e^{r(T-t)}\frac{\Gamma_t^h S_t^2}{2}[\sigma_t^2 - (\sigma_t^h)^2]\,\mathrm{d}t.\end{aligned}$$

At maturity T, both V_T^i and V_T^h become the terminal payoff, where

$$V_T^h = V_T^i = V(S_T).$$

Consequently, we have the following formula for the total P&L at maturity T (Carr and Madan, 1998):

$$\Pi_T = e^{rT}[V(S_0,0;\sigma_0^i) - V(S_0,0;\sigma_0^h)] + \int_0^T e^{r(T-t)}\frac{\Gamma_t^h S_t^2}{2}[(\sigma_t^h)^2 - \sigma_t^2]\,\mathrm{d}t. \quad (1.10)$$

Since the factor $\Gamma_t^h S_t^2$ appears as cash term in Π_T, it is commonly called the *cash gamma* or *dollar gamma*.

Remarks

1. The total P&L generated by delta hedging an option at hedge volatility σ_t^h can be decomposed into two parts. The first component is the future value of the time-0 price difference of the two options priced at the implied volatility σ_0^i and the hedge volatility σ_0^h, respectively. As shown in the second integral term in Π_T, the second component arises since the option is hedged at the hedge volatility σ_t^h instead of the realized volatility σ_t. It is equal to the future value of the weighted difference of the variance with reference to the hedge volatility and realized variance weighted by half of the cash gamma, $\frac{\Gamma_t^h S_t^2}{2}$ and integrated over the life of the option.

2. Suppose the trader is delta hedging the option at the implied volatility, where $\sigma_t^h = \sigma_t^i$, then the total P&L becomes

$$\Pi_T = \int_0^T e^{r(T-t)}\frac{\Gamma_t^i S_t^2}{2}[(\sigma_t^i)^2 - \sigma_t^2]\,\mathrm{d}t. \quad (1.11)$$

In this case, the total P&L is equal to the future value of the weighted sum of the difference between the implied variance and realized variance, where the weight factor is half of the cash gamma, $\frac{\Gamma_t^i S_t^2}{2}$. Usually, options are over priced, which means $\sigma_t^i > \sigma_t$. As most options observe the property $\Gamma_t^i > 0$, we then deduce that delta hedging strategy generates positive P&L. Note that the variance exposure associated with the delta hedged option is also dependent on the realized path of S_t. It is still possible to obtain

$$\int_0^T [(\sigma_t^i)^2 - \sigma_t^2]\,\mathrm{d}t > 0,$$

while the P&L can be negative due to the path-dependent factor $\frac{\Gamma_t^i S_t^2}{2}$ [see Exhibit 3 in Bossu (2006) for an illustrative example]. The use of variance derivatives whose payoffs depend only on realized variance would not have such undesirable outcome in P&L.

To achieve volatility exposure without such path dependency, one natural remedy is to form a portfolio of options with constant cash gamma. Indeed, this is possible if we choose a portfolio of options with their notional values in proportion to $1/K^2$, where K is their respective strike price. More precisely, assuming pure diffusion dynamics of S_t with no jump and suppose one owns a portfolio of options of all strikes ranging from zero to infinity with this choice of weighting scheme, one can obtain an exposure to variance that is independent of S_t. The issue of replication of variance swaps using traded options is explored in Sec. 1.4.

1.3 Derivatives on discrete realized variance

Volatility and variance derivatives are seen to be the more appropriate derivative instruments that provide pure volatility exposure. Volatility and variance derivatives can be classified into two categories. The first type is called the VIX-based derivatives, whose payoffs depend on the exchange-traded VIX, like VIX futures and options. The second type is the realized variance-based derivatives, whose payoffs depend on various forms of discrete realized variance calculated from daily monitored asset prices.

Trading of variance swaps started in the 1990s as a remedy to resolve the inherent difficulties in achieving pure volatility position using the straddle/strangle strategies or delta hedging options. Since then, various innovative realized variance-based derivatives have emerged in the equity derivatives markets. We discuss various types of derivative products with the discrete realized variance as the underlying state variable and the motivation behind the product designs of these realized variance products.

1.3.1 Swaps and options on realized variance and volatility

A variance swap is a swap contract between two parties who agree to exchange the realized variance over certain accrual period for a fixed strike at some future date (usually the end of the accrual period). The terminal payoff of a variance swap takes the form:

$$L[I(0,T;N)-K],$$

where L is some constant notional value and $I(0,T;N)$ is defined in (1.1). The fixed strike K, which is set at initiation of the contract, is usually chosen in such a way that it costs zero for both parties to enter into the swap contract. We commonly call K to be the fair strike of the variance swap.

One typical use of variance swap is the implied versus realized variance arbitrage. If an investor expects the future realized volatility to be above the implied volatility, then the investor can take a long position in a variance swap via buy-and-hold strategy. The strike of the variance swap is determined based on the implied volatility. The investor gains if his view of higher future realized variance does materialize. Unlike the inherent path-dependent volatility exposure created by a delta hedged option position, the variance swap provides the direct way of volatility trading by using the "pure" variance as the underlying. More detailed discussion on various uses of variance swaps can be found in the BNP Paribus research report on volatility investing (Mougeot, 2005).

For the seller of the variance swap contract (short position in the realized variance), the realized variance may spike due to market crash event. As a result, the loss on the short position may be too significant. A preset cap on the realized variance is imposed to protect the seller from potential catastrophic loss. The terminal payoff of the capped variance swap is modified as follows:

$$L[\max(I(0,T;N),C)-K],$$

where $C > K$ is the imposed cap on the realized variance. For single name variance swaps, the cap is set to be 2.5 times the variance swap rate as the usual market practice.

Investors who would like to seek full protection from uncertainty of the future volatility may buy an option on realized variance. For instance, the call option on the realized variance has the following terminal payoff:

$$L\max(I(0,T;N)-K,0).$$

Some investor may prefer the payoff to be dependent on the realized volatility $\sqrt{I(0,T;N)}$. The corresponding terminal payoff of the call option on the realized volatility is given by

$$L\max(\sqrt{I(0,T;N)}-K,0).$$

In our later discussion, we take the notional multiplier L to be unity for convenience and most often adopt the logarithm return specification in $I(0,T;N)$.

Barndorff-Nielsen and Shephard (2003) propose another measure of discrete realized volatility as defined by

$$\sqrt{\frac{\pi}{2NT}}\sum_{k=1}^{N}\left|\frac{S_{t_k}-S_{t_{k-1}}}{S_{t_k}}\right|,$$

where t_k, $k = 1,2,\ldots,N$, is the kth monitoring instant with $t_0 = 0$ and $t_N = T$. They performed empirical studies on this measure of discrete realized volatility and concluded that this is a more robust measure of discrete realized volatility. Also, this definition of discrete realized volatility provides better analytic tractability in pricing volatility swap (see Sec. 3.2.3). Mathematically, it is easier to deal with the absolute sign $\left|\frac{S_{t_k}-S_{t_{k-1}}}{S_{t_k}}\right|$ than the square root function $\sqrt{I(0,T;N)}$.

1.3.2 Generalized variance swaps

The generalized variance swaps, known as the third generation variance derivatives, provide sophisticated variance trading tools with more refined variance exposure. A generalized variance swap is a swap product on the modified discrete realized variance defined by

$$I_w(0,T;N) = \frac{A}{N}\sum_{k=1}^{N} w_k r_k^2, \tag{1.12}$$

where w_k is a weight process, which is usually chosen to be dependent on the path of the underlying asset price. By choosing an appropriate weight process w_k, one has the flexibility to construct customized variance exposure.

Gamma swaps and self-quantoed variance swaps

In a gamma swap, the weight process is chosen to be the ratio of the prevailing asset price over the initial price, where $w_k = S_{t_k}/S_{t_0}$. The terminal payoff of a gamma swap is given by

$$\frac{A}{N}\sum_{k=1}^{N} \frac{S_{t_k}}{S_{t_0}}\left(\ln\frac{S_{t_k}}{S_{t_{k-1}}}\right)^2 - K.$$

The motivation of choosing the weight to be proportional to the underlying level is to impose damping on the large downside variance when the stock price falls close to zero. This serves to protect the swap seller from paying substantial amount in a catastrophic crash.

A self-quantoed variance swap is quite similar to the gamma swap, where the weight process is fixed to be S_{t_N}/S_{t_0}. As a result, the terminal payoff of the self-quantoed swap is given by

$$\frac{A}{N}\sum_{k=1}^{N} \frac{S_{t_N}}{S_{t_0}}\left(\ln\frac{S_{t_k}}{S_{t_{k-1}}}\right)^2 - K.$$

Here, the damping effect upon market crash is determined by the terminal asset price only, rather than the asset prices over successive monitoring instants as in a gamma swap.

Corridor variance swaps

A corridor variance swap differs from the vanilla variance swap in that the underlying asset price must fall inside a specified corridor $(L,U]$, $L \geq 0$ and $U < \infty$, in order for its squared logarithm return to be included in the floating leg of the corridor variance swap. For a discretely sampled corridor variance swap with the tenor $t_0 < t_1 < t_2 < \cdots < t_N = T$, suppose the corridor is monitored on the underlying price at the old time level t_{k-1} for the kth squared logarithm return, the floating leg of the corridor variance swap with the corridor $(L,U]$ is given by

$$\frac{A}{N}\sum_{k=1}^{N} \left(\ln\frac{S_{t_k}}{S_{t_{k-1}}}\right)^2 \mathbf{1}_{\{L<S_{t_{k-1}}\leq U\}}.$$

Here, $\mathbf{1}_{\{\,\cdot\,\}}$ denotes the indicator function.

Corridor variance swaps with one-sided barrier are also widely traded in the financial markets. The downside variance swap and upside variance swap can be obtained by taking $L = 0$ and $U \to \infty$, respectively. Also, one can choose to have the corridor monitored on the underlying asset price at the new time level t_k (Sepp, 2007) or even in terms of asset prices at both time levels t_{k-1} and t_k (Carr and Lewis, 2004).

The corridor variance swaps allow the investors to take their view on the implied volatility skew. Suppose the implied volatility skew is expected to steepen at high value of moneyness of put options (low asset price), the investor may benefit from buying a downside variance swap or selling an upside variance swap if this view is realized. Also, investors seeking crash protection may buy the downside variance swap since it can provide almost the same level of crash protection as the vanilla variance swap but at a lower strike.

It is common to add the knock-out feature in a corridor variance swap that ends the swap contract earlier when the underlying asset price moves way out of the corridor range. The earlier knock-out helps to free the buyer from locking up margin in the trade when there is no longer any further accrual of realized variance. Also, the seller does not need to continue to pay for the hedge. More details on the uses of the corridor variance swaps can be found in Lee (2010).

Conditional variance swaps

A conditional variance swap is similar to a corridor variance swap, except for two modifications in the payoff structures.

(i) The accumulated sum of squared logarithm returns is divided by the number of daily observations D that the underlying asset price stays within the corridor instead of the total number of sampling observations N;

(ii) The final payoff to the holder is scaled by the ratio D/N.

The payoff of a conditional variance swap counts only the sampling dates at which the realized variance does accumulate (conditional on the underlying price lying within the corridor). Let K and K' be the strike of a conditional downside variance swap and its corridor variance swap counterpart, respectively. The holder's payoff of the conditional downside variance swap with the corridor's upper barrier U is given by

$$\frac{D}{N}\left[\frac{A}{D}\sum_{k=1}^{N}\left(\ln\frac{S_{t_k}}{S_{t_{k-1}}}\right)^2\mathbf{1}_{\{S_{t_{k-1}}\leq U\}}-K\right],$$

where $D = \sum_{k=1}^{N}\mathbf{1}_{\{S_{t_{k-1}}\leq U\}}$. By observing

$$\begin{aligned}&\frac{D}{N}\left[\frac{A}{D}\sum_{k=1}^{N}\left(\ln\frac{S_{t_k}}{S_{t_{k-1}}}\right)^2\mathbf{1}_{\{S_{t_{k-1}}\leq U\}}-K\right]\\&=\left[\frac{A}{N}\sum_{k=1}^{N}\left(\ln\frac{S_{t_k}}{S_{t_{k-1}}}\right)^2\mathbf{1}_{\{S_{t_{k-1}}\leq U\}}-K'\right]+\left(K'-\frac{D}{N}K\right),\end{aligned}\tag{1.13}$$

the conditional variance swap can be decomposed into a corridor variance swap with the same upper barrier plus a range accrual note of payoff $\left(K' - \frac{D}{N}K\right)$. Compared to the corridor variance swaps, the conditional variance swaps are structured specifically for investors who would like to be exposed *only* to volatility risk within a prespecified corridor.

In a corridor variance swap, the actual number of daily observations D that the underlying price falls within the corridor has a significant effect on the profit and loss to its holder. Much lower payoff is resulted in the corridor variance swap compared to that in the conditional variance swap when D is significantly small compared to N. The conditional variance swap helps reduce this risk since only the number of dates that the underlying asset price falls within the corridor matters. The decomposition formula (1.13) shows that the holder of a conditional variance swap receives high level of compensation from the range accrual note when D is small.

Switch corridor variance swaps

More complex variation of the corridor variance swaps have emerged in the Asian markets in recent years. For the innovative switch corridor variance swap, the standard structure has the variance underlying being the most liquid index SPX and the corridor condition underlying being a major Asian index, like Nikkei and HSI. Let S^c and S^v denote the corridor index and variance index, respectively. The variance S^v is accrued only when the corridor asset S^c falls within the corridor $[L,U]$. The terminal payoff of the switch corridor variance swap is typically given by

$$\frac{A}{N}\sum_{k=1}^{N}\left(\ln\frac{S^v_{t_k}}{S^v_{t_{k-1}}}\right)^2 \mathbf{1}_{\{L<S^c_{t_k}<U\}} - \frac{D}{N}K_{[L,U]},$$

where A is the annualized factor, $K_{[L,U]}$ is the strike of the swap, and D is the total number of days that S^c falls within $[L,U]$. The above payoff reduces to that of the single-asset corridor variance swap when the corridor index and variance index coincide. More details on the market uses of the product and its pricing methods can be found in Hong (2017).

1.3.3 Timer options

The level of implied volatility of a traded option is often higher than the realized volatility of the underlying asset price process, reflecting that options are usually overpriced. The additional option premium that is overpaid by the option buyer reflects the premium on the future market uncertainty, known as volatility risk premium. Based on empirical studies on traded options, Société Générale Corporate and Investment Banking discovered that 80% of three-month calls that have matured in-the-money were overpriced. As an attempt to minimize volatility risk premium in an option, Société Générale launched the first timer option in April 2007. Timer options are barrier style options in the volatility space. A typical timer option is similar to its European vanilla counterpart, except that the expiration date floats with the realized volatility.

In the contractual specification of a typical timer option, the investor can choose a target volatility level σ and investment horizon $[0,T]$ to establish a target variance budget. In terms of the actual number of trading days N corresponding to the time horizon T, assuming the day count convention of 252 trading days for one year, the variance budget B is given by

$$B = \sigma^2 \frac{N}{252}.$$

The discrete daily monitoring dates are denoted by $t_0, t_1, \ldots, t_N$, where t_0 is the initial time and $t_N = T$ is the mandatory expiration date of the timer option. The annualized realized variance over $[0, t_m]$ is given by

$$I(0,t_m;m) = \frac{252}{m} \sum_{k=1}^{m} \left(\ln \frac{S_{t_k}}{S_{t_{k-1}}} \right)^2, \quad m = 1, 2, \ldots, N. \tag{1.14a}$$

Provided that the mandated expiration date T has not been exceeded, the timer option is knocked out prematurely when $I(0,t_m;m)$ reaches the prespecified variance budget B for the first time. That is, an earlier knock out prior to T occurs at t_n, where the time index n is given by

$$n = \inf\{m : I(0,t_m;m) \geq B\}, \quad n < N. \tag{1.14b}$$

In summary, the finite maturity timer option expires either when the accumulated realized variance of the underlying asset has reached a pre-specified level or on the mandated expiration date, whichever comes earlier. The payoff at knock out or at expiry resembles the usual option payoff.

The buyer of a timer option chooses a reasonable volatility target so as to benefit from the reduction of the option premium. In one example of a timer call that was traded, the buyer set a target volatility level of 12% where the implied volatility on the plain vanilla call with the same strike and some target maturity was slightly above 15%. The variance budget set by the buyer is calculated based on the target volatility and maturity. The premium of the timer call had 20% discount compared to the vanilla call counterpart.

Uses of timer options

A timer option can be used to hedge a stock position and exploit the negative correlation that generally exists between stock return and realized volatility. For example, an investor wants to hedge against a drop of the stock price. To capture his view, the investor buys a timer put. Suppose the stock price goes down, the investor is compensated by the put payoff. Due to the negative correlation, the timer put expires earlier since the realized volatility increases. As a result, the compensation of the loss from the stock price decline arrives sooner in a more timely manner. On the contrary, in the bullish market where the stock price increases, the realized volatility decreases thus giving longer life of the time put. The insurance protection provided by the timer put lasts longer and the hedging cost associated with buying the put is lower.

In addition, the timer call can be used to exploit the negative correlation between stock return and realized volatility when the investor holds a bullish view of the stock. This can be achieved by longing a timer call and shorting a vanilla call, both with the same strike. The target volatility is set below the prevailing implied volatility of the vanilla call, which results in an upfront positive cash since the timer call is less expensive than the vanilla call. Suppose an increase of the stock price does occur, the realized volatility decreases. As a result, the timer call expires later than the vanilla call. A net gain is realized that captures the difference in the time value of the two call options, where the longer-lived timer call is more expensive. For more discussion on the uses of timer options, readers may refer to Sawyer (2007).

Exotic timer options

Timer options can be structured with more exotic forms of payoff. The timer out-performance was developed shortly after the first timer call option was sold in April 2007. In this two-asset out-performance timer option, the payoff upon knock-out or expiry is set to be dependent on the spread of two underlying asset prices on the termination date. The premature knock-out time is determined by the variance of the logarithm of the spread of the two asset prices. In a timer swap contract, the random first hitting time τ of exceeding the variance budget is used as the payoff variable and barrier variable. There is no mandatory expiration date and the swap expires at τ with payoff $L(\tau - T)$, where L is the notional and T is a preset fixed time horizon.

1.3.4 Target volatility options

A European target volatility option has its terminal payoff given by that of the vanilla option counterpart scaled by the ratio of the target volatility (or variance) and the realized volatility (or variance) over the life of the option. The option buyer receives higher payoff if the realized volatility (or variance) is lower than the target (predicted) level. For example, the European target volatility call option at maturity pays

$$\frac{\overline{\sigma}}{\sqrt{I(0,T;N)}}\max(S_T - K, 0),$$

where $\overline{\sigma}$ is the predicted volatility, $\sqrt{I(0,T;N)}$ is the realized volatility, S_T is the terminal asset price, and K is the strike price.

An extension of this class of options with joint payoff structure of terminal asset price and its realized volatility is the double digital call option, where the terminal payoff is given by

$$\mathbf{1}_{\{S_T \geq K_S, \sqrt{I(0,T;N)} \geq K_v\}}.$$

Here, K_S and K_v are the respective strike of the asset price and realized volatility.

Wang and Wang (2014) provide an interesting theoretical discussion on the implementation of variance-optimal hedge for target volatility options. There are numerous pricing models of the target volatility options based on different assumptions of joint dynamics of asset price and instantaneous variance. These include the stochastic volatility models (Di Graziano and Torricelli, 2012; Torricelli, 2013; Da Fonseca *et al.*, 2015), decoupled time-changed Lévy models (Torricelli, 2016), log-normal fractional SABR model (Alòs *et al.*, 2019), and affine GARCH models (Cao *et al.*, 2020).

1.4 Replication of variance swaps

Recall the path-dependent nature of the cash gamma $\Gamma_t^h S_t^2$ when a delta hedged option is used as a vehicle for trading volatility [see the last integral in (1.10)]. Can we remove this path dependency by delta hedging a well-designed portfolio of derivatives that has the property of constant cash gamma?

Let $h(S,t)$ be the price of a portfolio of derivatives that satisfies the property of constant cash gamma; that is,

$$\frac{\partial^2 h}{\partial S^2} S^2 = c,$$

where c is some constant. Solving the differential equation yields

$$h(S,t) = a(t) + b(t)S - c\ln S,$$

where $a(t)$ and $b(t)$ are two arbitrarily chosen functions in t. The term $\ln S$ relates to a log contract, which pays $\ln S_T$ at maturity time T. A portfolio of derivatives with constant cash gamma is supposed to be comprised of a static position of the log contract, plus a dynamic position of the underlying asset and money market account. An introductory discussion of the log contract as an instrument to trade on volatility can be found in Neuberger (1994).

Taylor expansion formula

By performing the Taylor expansion of a function up to the first power term, any twice differentiable payoff function $f(S_T)$ can be written as

$$\begin{aligned} f(S_T) &= f(\kappa) + f'(\kappa)[(S_T - \kappa)^+ - (\kappa - S_T)^+] \\ &\quad + \int_0^\kappa f''(K)(K - S_T)^+ \mathrm{d}K + \int_\kappa^\infty f''(K)(S_T - K)^+ \mathrm{d}K, \end{aligned} \tag{1.15}$$

where κ is an arbitrarily chosen positive real number (Carr and Madan, 1998). The last two integrals represent the remainder in the Taylor expansion. The proof of (1.15) is presented in Appendix. By applying (1.15) to the function $\ln S_T$ and rearranging the terms, we have

$$\ln\frac{\kappa}{S_T} + \frac{S_T}{\kappa} - 1 = \int_0^\kappa \frac{1}{K^2}(K - S_T)^+ \mathrm{d}K + \int_\kappa^\infty \frac{1}{K^2}(S_T - K)^+ \mathrm{d}K. \tag{1.16}$$

The right-hand side represents the terminal payoff of a portfolio of log contract, asset and money market account. One can check easily that the portfolio has unit cash gamma. The left-hand side represents the terminal payoff of a portfolio of continuum of options with their notional values that equal the reciprocal of the square of their respective strike prices. Both portfolios share the same cash gamma of unit value.

A log contract can be delta hedged without making any forecast of volatility. This is because the delta is $\frac{1}{S_t}$ so that $\Gamma_t^i = -\frac{1}{S_t^2}$. This corresponds to $1 worth of the underlying asset. The hedger adjusts the hedge position by keeping one dollar worth of the asset at all times, without the need to make forecast on the hedge volatility σ_t^h (see Sec. 1.2.2). According to (1.11), the corresponding P&L when $\Gamma_t^i = -\frac{1}{S_t^2}$ is given by

$$\Pi_T = \int_0^T \frac{e^{r(T-t)}}{2}[\sigma_t^2 - (\sigma_t^i)^2]\,\mathrm{d}t,$$

which is independent of the asset price path. The hedged position is a pure volatility play since Π_T is dependent on the difference of volatilities, $\sigma_t^2 - (\sigma_t^i)^2$.

1.4.1 Replication of continuous variance swaps

We would like to show that continuously sampled variance is related to the log contract. Suppose S_t follows the dynamics as depicted in (1.8) under a risk neutral measure Q. By applying Itô's lemma to

$$h(S_t) = \frac{2}{T}\left(\ln\frac{\kappa}{S_t} + \frac{S_t}{\kappa} - 1\right), \tag{1.17}$$

where $\kappa > 0$ is arbitrarily chosen, we obtain

$$\mathrm{d}h(S_t) = \frac{2}{T}\left[\left(\frac{1}{\kappa} - \frac{1}{S_t}\right)\mathrm{d}S_t + \frac{\sigma_t^2}{2}\,\mathrm{d}t\right].$$

Integrating the above stochastic differential equation from $t = 0$ to $t = T$, we have

$$\frac{1}{T}\int_0^T \sigma_t^2\,\mathrm{d}t = h(S_T) - \frac{2}{T}\int_0^T\left(\frac{1}{\kappa} - \frac{1}{S_t}\right)\mathrm{d}S_t - h(S_0). \tag{1.18a}$$

The first term $h(S_T)$ can be replicated by the terminal payoff of a portfolio of options as depicted in (1.16). The second term is related to the dynamic hedging of the log contract and underlying asset represented in $h(S_t)$.

It may be more convenient to express the continuously sampled integrated cumulative variance in terms of the futures price F_t, where $F_t = e^{(r-q)(T-t)}S_t$. Note that $F_T = S_T$, so $h(F_T) = h(S_T)$. With the choice of $\kappa = F_0$ and following similar procedure, we can establish

$$\begin{aligned}
&\frac{1}{T}\int_0^T \sigma_t^2\,\mathrm{d}t \\
&= h(F_T) - \frac{2}{T}\int_0^T\left(\frac{1}{F_0} - \frac{1}{F_t}\right)\mathrm{d}F_t \qquad (1.18\text{b}) \\
&= \frac{2}{T}\left[\int_0^{F_0}\frac{1}{K^2}(K - F_T)^+\mathrm{d}K + \int_{F_0}^{\infty}\frac{1}{K^2}(F_T - K)^+\mathrm{d}K - \int_0^T\left(\frac{1}{F_0} - \frac{1}{F_t}\right)\mathrm{d}F_t\right].
\end{aligned}$$

Under the assumption of no jump in the dynamics of S_t, the fair strike K_v of continuously sampled variance swap is given by

$$K_v = E_0^Q \left[\frac{1}{T} \int_0^T \sigma_t^2 \, dt \right].$$

The above replication argument reveals that K_v can be expressed in terms of the initial prices of the European call and put options, the details are presented in Proposition 1.1.

Proposition 1.1 *Under the assumption of no jump in the dynamics of S_t, the fair strike of the continuous variance swap is given by*

$$K_v = \frac{2e^{rT}}{T} \left[\int_0^{F_0} \frac{p(K)}{K^2} \, dK + \int_{F_0}^{\infty} \frac{c(K)}{K^2} \, dK \right], \tag{1.19}$$

where $p(K)$ and $c(K)$ are the respective price of the European T-maturity put and call option of strike K at initiation.

Proof *Under the assumption of no jump in the dynamics of F_t, we consider the dynamics $dF_t = F_t \sigma_t \, dW_t$ so that*

$$E_0^Q \left[\int_0^T \left(\frac{1}{F_0} - \frac{1}{F_t} \right) dF_t \right] = 0.$$

By taking expectation on both sides of (1.18b) *and observing*

$$E_0^Q[(K - F_T)^+] = e^{rT} p(K) \quad and \quad E_0^Q[(F_T - K)^+] = e^{rT} c(K),$$

we obtain the fair strike formula (1.19).

We conclude that the variance exposure of a portfolio of options becomes independent of the asset price when the weights of the options are chosen to be inversely proportional to the square of the respective strike. This result is illustrated by the numerical example shown in Exhibit 1 in Demeterfi *et al.* (1999). As a further remark, we may apply the put-call parity relation to eliminate the put option in (1.19). The resulting fair strike formula becomes (Britten-Jones and Neuberger, 2000)

$$K_v = \frac{2}{T} \left[\int_0^{\infty} \frac{e^{rT} c(K) - \max(F_0 - K, 0)}{K^2} \, dK \right]. \tag{1.20}$$

Weighted average of implied variances

Some practitioners may prefer to work with implied volatilities than option prices, so it is desirable to express K_v in terms of implied volatilities. Let K be the strike of an option and F_0 be the current forward price. Recall that the Black option price formula is expressed in terms of forward price. We define $x = \ln \frac{K}{F_0}$ and let

$\sigma^B_{imp}(x)$ denote the Black implied volatility. We define the moneyness of a call option as

$$y = \mathrm{d}_2^B(x) = \frac{\ln\frac{F_0}{K} - \frac{[\sigma^B_{imp}(x)]^2 T}{2}}{\sigma^B_{imp}(x)\sqrt{T}} = -\frac{x}{\sigma^B_{imp}(x)\sqrt{T}} - \frac{\sigma^B_{imp}(x)\sqrt{T}}{2}.$$

We prefer to express the Black implied volatility in terms of y instead of x. We write $x = (\mathrm{d}_2^B)^{-1}(y)$ so that $\sigma^B_{imp}(x)$ can be expressed as a function of y, where

$$\sigma^B_{imp}(y) = \sigma^B_{imp}(x) = \sigma^B_{imp}((d_2^B)^{-1}(y)).$$

Carr and Lee (2009) manage to show that

$$\begin{aligned} K_v &= \frac{2e^{rT}}{T}\left[\int_0^{F_0} \frac{p(K)}{K^2}\,\mathrm{d}K + \int_{F_0}^{\infty} \frac{c(K)}{K^2}\,\mathrm{d}K\right] \\ &= e^{rT}\int_{-\infty}^{\infty} n(y)[\sigma^B_{imp}(y)]^2\,\mathrm{d}y, \end{aligned} \tag{1.21}$$

where $n(y)$ is the standard normal density function. Interestingly, K_v can be expressed as the weighted average of the Black implied variance expressed in terms of moneyness y and the weight is the standard normal distribution.

To prove the result, we start with the relation:

$$\int_0^{F_0} \frac{p(K)}{K^2}\,\mathrm{d}K + \int_{F_0}^{\infty} \frac{c(K)}{K^2}\,\mathrm{d}K = \int_0^{F_0} \frac{p'(K)}{K}\,\mathrm{d}K + \int_{F_0}^{\infty} \frac{c'(K)}{K}\,\mathrm{d}K.$$

The derivatives of the Black option call and put price functions with respect to K are given by

$$\begin{aligned} c'(K) &= -N(d_2^B(K)) + \sqrt{T}n(d_2^B(K))K\frac{\mathrm{d}\sigma^B_{imp}(K)}{\mathrm{d}K}, \\ p'(K) &= N(-d_2^B(K)) + \sqrt{T}n(d_2^B(K))K\frac{\mathrm{d}\sigma^B_{imp}(K)}{\mathrm{d}K}, \end{aligned}$$

where $n(x)$ is the standard normal density function and $N(x)$ is the standard normal distribution. By performing tedious calculus exercise of simplifying the two integrals involving $N(d_2^B(K))$, $n(d_2^B(K))$ and $\frac{\mathrm{d}\sigma^B_{imp}(K)}{\mathrm{d}K}$ (Carr and Lee, 2009), we obtain (1.21).

The appropriate choice of $h(F,t)$ in (1.17) dictates constant cash gamma of the underlying derivative and it has close relation with the continuously sampled variance [see (1.18a)]. Next, we would like to show how to choose suitable functions $h_g(F_t)$ and $h_c(F_t)$ that are related to the continuously sampled gamma swap and corridor variance swap, respectively. We then use these results to derive the corresponding replicating portfolios of options with continuum of strikes for these two exotic variance swaps.

Gamma swaps

We would like to derive similar replication formula for the continuous gamma swap. Again, it is convenient to work with the futures price F_t and modify the floating log payoff of the continuous gamma swap to become

$$\frac{1}{T}\int_0^T \frac{F_t}{F_0}\sigma_t^2\, dt = \frac{1}{T}\int_0^T e^{-(r-q)t}\frac{S_t}{S_0}\sigma_t^2\, dt.$$

Consider the following function

$$h_g(F_t) = \frac{2}{T}\left[F_t\left(\ln\frac{F_t}{\kappa} - 1\right) + \kappa\right], \tag{1.22}$$

by applying Itô's lemma, the differential of $h_g(F_t)$ is given by

$$dh_g(F_t) = \frac{2}{T}\left(\ln\frac{F_t}{\kappa}\, dF_t + F_t\frac{\sigma_t^2}{2}\, dt\right).$$

Integrating from $t = 0$ to $t = T$, we obtain

$$\frac{1}{T}\int_0^T \frac{F_t}{F_0}\sigma_t^2\, dt = \frac{h_g(F_T)}{F_0} - \frac{2}{TF_0}\int_0^T \ln\frac{F_t}{\kappa}\, dF_t - \frac{h_g(F_0)}{F_0}. \tag{1.23a}$$

By setting $f(F_T)$ to be $\frac{h_g(F_T)}{F_0}$ in (1.15) with $\kappa = F_0$ and noting $F_T = S_T$, the first term on the right-hand side of (1.23a) can be written as

$$\frac{h_g(F_T)}{F_0} = \frac{2}{TF_0}\left[\int_0^\kappa \frac{(K - S_T)^+}{K}\, dK + \int_\kappa^\infty \frac{(S_T - K)^+}{K}\, dK\right]. \tag{1.23b}$$

Under continuous sampling, the fair strike of the continuous gamma swap is defined by

$$K_g = E_0^Q\left[\frac{1}{T}\int_0^T \frac{F_t}{F_0}\sigma_t^2\, dt\right],$$

where the continuous weight process is $w_t = \frac{F_t}{F_0} = e^{-(r-q)t}\frac{S_t}{S_0}$. Combining (1.23a) and (1.23b), and noting

$$E_0^Q\left[\int_0^T \ln\frac{F_t}{F_0}\, dF_t\right] = 0,$$

we take expectation on both sides of the equation to obtain the following replication formula (1.24) in Proposition 1.2.

Proposition 1.2 *The fair strike of the continuous gamma swap is given by*

$$K_g = \frac{2e^{rT}}{TF_0}\left[\int_0^{F_0} \frac{p(K)}{K}\, dK + \int_{F_0}^\infty \frac{c(K)}{K}\, dK\right], \tag{1.24}$$

where $p(K)$ and $c(K)$ are the respective price of the European T-maturity put and call option with strike K.

Corridor variance swaps

For the corridor variance swap with corridor $[L,U]$, we consider

$$h_c(F_t) = \frac{2}{T}\left(\ln\frac{F_0}{F_t} + \frac{F_t}{F_0} - 1\right)\mathbf{1}_{\{L<S_t\leq U\}}. \tag{1.25}$$

It is reasonable to choose the corridor so that F_0 lies within the corridor. Following similar procedures as above and observing $F_T = S_T$, we obtain

$$\begin{aligned}\frac{1}{T}\int_0^T \sigma_t^2 \mathbf{1}_{\{L<S_t\leq U\}}\,\mathrm{d}t &= \frac{2}{T}\left(\ln\frac{F_0}{F_T} + \frac{F_T}{F_0} - 1\right)\mathbf{1}_{\{L<S_T\leq U\}} \\ &\quad - \frac{2}{T}\int_0^T \left(\frac{1}{F_0} - \frac{1}{F_t}\right)\mathbf{1}_{\{L<S_t\leq U\}}\,\mathrm{d}F_t.\end{aligned}$$

The first term involves $\ln\frac{F_0}{S_T} + \frac{S_T}{F_0} - 1$ and $\mathbf{1}_{\{L<S_T\leq U\}}$. By applying the Taylor formula (1.15), the product of these two quantities can be represented as the terminal payoff of a portfolio of options with continuum of strikes lying within the corridor $[L,U]$, where

$$\begin{aligned}&\left(\ln\frac{F_0}{S_T} + \frac{S_T}{F_0} - 1\right)\mathbf{1}_{\{L<S_T\leq U\}} \\ &= \int_L^{F_0} \frac{1}{K^2}(K-S_T)^+\,\mathrm{d}K + \int_{F_0}^U \frac{1}{K^2}(S_T-K)^+\,\mathrm{d}K.\end{aligned}$$

Compared to the variance swap, the replication of the corridor variance swap only requires options with strikes lying within the corridor $(L,U]$. Following the same procedure as in the proof of Proposition 1.1, the fair strike of the continuous corridor swap is given by

$$K_v = \frac{2e^{rT}}{T}\left[\int_L^{F_0} \frac{p(K)}{K^2}\,\mathrm{d}K + \int_{F_0}^U \frac{c(K)}{K^2}\,\mathrm{d}K\right]. \tag{1.26}$$

It is interesting to observe that an approximate replication of the variance swap with truncation on the range of strikes indeed resembles the replication of a corridor variance swap.

1.5 Practical implementation of replication: Finite strikes and discrete monitoring

We have discussed the replication of the generalized continuously sampled realized variance using a portfolio of options with continuum of strikes spanning the full range of values. In actual implementation of replication using a portfolio of options, we face with several practical issues. First, only options with finite set of strikes within a bounded range are available in the market. Second, most variance derivative contracts specify the payoff in terms of discrete realized variance since asset prices can only be monitored at discrete time instants.

In this section, we explore the practical implementation of replication under the realistic market conditions of options limited to finite strikes and discrete monitoring of asset prices. We start with the discussion of replication of continuously sampled variance using options of finite strikes within a bounded range. We then apply similar concepts to examine how the Chicago Board of Options Exchange (CBOE) structures the CBOE volatility index (VIX) that aims to extract model-free volatility from the market prices of S&P 500 index options. The errors of replication of the realized volatility of S&P 500 index are discussed. Finally, we consider how to construct various replicating portfolios of generalized swaps on discrete realized variance and examine the errors incurred in these replication procedures.

1.5.1 Continuously sampled realized variance replicated by options of finite strikes

Recall that replication of different types of variance swaps amounts to replication of the corresponding payoff function $h(S_T;\kappa)$ using an appropriate portfolio of options with common maturity T. Note that $h(S_T;\kappa)$ is typically increasing on (κ,∞) and decreasing on $(0,\kappa)$ with $h=0$ at $S_T=\kappa$ [see (1.17), (1.22) and (1.25)]. For $S_T \geq \kappa$, the curve $h(S_T;\kappa)$ is approximated by the discrete set of line segments that connect the points: $(K_0,h(K_0))$, $(K_{1c},h(K_{1c}))$, $\cdots$, $(K_{nc},h(K_{nc}))$ (see Fig. 1.3). The approximate payoff can be replicated exactly by the portfolio of call options with a discrete set of strike prices $\kappa = K_0 < K_{1c} < \cdots < K_{nc}$ (Demeterfi *et al.*, 1999).

The first segment on the right-hand side with an upward slope can be replicated by the call option with strike K_0 and notional amount

$$w_c(K_0) = \frac{h(K_{1c})-h(K_0)}{K_{1c}-K_0},$$

which is the slope of the first segment. The second line segment is seen to be related to the two call options with strikes K_0 and K_{1c}. Given that we already hold $w_c(K_0)$ units of call option with strike K_0, we need to hold $w_c(K_{1c})$ units of call option with strike K_{1c}. Here, $w_c(K_{1c})$ is given by the difference between the slopes of the first segment and the second one, where

$$w_c(K_{1c}) = \frac{h(K_{2c})-h(K_{1c})}{K_{2c}-K_{1c}} - w_c(K_0).$$

In general, the recursive formula for $w_c(K_{ic})$ is given by

$$w_c(K_{ic}) = \frac{h(K_{i+1,c})-h(K_{ic})}{K_{i+1,c}-K_{ic}} - \sum_{j=0}^{i-1} w_c(K_{jc}),\ i=1,2,\cdots,n-1. \tag{1.27}$$

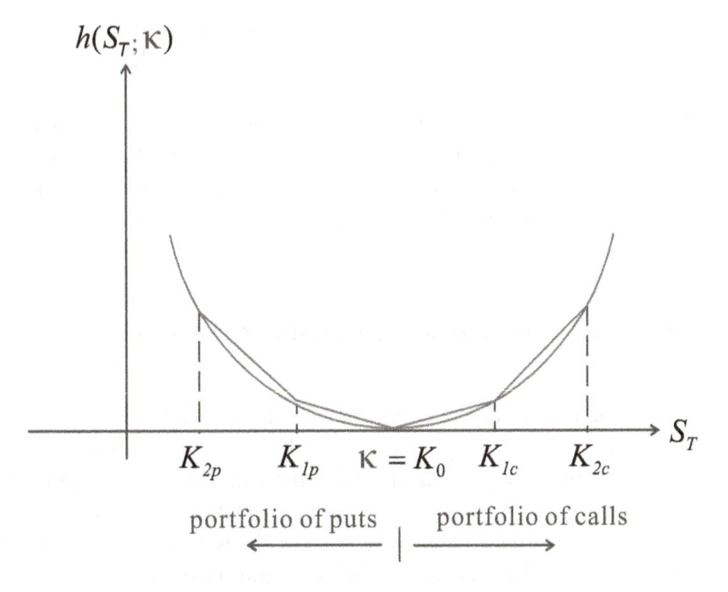

FIGURE 1.3 Approximation of $h(S_T;\kappa)$ by discrete line segments. The line segments on the right (left) can be replicated by portfolio of calls (puts) with discrete set of strikes at and above (below) κ.

Mathematically, we would like to show that when $\max|K_{ic} - K_{i-1,c}| \to 0$ and $n \to \infty$, the weight $w_c(K_{ic})$ defined by (1.27) indeed converges to $\frac{2}{T}\frac{\mathrm{d}K}{K^2}$. First, we express $w_c(K_{ic})$ as the difference of two gradient terms, where

$$w_c(K_{ic}) = \frac{h(K_{i+1,c}) - h(K_{ic})}{K_{i+1,c} - K_{ic}} - \frac{h(K_{ic}) - h(K_{i-1,c})}{K_{ic} - K_{i-1,c}}.$$

In the limit of taking differential small change in K, the above difference of the two gradient terms is related to the second-order derivative $h''(K)$. By noting

$$h''(K) = \frac{2}{T}\frac{1}{K^2},$$

we obtain

$$w_c(K) \to h''(K)\,\mathrm{d}K = \frac{2}{T}\frac{\mathrm{d}K}{K^2}.$$

The left side of the curve $h(S_T;\kappa)$ can be approximated by using put options with discrete strikes: $K_{1p}, K_{2p}, \ldots, K_{np}$. The notional amount of the put option with strike K_{ip} is given by

$$w_p(K_{ip}) = -\frac{h(K_{i+1,p}) - h(K_{ip})}{K_{i+1,p} - K_{ip}} - \sum_{j=0}^{i-1} w_p(K_{jp}), \quad i = 1,2,\ldots,n-1. \tag{1.28}$$

The above procedure uses the line segments to approximate the function $h(S_T;\kappa)$. One may use other piecewise approximation (trapezoidal method) or go further to the Simpson quadrature method to achieve higher accuracy of approximation. Leung and

Lorig (2016) solve the optimal hedge of the log contract using discrete sets of call options and put options by minimizing the expected squared hedging error subject to a cost constraint. Le Floc'h (2018) studies the errors in variance swap replication using different approximation methods, and also assesses the performance of the optimal quadratic hedge method of Leung and Lorig (2016). His results show that the Simpson quadrature has the best performance among the different approximation methods while the optimal quadratic hedge method may result in a worse static hedge in some scenarios.

1.5.2 VIX: Extracting model-free volatility from S&P 500 index options

The volatility of an asset price process or stock index process is a latent stochastic process on its own, which changes stochastically over time. Volatility appears to revert to the mean. After a large volatility spike, the volatility can potentially decrease rapidly; and it may start to decrease slowly after a low volatility period. Volatility is often negatively correlated with the stock price or index level, commonly called the negative leverage effect. The volatility of some common stock index, like S&P 500 index, may serve as a proxy of market confidence or fear, so it is nicknamed "fear gauge". We discuss how to extract model-free expected variance of S&P 500 index from traded prices of index options. A comprehensive survey of theoretical studies on the model-free VIX can be found in Gonzalez-Perez (2015).

The VIX is a volatility index calculated by the Chicago Board Options Exchange (CBOE). It informs investors about the expected market volatility of S&P 500 index in the next 30 calendar days. The old volatility index (now called VXO) was first computed in 1993 as a linear combination of 8 at-the-money implied volatilities on the S&P 100 options with maturities closest to 30 calendar days. The old index is model dependent since its calculation is based on the Black-Scholes pricing model. The new VIX was launched in 2003, which is the model-free volatility measure of the S&P 500 index (ticker symbol SPX). The use of the model-free volatility measure would ease the pricing procedures of VIX futures and options. Carr and Wu (2006) present an interesting and detailed account on the shortcomings of the old index and why the new index is adopted. A more recent expository article on the CBOE VIX can be found in Cboe (2019). Jiang and Tian (2005) conclude from their empirical studies that the model-free implied volatility is an efficient forecast for future realized volatility and subsumes all information contained in the Black-Scholes implied volatility and historical realized volatility. This is because the model-free implied volatility extracts information from all relevant option prices instead of a few Black-Scholes implied volatilities as in VXO. Also, empirical tests on the model-free implied volatility is a direct test of market efficiency rather than a joint test of market efficiency and the assumed Black-Scholes option pricing model, so there is no contamination by the model misspecification errors. The theoretical underpinning is the success of replication of expected integrated variance using portfolio of options as discussed in the last section. The VIX formula is expressed as the weighted average of out-of-the-money S&P 500 options within a bounded range and finite set of strikes, which is an approximation to the replication formula of integrated variance within the limits of practical implementation.

Some researchers may debate the "model free" property of the index since the VIX formula is derived based on pure diffusion process with no jump of the underlying S&P 500 index. The weighted schemes of the out-of-the-money S&P index options would differ under different assumptions of the underlying process, say with inclusion of jumps in the underlying index value process and instantaneous variance process. More detailed discussion on the VIX formula with inclusion of jumps in the underlying processes can be found in Sec. 3.4.

Derivation of VIX formula

VIX expresses volatility in percentage points. It is calculated as 100 times the square root of the expected 30-day variance of the rate of return of the forward price of the S&P 500 index. The following derivation of VIX formula can be considered as an alternative but more direct proof of the fair strike formula (1.19) in Proposition 1.1.

The VIX formula is derived under the assumption of zero jump on the process of the forward price F_t of the S&P 500 index. Under a risk neutral measure Q, the dynamics of F_t is governed by

$$\frac{\mathrm{d}F_t}{F_t} = \sigma_t\,\mathrm{d}W_t,$$

where σ_t is the stochastic volatility and W_t is the standard Q-Brownian motion. We define VIX^2 to be

$$\mathrm{VIX}^2 = E_0^Q\left[\frac{1}{T}\int_0^T \sigma_t^2\,\mathrm{d}t\right]. \tag{1.29}$$

By Itô's lemma, we obtain

$$\mathrm{d}\ln F_t = -\frac{\sigma_t^2}{2}\,\mathrm{d}t + \sigma_t\,\mathrm{d}W_t$$

so that

$$\frac{\mathrm{d}F_t}{F_t} - \mathrm{d}\ln F_t = \frac{\sigma_t^2}{2}\,\mathrm{d}t$$

or

$$\frac{1}{T}\int_0^T \sigma_t^2\,\mathrm{d}t = \frac{2}{T}\left[\int_0^T \frac{\mathrm{d}F_t}{F_t} - \ln\frac{F_T}{F_0}\right].$$

Since F_t is an Itô process with zero drift, by taking expectation under Q, we obtain

$$E_0^Q\left[\int_0^T \frac{\mathrm{d}F_t}{F_t}\right] = E_0^Q\left[\int_0^T \sigma_t\,\mathrm{d}W_t\right] = 0.$$

We then deduce that

$$\mathrm{VIX}^2 = \frac{2}{T}E_0^Q\left[\int_0^T \frac{\mathrm{d}F_t}{F_t} - \mathrm{d}\ln F_t\right] = -\frac{2}{T}E_0^Q\left[\ln\frac{F_T}{F_0}\right]. \tag{1.30}$$

This establishes the link between VIX^2 with the log contract. The CBOE procedure of calculating VIX^2 relates the log contract with the usual vanilla call and put options on S&P 500 index.

Applying the Taylor formula (1.15) to the function $f(F_T) = \ln \frac{F_T}{F_0}$ gives

$$\ln \frac{F_T}{F_0} = \frac{F_T - F_0}{F_0} - \int_0^{F_0} \frac{(K - F_T)^+}{K^2}\, \mathrm{d}K - \int_{F_0}^{\infty} \frac{(F_T - K)^+}{K^2}\, \mathrm{d}K.$$

Taking expectation on both sides and combining results in (1.30), we obtain

$$\begin{aligned} \text{VIX}^2 &= -\frac{2}{T} E_0^Q \left[\ln \frac{F_T}{F_0} \right] \\ &= -\frac{2}{T} E_0^Q \left[\frac{F_T - F_0}{F_0} \right] + \frac{2}{T} E_0^Q \left[\int_0^{F_0} \frac{(K - F_T)^+}{K^2}\, \mathrm{d}K + \int_{F_0}^{\infty} \frac{(F_T - K)^+}{K^2}\, \mathrm{d}K \right]. \end{aligned}$$

Note that $E_0^Q \left[\frac{F_T - F_0}{F_0} \right] = 0$ since $E_0^Q[F_T] = F_0$. Furthermore, we visualize the following put price and call price formulas:

$$E_0^Q[(K - F_T)^+] = e^{rT} p(K) \quad \text{and} \quad E_0^Q[(F_T - K)^+] = e^{rT} c(K),$$

where $p(K)$ and $c(K)$ denote the respective T-maturity put price and call price with strike K. The theoretical VIX^2 formula then takes the form:

$$\begin{aligned} \text{VIX}^2 = E_0^Q \left[\frac{1}{T} \int_0^T \sigma_t^2\, \mathrm{d}t \right] &= -\frac{2}{T} E_0^Q \left[\frac{F_T}{F_0} \right] \\ &= \frac{2e^{rT}}{T} \left[\int_0^{F_0} \frac{p(K)}{K^2}\, \mathrm{d}K + \int_{F_0}^{\infty} \frac{c(K)}{K^2}\, \mathrm{d}K \right]. \end{aligned} \tag{1.31}$$

The VIX^2 formula represents replication of the expected integrated continuously sampled variance by a continuum of out-of-the-money S&P 500 index options.

Actual implementation of VIX formula

In the actual implementation of the replication formula (1.31), one has to face with options that are available only with discrete number of strikes. Also, only a finite range of strikes of traded options are available. In addition, the option with strike that exactly equals F_0 is not available in general. In the CBOE calculation procedure, we choose the out-of-the-money options within the bounded interval $[K_L, K_U]$, and set K_0 to be the closest listed strike below F_0. The out-of-the-money options include all listed put options with strikes at or below K_0, and all listed call options with strike at or above K_0, far in the moneyness range until two consecutive zero bid prices are found.

Applying the Taylor formula (1.15) to $\ln \frac{F_T}{K_0}$, we obtain

$$\ln \frac{F_T}{K_0} = \frac{F_T - K_0}{K_0} - \int_0^{K_0} \frac{(K - F_T)^+}{K^2}\, \mathrm{d}K - \int_{K_0}^{\infty} \frac{(F_T - K)^+}{K^2}\, \mathrm{d}K$$

so that

$$-\frac{2}{T}E_Q\left[\ln\frac{F_T}{K_0}\right] = -\frac{2}{T}\left(\frac{F_0}{K_0}-1\right)+\frac{2e^{rT}}{T}\left[\int_0^{K_0}\frac{p(K)}{K^2}\,\mathrm{d}K+\int_{K_0}^{\infty}\frac{c(K)}{K^2}\,\mathrm{d}K\right].$$

Furthermore, by observing

$$\ln\frac{F_T}{F_0} = \ln\frac{F_T}{K_0} - \ln\frac{F_0}{K_0}$$

and using the approximation

$$\ln\frac{F_0}{K_0} \approx \left(\frac{F_0}{K_0}-1\right)-\frac{1}{2}\left(\frac{F_0}{K_0}-1\right)^2,$$

we obtain

$$\begin{aligned}&\frac{2e^{rT}}{T}\left\{\left[\int_0^{F_0}\frac{p(K)}{K^2}\,\mathrm{d}K+\int_{F_0}^{\infty}\frac{c(K)}{K^2}\,\mathrm{d}K\right]-\left[\int_0^{K_0}\frac{p(K)}{K^2}\,\mathrm{d}K+\int_{K_0}^{\infty}\frac{c(K)}{K^2}\,\mathrm{d}K\right]\right\}\\ =\;&\frac{2}{T}\left[\ln\frac{F_0}{K_0}-\left(\frac{F_0}{K_0}-1\right)\right]\approx -\frac{1}{T}\left(\frac{F_0}{K_0}-1\right)^2.\end{aligned}$$

Based on these observations, the CBOE procedure of calculating VIX^2 is given by

$$\mathrm{VIX}^2 = \frac{2e^{rT}}{T}\left[\sum_{K_L}^{K_0}\frac{\Delta K}{K^2}p(K)+\sum_{K_0}^{K_U}\frac{\Delta K}{K^2}c(K)\right]-\frac{1}{T}\left(\frac{F_0}{K_0}-1\right)^2. \tag{1.32}$$

The option prices are taken to be the midpoint of the latest available bid and ask prices. To express as percentage point, VIX is calculated as 100 times the square root of the right-hand side of (1.32).

Another approximation error arises from the linear maturity interpolation. The VIX is calculated based on options with a fixed 30-day maturity. However, there are generally no options that expire exactly on 30-day maturity. The CBOE calculation procedure finds two maturities: near-term T_1 and next-term T_2 that are closest to the required 30-day maturity T_0. According to the CBOE calculation procedures [see Cboe White Paper (2019) for details], the near-term maturity and next-term maturity are set to be more than 23 days and not more than 37 days to expiration, respectively. The forward S&P index level F_0 corresponding to the near-term and next-term is determined by identifying the strike price at which the absolute difference between the call and put prices is smallest at that maturity term T, where

$$F_0 = \text{strike price} + e^{RT}(\text{call price} - \text{put price}).$$

Once F_0 is found, K_0 is the first strike below F_0. We label the variance measures for maturities T_1 and T_2 as $\widehat{\mathrm{VIX}}^2_{T_1}$ and $\widehat{\mathrm{VIX}}^2_{T_2}$, respectively. The VIX^2 is computed by applying the linear maturity interpolation as follows:

$$\mathrm{VIX}^2 = \frac{1}{30}\left[\alpha T_1\widehat{\mathrm{VIX}}^2_{T_1}+(1-\alpha)T_2\widehat{\mathrm{VIX}}^2_{T_2}\right], \tag{1.33}$$

where the weight α is given by

$$\alpha = \frac{T_2 - T_0}{T_2 - T_1}.$$

As a summary, we observe the following sources of errors in the CBOE procedure of calculating VIX.

1. Truncation error

   We choose a bounded truncation interval $[K_L, K_U]$ of strikes instead of the theoretical interval $[0, \infty]$. Note that the CBOE may add new strikes as the underlying S&P 500 index moves. The added strikes expand the truncation interval. During the period of frequent spikes, the expansion of interval can be frequent and significant.

2. Discretization error

   The continuous integration of option prices with respect to continuum of strikes is approximated by the sum of weighted out-of-the-money option prices.

3. Approximation error arising from replacing F_0 by K_0

   The logarithm term $\ln \frac{F_0}{K_0}$ is approximated by the Taylor expansion in powers of $\frac{F_0}{K_0} - 1$ up to the quadratic term.

4. Linear maturity interpolation

   Errors are introduced since the model-free volatility is a nonlinear function of maturity.

Jiang and Tian (2007) analyze the errors in the CBOE's implementation of the VIX formula. Using option prices simulated under typical market conditions, they discover that the CBOE procedure may exhibit estimation error of the true volatility by several percents. More importantly, these errors show predictable patterns in relation to volatility levels. They propose a simple fix of the problems using a smooth interpolation-extrapolation of the implied volatility function.

1.5.3 Replication of swaps on discrete realized variance

The replication discussed in the last two subsections reproduces continuously sampled realized variance. In reality, the realized variance is sampled discretely (typically daily sampled). In this subsection, we consider the replication of discrete realized variance using portfolio of options (Carr and Lee, 2009).

Let the simple rate of return of the asset price process S_k over (t_{k-1}, t_k) be defined by

$$r_k = \frac{S_{t_k} - S_{t_{k-1}}}{S_{t_{k-1}}}.$$

Recall the following relations between the logarithm return and simple rate of return:

$$\ln \frac{S_{t_k}}{S_{t_{k-1}}} = r_k - \frac{r_k^2}{2} + \frac{r_k^3}{3} + O(r_k^4), \quad k = 1,2,\cdots,N; \tag{1.34a}$$

and

$$\left(\ln \frac{S_{t_k}}{S_{t_{k-1}}}\right)^2 = r_k^2 - r_k^3 + O(r_k^4), \quad k = 1,2,\cdots,N. \tag{1.34b}$$

Adding two times (1.34a) to (1.34b) and rearranging the terms, we obtain

$$\left(\ln \frac{S_{t_k}}{S_{t_{k-1}}}\right)^2 = 2r_k - 2\ln \frac{S_{t_k}}{S_{t_{k-1}}} - \frac{r_k^3}{3} + O(r_k^4), \quad k = 1,2,\cdots,N.$$

Summing over $k = 1,2,\cdots,N$, and observing the relation:

$$\sum_{k=1}^{N} \ln \frac{S_{t_k}}{S_{t_{k-1}}} = \ln \frac{S_T}{S_0},$$

we obtain

$$\sum_{k=1}^{N} \left(\ln \frac{S_{t_k}}{S_{t_{k-1}}}\right)^2 = 2\left[\sum_{k=1}^{N} \frac{\Delta S_{t_k}}{S_{t_{k-1}}} - \ln \frac{S_T}{S_0}\right] - \frac{1}{3}\sum_{k=1}^{N} r_k^3 + \sum_{k=1}^{N} O(r_k^4), \tag{1.35}$$

where $\Delta S_{t_k} = S_{t_k} - S_{t_{k-1}}$. By expressing the log contract in terms of a static position of options [see (1.16)] and setting $\kappa = S_0$, we obtain the replication strategy of the unannualized floating leg of the discrete variance swap based on logarithm returns. The details are summarized in Proposition 1.3.

Proposition 1.3 *The unannualized floating leg of the discrete variance swap based on logarithm returns can be written as*

$$\begin{aligned}
&\sum_{k=1}^{N} \left(\ln \frac{S_{t_k}}{S_{t_{k-1}}}\right)^2 \\
&= 2\left[\sum_{k=1}^{N} \left(\frac{1}{S_{t_{k-1}}} - \frac{1}{S_0}\right)\Delta S_{t_k} + \int_0^{S_0} \frac{(K - S_T)^+}{K^2}\,\mathrm{d}K + \int_{S_0}^{\infty} \frac{(S_T - K)^+}{K^2}\,\mathrm{d}K\right] \\
&\quad - \frac{1}{3}\sum_{k=1}^{N} r_k^3 + \sum_{k=1}^{N} O(r_k^4).
\end{aligned} \tag{1.36}$$

The above formula reveals that the discrete realized variance based on logarithm returns can be approximated by a static position of out-of-the-money calls and puts and a dynamically rebalanced position of the underlying asset with number of units equals $\frac{1}{S_{t_{k-1}}} - \frac{1}{S_0}$. The leading order of the replication error is the accumulated cubed returns. If the distribution of the return under the risk neutral measure is skewed to the left, since the expectation of the cubed returns tends to be negative, the replicating portfolio tends to underestimate the fair strike of the discrete variance swap.

Suppose that the simple rate of return is used in calculating the discrete realized variance. By using the Taylor expansion of $[\ln(1+r_k)]^2$ [see (1.2)], we deduce that

$$\sum_{k=1}^{N}\left(\ln\frac{S_{t_k}}{S_{t_{k-1}}}\right)^2 - \sum_{k=1}^{N} r_k^2 = -\sum_{k=1}^{N} r_k^3 + \sum_{k=1}^{N} O(r_k^4).$$

Combining the above equation with (1.36) yields the following proposition.

Proposition 1.4 *The unannualized floating leg of the discrete variance swap based on simple returns can be expressed as*

$$\begin{aligned}
&\sum_{k=1}^{N}\left(\frac{S_{t_k}-S_{t_{k-1}}}{S_{t_{k-1}}}\right)^2 \\
&= 2\left[\sum_{k=1}^{N}\left(\frac{1}{S_{t_{k-1}}}-\frac{1}{S_0}\right)\Delta S_{t_k} + \int_0^{S_0}\frac{(K-S_T)^+}{K^2}\,\mathrm{d}K + \int_{S_0}^{\infty}\frac{(S_T-K)^+}{K^2}\,\mathrm{d}K\right] \\
&\quad + \frac{2}{3}\sum_{k=1}^{N} r_k^3 + \sum_{k=1}^{N} O(r_k^4).
\end{aligned} \tag{1.37}$$

Jump effect

The replication approach discussed so far assumes continuous path of the underlying asset price process. The actual asset price process may have occurrences of jumps. We would like to examine the replication error due to presence of potential jumps.

As an illustration, we assume that only one jump occurs during the whole monitoring period of the realized variance and S jumps to Se^J. Ignoring the higher order terms in (1.35), we have

$$\sum_{k=1}^{N}\left(\ln\frac{S_{t_k}}{S_{t_{k-1}}}\right)^2 \approx 2\sum_{k=1}^{N}\left(\frac{\Delta S_{t_k}}{S_{t_{k-1}}} - \ln\frac{S_{t_k}}{S_{t_{k-1}}}\right). \tag{1.38}$$

Suppose within the period (t_{k-1},t_k), the kth logarithm return $\ln\frac{S_{t_k}}{S_{t_{k-1}}}$ has the additional jump contribution J. As a result, the left-hand side of (1.38) increases by J^2. On the other hand, the term $\frac{\Delta S_{t_k}}{S_{t_{k-1}}}$ approximately changes by e^J-1, and the term $\ln\frac{S_{t_k}}{S_{t_{k-1}}}$ changes by J. By using the Taylor expansion

$$e^J - 1 - J = \frac{J^2}{2} + \frac{J^3}{6} + O(J^4),$$

the difference of the responses of the two sides of (1.38) due to the single jump is seen to be

$$2(e^J - J - 1) - J^2 = \frac{J^3}{3} + O(J^4). \tag{1.39}$$

If J is small and negative, corresponding to a small downward jump, the replicating strategy that is used under the continuous path assumption tends to underestimate the realized variance. The error is of order J^3.

Futures price based variance swaps

In some variance swap contracts, the futures price is used to compute the discrete realized variance. The unannualized floating leg of a futures price based variance swap is defined by

$$\sum_{k=1}^{N}\left(\ln\frac{F_{t_k}}{F_{t_{k-1}}}\right)^2,$$

where $F_t = S_t e^{(r-q)(T-t)}$ and $\Delta t_k = t_k - t_{k-1}$. Since

$$\left(\ln\frac{F_{t_k}}{F_{t_{k-1}}}\right)^2 = \left(\ln\frac{S_{t_k}}{S_{t_{k-1}}}\right)^2 - 2\ln\frac{S_{t_k}}{S_{t_{k-1}}}(r-q)\Delta t_k + (r-q)^2\Delta t_k^2,$$

the difference between the squared logarithm returns calculated from the spot prices and futures prices is $O(\Delta t_k)$. Hence, the futures price based realized variance is a close proxy of the spot price based discrete realized variance, where the difference converges to zero when the continuous sampling limit is taken.

Replacing the spot price S_t by the futures price F_t in (1.36), we obtain the following replication of the futures price based variance swap as stated in Proposition 1.5.

Proposition 1.5 *The unannualized floating leg of the futures price based variance swap can be expressed as*

$$\begin{aligned}
&\sum_{k=1}^{N}\left(\ln\frac{F_{t_k}}{F_{t_{k-1}}}\right)^2 \\
&= 2\left[\sum_{k=1}^{N}\left(\frac{1}{F_{t_{k-1}}}-\frac{1}{S_0}\right)\Delta F_{t_k} + \int_0^{S_0}\frac{1}{K^2}(K-F_T)^+\mathrm{d}K + \int_{S_0}^{\infty}\frac{1}{K^2}(F_T-K)^+\mathrm{d}K\right] \\
&\quad -\frac{1}{3}\sum_{k=1}^{N}\tilde{r}_k^3 + \sum_{k=1}^{N}O(\tilde{r}_k^4),
\end{aligned} \tag{1.40}$$

where

$$\tilde{r}_k = \frac{F_{t_k}-F_{t_{k-1}}}{F_{t_{k-1}}}.$$

Replication of the corridor variance swaps

The extension of the replication of discrete realized variance swaps with corridor feature using a portfolio of options is quite tedious. Carr and Lewis (2004) propose an approximate replication method for the futures price based corridor variance swaps under a modified asymmetric specification of the discrete realized variance on the entry and exit of the corridor. Unlike the traditional definition of the corridor variance, Carr and Lewis (2004) define a typical term over the period (t_{k-1}, t_k) of the downside variance with an upper barrier U as follows:

$$\begin{aligned} R_k^d = & \mathbf{1}_{\{F_{t_{k-1}} \leq U,\, F_{t_k} \leq U\}} \left(\ln \frac{F_{t_k}}{F_{t_{k-1}}} \right)^2 + \mathbf{1}_{\{F_{t_{k-1}} > U,\, F_{t_k} \leq U\}} \left(\ln \frac{F_{t_k}}{U} \right)^2 \\ & + \mathbf{1}_{\{F_{t_{k-1}} \leq U,\, F_{t_k} > U\}} \left[\left(\ln \frac{F_{t_k}}{F_{t_{k-1}}} \right)^2 - \left(\ln \frac{F_{t_k}}{U} \right)^2 \right]. \end{aligned} \tag{1.41}$$

There are 3 scenarios to be considered when one updates the discrete realized variance at t_k. When $F_{t_{k-1}} \leq U$ and $F_{t_k} \leq U$, the normal squared logarithm return is used to update the realized downside variance. If F_{t_k} is in the corridor while $F_{t_{k-1}}$ is outside, then $F_{t_{k-1}}$ is replaced by U in calculating the logarithm return for the period (t_{k-1}, t_k). Lastly, if $F_{t_{k-1}}$ is in the corridor while F_{t_k} is outside, we subtract the exceeded squared logarithm return from the normal squared logarithm return. Under this formulation, there exists a model-free hedging strategy whose error is only third order, the details are stated in Proposition 1.6. The reason for choosing this asymmetric treatment of entry and exit in (1.41) is that the sum of the payoffs of an upside variance contract and a downside variance contract with the same barrier U remains equal to the payoff of a standard variance contract. A symmetric treatment of entry and exit payoffs will lead to the violation of this nice property.

The asymmetry of the payoff prescription of entry and exit vanishes if one assumes continuous price process, continuous sampling of the realized variance and continuous rebalancing of the underlying position. Under these idealized conditions, the replication error can be kept to the third order. These results are summarized in Proposition 1.6.

Proposition 1.6 *For R_k^d defined in (1.41), the unannualized floating leg of the downside variance swap with upper barrier U can be expressed as*

$$\begin{aligned} \sum_{i=1}^N R_k^d = & 2 \left[\int_0^U \frac{(K - F_T)^+}{K^2} \, dK - \left(\ln \frac{U}{F_0} + \frac{F_0}{U} - 1 \right) \mathbf{1}_{\{F_0 \leq U\}} \right. \\ & \left. + \sum_{k=1}^N \left(\frac{1}{F_{t_{k-1}}} - \frac{1}{U} \right)^+ \Delta F_{t_k} \right] + \sum_{k=1}^N O(\tilde{r}_k^3). \end{aligned} \tag{1.42}$$

The proof of Proposition 1.6 can be found in Carr and Lewis (2004).

Final remarks

There have been several research works on deriving no-arbitrage bounds on the fair strikes of variance swaps and constructing the relevant replicating strategies using other traded instruments (Hobson and Klimmek, 2012; Davis *et al.*, 2014; Kahalé, 2016). In these papers, the authors show how to construct model-independent hedging strategies that super-replicate and/or sub-replicate the payoff of a variance swap for any price path of the underlying asset, including jumps.

The replication of nonlinear payoffs on discrete realized variance, such as volatility swap and variance options, remains a challenging task. Carr and Lee (2008) manage to replicate the volatility swap using the so-called correlation-robust synthetic volatility swap via decomposition into a portfolio of straddles, calls and puts. The weights of the calls and puts in the portfolio are expressed implicitly in terms of the Bessel functions.

More recently, Burgard and Torné (2018) consider the super-replication of corridor variance swaps and related products. They show that the hedge involves two components: (i) static hedge that consists of a strip of appropriately weighted European vanilla options with expiration dates coinciding with the payout date of the variance swap contract, (ii) dynamic hedge that consists of a continuously rebalanced position in the underlying asset. In a later work, Ahallal and Torné (2019) derive a replication formula that expresses the knock-out corridor variance swap in terms of barrier options. Adding the knock-out feature into a corridor variance swap is sensible since a corridor variance swap that is deeply out-of-the-money will no longer accumulate realized variance while it may continue to consume margin until the expiration date.

Appendix

Proof of Dupire formula (1.6)

The original proof presented by Dupire (1994) assumes deterministic local volatility $\sigma(S_t,t)$. Let $\psi(K,T;S_t,t)$ denote the transition density function of the asset price process. Starting from the call price function:

$$c(S_t,t;K,T) = e^{-r(T-t)} \int_K^\infty (S_T - K)\psi(S_T,T;S_t,t)\, \mathrm{d}S_T,$$

we differentiate c with respect to K twice to obtain

$$\psi(K,T;S_t,t) = e^{r(T-t)} \frac{\partial^2 c(K,T)}{\partial K^2}.$$

Recall that the forward Fokker-Planck equation of $\psi(K,T;S_t,t)$ is given by

$$\frac{\partial \psi}{\partial T} = \frac{\partial^2}{\partial K^2}\left[\frac{\sigma^2(K,T)}{2} K^2 \psi\right] - \frac{\partial}{\partial K}[(r-q)K\psi].$$

By eliminating ψ in the above two equations, we obtain the following fourth-order differential equation for $c(K,T)$:

$$\begin{aligned} &\frac{\partial^2}{\partial K^2}\frac{\partial c}{\partial T} + r\frac{\partial^2 c}{\partial K^2} \\ = \;& \frac{\partial^2}{\partial K^2}\left[\frac{\sigma^2(K,T)}{2}K^2\frac{\partial^2 c}{\partial K^2}\right] - \frac{\partial}{\partial K}\left[(r-q)K\frac{\partial^2 c}{\partial K^2}\right]. \end{aligned}$$

By integrating the differential equation with respect to K twice and incorporating the two far field boundary conditions: $\lim_{K\to\infty} K\frac{\partial c}{\partial K} = 0$ and $\lim_{K\to\infty} K^2\frac{\partial^2 c}{\partial K^2} = 0$, we obtain the Dupire equation:

$$\frac{\partial c}{\partial T} - \frac{\sigma^2(K,T)}{2}K^2\frac{\partial^2 c}{\partial K^2} + (r-q)K\frac{\partial c}{\partial K} + qc = 0.$$

Indeed the Dupire equation can be interpreted as the forward equation of the Black-Scholes equation. The Dupire formula is obtained by expressing $\sigma^2(K,T)$ in terms of various derivatives of $c(K,T)$.

In an alternative proof presented below, we assume the volatility process σ_t to be stochastic in order to derive the more general result. The Dupire formula is then deduced as a special case when the stochastic volatility process σ_t is reduced to the deterministic local volatility $\sigma(S_t,t)$.

Let r and q be constant interest rate and dividend yield. Let $g(S_T,T)$ denote $e^{-r(T-t)}(S_T-K)^+$, the discounted terminal payoff of a call option with strike K at maturity T. We assume that S_t follows the following diffusion process under a risk neutral measure Q

$$\frac{dS_t}{S_t} = (r-q)\mathrm{d}t + \sigma_t\,\mathrm{d}W_t,$$

where W_t is the standard Brownian motion under Q and σ_t is the stochastic volatility process. The dynamics of S_t over $[T, T+\mathrm{d}T]$ determines the change of the call option price when maturity changes from T to $T+dT$. By Itô's lemma, the differential of g is given by

$$dg = \left[\frac{\partial g}{\partial T} + (r-q)S_T\frac{\partial g}{\partial S_T} + \frac{\sigma_T^2}{2}S_T^2\frac{\partial^2 g}{\partial S_T^2}\right]\mathrm{d}T + \sigma_T S_T\frac{\partial g}{\partial S_T}\mathrm{d}W_T.$$

Recall the following identities:

$$\frac{\partial}{\partial S_T}(S_T-K)^+ = \mathbf{1}_{\{S_T>K\}},\quad \frac{\partial^2}{\partial S_T^2}(S_T-K)^+ = \frac{\partial}{\partial S_T}\mathbf{1}_{\{S_T>K\}} = \delta(S_T-K),$$

we then have

$$\begin{aligned}\mathrm{d}g \;&=\; e^{-r(T-t)}[-r(S_T-K)^+ + (r-q)S_T\mathbf{1}_{\{S_T>K\}} + \frac{\sigma_T^2}{2}S_T^2\delta(S_T-K)]\mathrm{d}T \\ &\quad + e^{-r(T-t)}\sigma_T S_T\mathbf{1}_{\{S_T>K\}}\mathrm{d}W_T.\end{aligned}$$

Note that

$$c = E_t^Q[g] \quad\text{and}\quad \frac{\partial c}{\partial T}\mathrm{d}T = E_t^Q[\mathrm{d}g],$$

since $g(S_T,T)$ is defined for a fixed strike K and the change of dg over $[T, T+\mathrm{d}T]$ arises from the change in the asset price. Putting all the above relations together and observing $E_t^Q[\mathrm{d}W_T] = 0$, we obtain

$$\begin{aligned}\frac{\partial c}{\partial T}\mathrm{d}T &= E_t^Q[\mathrm{d}g] \\ &= e^{-r(T-t)}E_t^Q\left[rK\mathbf{1}_{\{S_T>K\}} - qS_T\mathbf{1}_{\{S_T>K\}} + \frac{\sigma_T^2}{2}S_T^2\delta(S_T-K)\right]\mathrm{d}T.\end{aligned}$$

Furthermore, we observe

$$\frac{\partial c}{\partial K} = -e^{-r(T-t)}E_t^Q[\mathbf{1}_{\{S>K\}}], \quad \frac{\partial^2 c}{\partial K^2} = e^{-r(T-t)}E_t^Q[\delta(S_T-K)],$$
$$e^{-r(T-t)}E_t^Q[S_T\mathbf{1}_{\{S_T>K\}}] = c + e^{-r(T-t)}KE_t^Q[\mathbf{1}_{\{S_T>K\}}].$$

Combining all the results together, we obtain

$$\begin{aligned}\frac{\partial c}{\partial T} &= e^{-r(T-t)}rKE_t^Q[\mathbf{1}_{\{S_T>K\}}] - q\left\{c + e^{-r(T-t)}KE_t^Q\left[\mathbf{1}_{\{S_T>K\}}\right]\right\} \\ &\quad + e^{-r(T-t)}E_t^Q\left[\frac{\sigma_T^2}{2}S_T^2\delta(S_T-K)\right] \\ &= -(r-q)K\frac{\partial c}{\partial K} - qc + e^{-r(T-t)}E_t^Q\left[\frac{\sigma_T^2}{2}S_T^2\delta(S_T-K)\right] \\ &= -(r-q)K\frac{\partial c}{\partial K} - qc + e^{-r(T-t)}E_t^Q\left[\frac{\sigma_T^2}{2}S_T^2\middle| S_T=K\right]E_t^Q[\delta(S_T-K)] \\ &= E_t^Q\left[\frac{\sigma_T^2}{2}\middle| S_T=K\right]K^2\frac{\partial^2 c}{\partial K^2} - (r-q)K\frac{\partial c}{\partial K} - qc.\end{aligned}$$

Rearranging the terms, we obtain

$$E_t^Q[\sigma_T^2|S_T=K] = 2\frac{\frac{\partial c}{\partial T} + qc + (r-q)K\frac{\partial c}{\partial K}}{K^2\frac{\partial^2 c}{\partial K^2}}.$$

The right-hand side term is visualized as the square of the *local volatility function* $\sigma(K,T)$, which can be determined by computing the derivatives of call prices with respect to K and T.

Suppose the volatility process σ_t is a deterministic function of S and t as specified by the local volatility $\sigma(S_t,t)$ in (1.4), the square of local volatility $\sigma^2(K,T)$ then equals $E_t^Q[\sigma_T^2|S_T=K]$. We then obtain the Dupire formula (1.6).

Proof of Taylor formula (1.15)

Assuming $x_0 > 0$, we have

$$\begin{aligned}\int_0^\xi \delta(x-x_0)\,\mathrm{d}x &= \mathbf{1}_{\{x_0<\xi\}} = \begin{cases}0 & \text{if } \xi < x_0 \\ 1 & \text{if } \xi > x_0\end{cases}; \\ \int_\xi^\infty \delta(x-x_0)\,\mathrm{d}x &= \mathbf{1}_{\{x_0>\xi\}} = \begin{cases}1 & \text{if } \xi < x_0 \\ 0 & \text{if } \xi > x_0\end{cases}.\end{aligned}$$

We perform repeated integration by parts in order to generate the option payoff terms: $(S_T - K)^+$ and $(K - S_T)^+$. For any choice of $\kappa > 0$, we have

$$\begin{aligned}
f(S_T) &= \int_0^{\kappa} f(K)\delta(S_T - K)\,\mathrm{d}K + \int_{\kappa}^{\infty} f(K)\delta(S_T - K)\,\mathrm{d}K \\
&= f(K)\mathbf{1}_{\{S_T<K\}}\Big]_0^{\kappa} - \int_0^{\kappa} f'(K)\mathbf{1}_{\{S_T<K\}}\,\mathrm{d}K \\
&\quad - f(K)\mathbf{1}_{\{S_T\geq K\}}\Big]_{\kappa}^{\infty} + \int_{\kappa}^{\infty} f'(K)\mathbf{1}_{\{S_T\geq K\}}\,\mathrm{d}K
\end{aligned}$$

$$\begin{aligned}
&= f(\kappa)\mathbf{1}_{\{S_T<\kappa\}} - [f'(K)(K - S_T)^+]_0^{\kappa} + \int_0^{\kappa} f''(K)(K - S_T)^+\,\mathrm{d}K \\
&\quad + f(\kappa)\mathbf{1}_{\{S_T\geq\kappa\}} - [f'(K)(S_T - K)^+]_{\kappa}^{\infty} + \int_{\kappa}^{\infty} f''(K)(S_T - K)^+\,\mathrm{d}K \\
&= f(\kappa) + f'(\kappa)[(S_T - \kappa)^+ - (\kappa - S_T)^+] \\
&\quad + \int_0^{\kappa} f''(K)(K - S_T)^+\,\mathrm{d}K + \int_{\kappa}^{\infty} f''(K)(S_T - K)^+\,\mathrm{d}K.
\end{aligned}$$

Chapter 2

Lévy Processes and Stochastic Volatility Models

The local volatility model for the asset price process is the most direct extension of the Black-Scholes model of Geometric Brownian motion with constant drift and volatility. In the local volatility model, the instantaneous volatility is a function of asset price and time. This serves as an attempt to model the skews and smiles that exist in implied volatility surfaces derived from traded option prices. The Dupire formula (1.6) and Andersen and Brotherton-Ratcliffe formula (1.7) relate the local volatilities to traded option prices and implied volatilities, respectively. The local volatility model is a self-consistent model that is capable of producing the implied volatility surface observed in the market place, thus providing an effective and direct solution to the calibration problem. However, the local volatility model has limited scope and exhibits several significant weaknesses. It leads to unreasonable skew dynamics of the asset price process. Also, Greek values calculated under the local volatility framework are generally not consistent with what are empirically observed. Indeed, volatility is widely accepted to be a latent (hidden) stochastic process itself. Therefore, the local volatility model tends to underestimate the volatility of volatility.

Asset return time series data invariably exhibit volatility clustering phenomena, where large changes in prices tend to cluster together while small changes tend to be followed by small changes. The variance of returns is stochastic, displacing alternative turbulent and quiet periods of varying time lengths and fluctuation magnitudes. In diffusion-based stochastic volatility models, asset price volatility is assumed to be driven by a Brownian motion, which is correlated with the asset price process. The stochastic volatility model can offer enough distribution structures such that the negative correlation (so-called leverage effect) between volatility innovations and underlying asset returns serves to control the level of skewness. Also, the volatility variation impacts the level of kurtosis. Heston (1993) pioneers the inclusion of the CIR/square root process for the instantaneous variance process, where positiveness of variance is guaranteed. Under the stochastic volatility framework, only the joint process of asset price and volatility is a Markov process while the asset price itself is no longer Markovian. The diffusion-based stochastic volatility models can produce reasonable profile of implied volatilities at a given maturity. However, they may fail to yield a realistic term structure of implied volatilities.

The proper modeling of asset price dynamics should include jumps since asset prices are seen to have occasions of large and rapid price movements resembling jumps. It has been common to add jumps to diffusion type asset price models to

DOI: 10.1201/9781003263524-2

model the jump feature. Merton (1976) initiates the addition of a jump component with Gaussian jump size distribution to the classical geometric Brownian motion in the asset price process. The later versions of stochastic volatility models add jumps in either the asset price process or both processes of the asset price and volatility. Bakshi *et al.* (1997) show that adding the jump feature to the stochastic volatility model improves pricing and hedging of equity options, especially for short-term options and relatively long-term options. Discontinuous jumps cause the negative skewness and excess kurtosis to exist in option prices. The presence of a downward skew is attributed to the fear of large negative jumps in equity prices by market participants. In currency options, symmetric jumps are more often to occur so smile patterns prevail. By analyzing S&P 500 and Nasdaq 100 index options, Eraker *et al.* (2003) performed empirical studies on the impact of jumps in volatility and return. They found that models with only diffusive stochastic volatility and jumps in returns are misspecified since they do not have a component that drives the conditional volatility of returns, which is rapidly moving. Jumps in returns can generate large price moves but the impact of a jump is transient. A jump in returns at the current time has no impact on the future distribution of returns. Diffusive stochastic volatility can only increase gradually. Jumps in volatility can provide a rapidly moving but persistent factor that drives the conditional volatility of returns.

In our later discussion on pricing variance derivatives, we focus on the use of stochastic volatility models with jumps. The addition of stochastic volatility as an additional state variable leads to market incompleteness in the option pricing model. The market price of risk under the risk neutral valuation framework must be calibrated to market option prices.

The jump-diffusion stochastic volatility models exclusively use the Compound Poisson processes to model jumps. This is referred to as finite activity jump processes since a finite number of jumps occur over a finite time period. However, empirical observations on asset prices reveal many small jump on a finite time scale. This motivates another approach that employs the Lévy processes to model asset price dynamics since the Lévy framework allows more general jump structures. In particular, infinite activity jumps are allowed to occur within any finite time interval. Geman *et al.* (2001) even argue that price processes must have a jump component but not necessary to include diffusion component. The homogeneous Lévy processes observe independence of increments, thus imposing restrictions on capturing the term structure of the risk neutral variance, skewness, and kurtosis of asset returns observed in the market. By employing time change on common Lévy processes with stochastic time clock, it is possible to incorporate stochastic and mean reverting volatilities as well as leverage effects on asset returns.

In summary, feasible and reliable asset returns models should reflect the following three basic features. First, volatility of returns varies stochastically over time. Second, returns and their volatilities are correlated, typically negative correlation for equity returns. Third, asset prices and volatilities exhibit jumps. Both the stochastic volatility and jump features reflect the leptokurtic feature in asset returns. The negative correlation between asset returns and volatility explains the skewness feature in implied volatilities.

In this chapter, we start with review of the compound Poisson processes, jump-diffusion models and basic properties of Lévy processes. The classical Merton's Gaussian jump-diffusion model (Merton, 1976) and Kou's double exponential jump model (Kou, 2002) are seen to be finite activity Lévy jump models. We show the relation between infinitively divisible distribution and Lévy process. The key theoretical properties are the Lévy-Khintchine representation and Lévy-Itô decomposition theorem. We then introduce the notion of time change in infinite activity Lévy processes and show how to obtain several prototype Lévy models via subordination of a Brownian motion, like the Variance Gamma model and Normal Inverse Gaussian model. We illustrate the versatility of applying time change via stochastic activity rate to generate stochastic volatility in exponential Lévy models. Lastly, we present formulation and analytic properties of stochastic volatility models with jumps. In particular, we derive analytic formulas for the joint moment generating functions for the affine stochastic volatility models and 3/2-model.

2.1 Compound Poisson process

In this section, we consider a random process with pure jumps. Combination of jumps and diffusion in asset price processes will be discussed in later sections. We start our discussion with modeling of the random jump times, followed by random jump sizes.

Let $(\tau_i)_{i\geq 1}$ be a sequence of independent and identically distributed (iid) exponential random variables with parameter $\lambda > 0$, where $\tau_i \sim \text{Exp}(\lambda)$. We consider a jump model whose jumps occur randomly and τ_i is the time interval of the ith jump from the previous one. The random time T_n of the nth jump is then given by

$$T_n = \sum_{i=1}^{n} \tau_i, \tag{2.1}$$

with the convention $T_0 = 0$. In other words, $(T_n - T_{n-1})_{n\geq 1}$ is an independent and identically distributed (iid) sequence of exponential variables, where

$$E[T_n - T_{n-1}] = E[\tau_n] = \frac{1}{\lambda}, \quad n = 1, 2, \ldots. \tag{2.2}$$

The exponential distribution $\tau \sim \text{Exp}(\lambda)$, $\lambda > 0$, has the important "memoryless" property. For any $t > 0$ and $s > 0$, we observe

$$P[\tau > t+s | \tau > t] = \frac{\int_{t+s}^{\infty} \lambda e^{-\lambda y}\, dy}{\int_t^{\infty} \lambda e^{-\lambda y}\, dy} = \frac{e^{-\lambda(t+s)}}{e^{-\lambda t}} = e^{-\lambda s} = P[\tau > s]. \tag{2.3}$$

Treating τ as random time, the distribution of $\tau - t$ given $\tau > t$ is the same as the distribution of τ itself.

2.1.1 Poisson process

The Poisson process N_t is defined by

$$N_t = \sum_{n\geq 1} \mathbf{1}_{\{t\geq T_n\}}, \tag{2.4}$$

which is a counting process that counts the number of random jumps occurring between 0 and t. Note that N_t is a non-negative integer-valued process that is piecewise constant with jump magnitude of one.

With reference to (2.1) and (2.2), $\lambda > 0$ is called the intensity parameter of the Poisson process N_t. This would mean that on average there are λ jumps over one unit of time. Alternatively, the mean number of jumps over time t is λt. One may extend the Poisson process to allow time-dependent intensity or even random intensity (known as the Cox process).

We observe the following properties of N_t:

1. The sample paths $t \mapsto N_t$ are right continuous with left limit (commonly called the càdlàg process).

2. The discontinuities of N_t occur at the jump times T_n, $n = 1,2,\ldots$. For $t > 0$, we observe $P[T_n = t] = 0$, so almost all trajectories of N_t are continuous at t.

3. For any $t > 0$, N_t follows a Poisson distribution with parameter λt, where

$$P[N_t = n] = e^{-\lambda t}\frac{(\lambda t)^n}{n!}, \quad n = 0,1,2,\ldots. \tag{2.5}$$

   As a result, we deduce that $E[N_t] = \lambda t$ and $\text{var}(N_t) = \lambda t$.

4. N_t has independent increments: for any $t_1 < t_2 < \cdots < t_{n-1} < t_n$, the increments $N_{t_n} - N_{t_{n-1}}, \ldots, N_{t_2} - N_{t_1}, N_{t_1}$ are independent random variables.

5. The increments of N_t are stationary: for any $t > s$, $N_t - N_s$ has the same distribution as N_{t-s}.

Both properties (4) and (5) are consequences of the memoryless property of the exponential random variable $\text{Exp}(\lambda)$ [see (2.3)]. Conversely, suppose a counting process has independent and stationary increments, then it must be a Poisson process. In other words, the only counting processes that have independent and stationary increments are Poisson processes. A proof of the above statement can be found in Lemma 2.1 in Cont and Tankov (2003).

Compensated Poisson process

Let $\mathcal{F}_t$ be the filtration generated by the Poisson process N_t with intensity λ. By virtue of independence of increments, for $t > s \geq 0$, we observe that

$$E[N_t|\mathcal{F}_s] = E[N_t - N_s] + N_s = \lambda(t-s) + N_s. \tag{2.6}$$

As a result, we deduce that $N_t - \lambda t$ is a martingale, and it is called the compensated Poisson process. Note that $N_t - \lambda t$ is no longer an integer-valued process.

2.1.2 Random jump sizes

Asset price processes exhibit random jump sizes, not limited to jump of unit magnitude. To model an asset price process with jumps, we generalize a Poisson jump process that allows random jump sizes instead of unit jump magnitude in the standard Poisson process. We let the jump sizes $(Y_i)_{i\geq 1}$ be iid sequence of square integrable random variables with jump size distribution $f(\mathrm{d}y)$ on $\mathbb{R}$ and independent of the Poisson process N_t, where

$$P[Y_i \in [a,b]] = f([a,b]) = \int_a^b f(\mathrm{d}y), \quad i = 1,2,\ldots. \tag{2.7}$$

A compound Poisson process X_t with intensity $\lambda > 0$ and jump size distribution $f(\mathrm{d}y)$ is defined by

$$X_t = \sum_{i=1}^{N_t} Y_i, \tag{2.8}$$

with the convention $X_0 = 0$. At the jump time T_i, we have

$$X_{T_i} = X_{T_i^-} + Y_i, \quad i = 1,2,\ldots,N_t. \tag{2.9}$$

It is seen that the sample paths of X_t are càdlàg piecewise constant and the jump times of X_t have the same law as the jump times of the Poisson process N_t. Let $M_{X_t}(\alpha)$ denote the moment generating function (mgf) of X_t, which can be shown to admit the following integral representation:

$$M_{X_t}(\alpha) = E[e^{\alpha X_t}] = \exp\left(\lambda t \int_{-\infty}^{\infty} (e^{\alpha y} - 1) f(\mathrm{d}y)\right), \quad \alpha \in \mathbb{R}. \tag{2.10a}$$

To show the above result, we consider

$$\begin{aligned} E[e^{\alpha X_t}] &= E\left[\exp\left(\alpha \sum_{i=1}^{N_t} Y_i\right)\right] \\ &= \sum_{n=0}^{\infty} E\left[\exp\left(\alpha \sum_{i=1}^{N_t} Y_i\right) \middle| N_t = n\right] P[N_t = n] \\ &= e^{-\lambda t} \sum_{n=0}^{\infty} \frac{(\lambda t)^n}{n!} \prod_{i=1}^{n} E[e^{\alpha Y_i}]. \end{aligned}$$

Since Y_i are iid jumps and expressing $e^{-\lambda t} = \exp\left(-\lambda t \int_{-\infty}^{\infty} f(\mathrm{d}y)\right)$, we have

$$\begin{aligned} E[e^{\alpha X_t}] &= e^{-\lambda t} \sum_{n=0}^{\infty} \frac{(\lambda t)^n}{n!} \left[\int_{-\infty}^{\infty} e^{\alpha y} f(\mathrm{d}y)\right]^n \\ &= \exp\left(\lambda t \left[\int_{-\infty}^{\infty} e^{\alpha y} f(\mathrm{d}y) - \int_{-\infty}^{\infty} f(\mathrm{d}y)\right]\right) \\ &= \exp\left(\lambda t \int_{-\infty}^{\infty} (e^{\alpha y} - 1) f(\mathrm{d}y)\right). \end{aligned}$$

By virtue of the stationary property of N_t, for any $t \in [0,T]$, we can obtain the generalized version of (2.10a) as follows:

$$E\left[e^{\alpha(X_T - X_t)}\right] = \exp\left(\lambda(T-t)\int_{-\infty}^{\infty}(e^{\alpha y}-1)f(\mathrm{d}y)\right). \tag{2.10b}$$

Let $M_Y(\alpha)$ denote the mgf of the random jump distribution Y, where

$$M_Y(\alpha) = \int_{-\infty}^{\infty} e^{\alpha y} f(\mathrm{d}y).$$

It is seen that $M_{X_t}(\alpha)$ and $M_Y(\alpha)$ are related by

$$M_{X_t}(\alpha) = \exp(\lambda t[M_Y(\alpha)-1]). \tag{2.10c}$$

Using the mgf of X_t, the mean of X_t can be found to be

$$E[X_t] = \frac{\mathrm{d}}{\mathrm{d}\alpha}E[e^{\alpha X_t}]\bigg|_{\alpha=0} = \lambda t \int_{-\infty}^{\infty} y\, f(\mathrm{d}y) = \lambda t E[Y]. \tag{2.11}$$

The compensated compound Poisson process is defined by

$$X_t - E[X_t] = X_t - \lambda t E[Y],$$

which is a martingale with respect to the filtration generated by N_t and Y.

To calculate the variance of X_t, we consider

$$\begin{aligned}\mathrm{var}(X_t) &= E[\mathrm{var}(X_t|N_t)] + \mathrm{var}(E[X_t|N_t])\\ &= E[N_t\mathrm{var}(Y)] + \mathrm{var}(N_t E[Y])\\ &= E[N_t]\mathrm{var}(Y) + \mathrm{var}(N_t)E[Y]^2.\end{aligned}$$

Since $\mathrm{var}(N_t) = E[N_t]$ for the Poisson process N_t, so

$$\mathrm{var}(X_t) = E[N_t][\mathrm{var}(Y) + E[Y]^2] = E[N_t]E[Y^2].$$

In terms of the order moments of Y, the ratio of $\mathrm{var}(X_t)$ to $E[X_t]$ is given by

$$\frac{\mathrm{var}(X_t)}{E[X_t]} = \frac{E[N_t]E[Y^2]}{E[N_t]E[Y]} = \frac{E[Y^2]}{E[Y]}. \tag{2.12}$$

Monte Carlo simulation

Simulation of the compound Poisson process can be performed by first simulating the random arrival times of the Poisson jumps over $[0,t]$. This is based on the exponential distribution of the next arrival time interval τ of successive jumps. The simulated value of N_t is the number of jumps realized in simulation. It is then followed by the simulation of N_t random jumps with iid jump distribution Y. Based on (2.8), we add these simulated jumps together to obtain the simulated value of X_t.

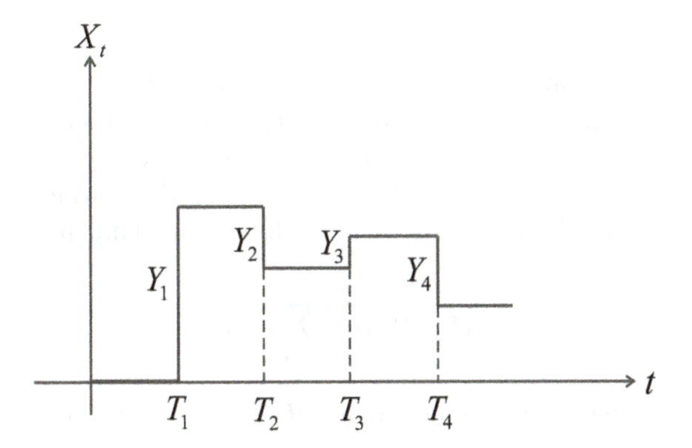

FIGURE 2.1: Simulation path of $X_t = \sum_{i=1}^{N_t} Y_i$, $X_0 = 0$. The random jump times $T_1, T_2, \ldots$ are obtained by simulation of the Poisson process N_t with intensity λ. The iid jump sizes Y_i, $i = 1, 2, \ldots, N_t$ are simulated based on the jump distribution $f(\mathrm{d}y)$.

2.1.3 Stochastic integration

Let X_t be a compound Poisson process with intensity $\lambda > 0$ as defined in (2.8). Let ϕ_t be a stochastic process adapted to the natural filtration generated by X_t, observing the technical condition

$$E\left[\int_0^T |\phi_t| \, \mathrm{d}t\right] < \infty, \quad T > 0.$$

By observing $\Delta X_t = Y_{N_t} \Delta N_t$, the stochastic integral of ϕ_t with respect to X_t is defined by

$$\int_0^T \phi_t \, \mathrm{d}X_t = \int_0^T \phi_t Y_{N_t} \, \mathrm{d}N_t = \sum_{i=1}^{N_t} \phi_{T_i} Y_{T_i}, \tag{2.13}$$

where T_i is the increasing family of jump times of N_t. Taking $\phi_t = 1$, the stochastic integral representation of X_t can be expressed as

$$X_t = X_0 + \sum_{i=1}^{N_t} Y_{T_i} = X_0 + \int_0^t Y_{N_s} \, \mathrm{d}N_s. \tag{2.14}$$

We may formally write $\mathrm{d}X_t = Y_t \, \mathrm{d}N_t$.

Let $g(t)$ be a function adapted to X_t. Similar to the derivation of (2.10a), we can establish

$$E\left[\exp \int_0^T g(t) \, \mathrm{d}X_t\right] = \exp\left(\lambda \int_0^T \int_{-\infty}^{\infty} \left[e^{yg(t)} - 1\right] f(\mathrm{d}y) \, \mathrm{d}t\right), \tag{2.15}$$

where λ is the intensity of X_t and $f(\mathrm{d}y)$ is the jump size distribution of Y.

2.1.4 Jump measure and Lévy measure

Recall that the compound Poisson process is defined as summation of iid Y_i, $i = 1, 2, \ldots, N_t$ [see (2.8)]. The jump measure $J_X(A, [t_1, t_2])$ of a compound Poisson process X counts the number of jumps of X within the time interval $[t_1, t_2]$ such that their jump sizes are in A. We define the Dirac function $\delta_{Y_i(A)}$, which equals 1 when $Y_i \in A$ and equals 0 if otherwise. It is seen that the random jump measure is given by

$$J_X(A, [0,t]) = \sum_{i=1}^{N_t} \delta_{Y_i(A)}. \tag{2.16}$$

The expectation of the random jump measure $J_X([0,t], A)$ is given by

$$\begin{aligned} E[J_X(A, [0,t])] &= E\left[\sum_{i=1}^{N_t} \delta_{Y_i(A)}\right] = \sum_{n=1}^{\infty} P[N_t = n] \sum_{i=1}^{n} P[\delta_{Y_i(A)}] \\ &= e^{-\lambda t} \sum_{n=1}^{\infty} \frac{(\lambda t)^n}{n!} n P[Y_i \in A] = \lambda t \int_A f(\mathrm{d}y), \end{aligned} \tag{2.17}$$

where $f(\mathrm{d}y)$ is the jump size distribution of Y_i.

The Lévy measure $\Pi(A)$ of X is defined to be the expected number of jumps over $[0,1]$ (equivalently, per unit time due to stationary property) whose jump size lies within A. This definition of Lévy measure can be generalized to Lévy processes (to be discussed in later sections). As deduced from (2.17), the differential form of the Lévy measure can be expressed as

$$\Pi(\mathrm{d}x) = \lambda f(\mathrm{d}x), \tag{2.18}$$

where $\Pi(\mathrm{d}x)$ is the expected number of jumps per unit time such that $\Delta X \in [x, x + \mathrm{d}x]$. Note that the Lévy measure is not a probability measure since

$$\int_{\mathbb{R}} \Pi(\mathrm{d}x) = \lambda \int_{\mathbb{R}} f(\mathrm{d}x) = \lambda \neq 1.$$

In terms of $J_X(\mathrm{d}x, \mathrm{d}t)$, one can express the compound Poisson process X_t in the following integral formula:

$$X_t = \sum_{\substack{s \in [0,t] \\ \Delta X_s \neq 0}} \Delta X_s = \int_0^t \int_{\mathbb{R}} x \, J_X(\mathrm{d}x, \mathrm{d}s). \tag{2.19}$$

We may visualize the jump measure $J_X(\mathrm{d}x, \mathrm{d}s)$ as a Poisson random measure with intensity measure

$$\mu(\mathrm{d}x, \mathrm{d}t) = \Pi(\mathrm{d}x)\, \mathrm{d}t = \lambda f(\mathrm{d}x)\, \mathrm{d}t. \tag{2.20}$$

The formal definition and properties of the Poisson random measure can be found in

Cont and Tankov (2003). For a function $g(X_t, t)$ that satisfies some mild regularity condition, the integral formula (2.19) can be generalized as follows:

$$\sum_{\substack{s \in [0,t] \\ \Delta X_s \neq 0}} g(\Delta X_s, s) = \int_0^t \int_{\mathbb{R}} g(x,s)\, J_X(\mathrm{d}x, \mathrm{d}s). \tag{2.21a}$$

For example, suppose $g(x,s) = x^2$, then

$$\int_0^t \int_{\mathbb{R}} x^2\, J_X(\mathrm{d}x, \mathrm{d}s) = \sum_{\substack{s \in [0,t] \\ \Delta X_s \neq 0}} (\Delta X_s)^2, \tag{2.21b}$$

which gives the sum of squares of the jumps of X over $[0,t]$.

We define the compensated jump measure $\widetilde{J}_X(\mathrm{d}x, \mathrm{d}t)$ by

$$\widetilde{J}_X(\mathrm{d}x, \mathrm{d}t) = J_X(\mathrm{d}x, \mathrm{d}t) - \Pi(\mathrm{d}x)\, \mathrm{d}t. \tag{2.22}$$

It can be shown that

$$\begin{aligned} M_t &= \int_0^t \int_{\mathbb{R}} g(x,s)\, \widetilde{J}_X(\mathrm{d}x, \mathrm{d}s) \\ &= \int_0^t \int_{\mathbb{R}} g(x,s)\, J_X(\mathrm{d}x, \mathrm{d}s) - \int_0^t \int_{\mathbb{R}} g(x,s)\, \Pi(\mathrm{d}x)\, \mathrm{d}s \end{aligned}$$

observes $E[M_t] = 0$, so M_t is a martingale.

2.2 Jump-diffusion models

Asset price processes and exchange rate processes are seen to exhibit jumps due to frequent arrivals of market shocks. Under the diffusion-type models, the chance of a significant move in asset price over a short time period is very small. Empirical studies show that options with short maturities are traded at higher prices compared to option prices derived under diffusion models. This is because practitioners have already priced in the possibility of finite jump over a short time period. It is now widely accepted that asset price models should include jump components on top of the diffusion mechanism.

There are various financial issues and technical tools to be considered in option pricing under the jump-diffusion models. First, we derive the Itô formula with the inclusion of jumps. Market completeness will not hold with additional jump risk when hedging is performed using only the underlying asset, so risk neutral measures would not be unique. We derive the martingale condition for the asset price process under a risk neutral measure. We apply these techniques to analyze the model formulation and pricing procedure of two pioneering jump-diffusion models: Merton's (1976) model with Gaussian jumps and Kou's (2002) model with double exponential jumps.

A jump-diffusion process X_t is a combination of Brownian motion with drift and compound Poisson process, which can be represented by

$$X_t = \mu t + \sigma W_t + \sum_{i=1}^{N_t} Y_i, \tag{2.23}$$

where μ is the drift, σ is the volatility of the Brownian diffusion, W_t is the standard Brownian motion, N_t is the Poisson process counting the number of jumps of X_t, and Y_i are jump sizes (assumed to be iid). The Poisson process N_t and Brownian motion W_t are taken to be independent processes.

The characteristic function of the jump-diffusion process X_t is defined as the Fourier transform of the density function of X_t, where

$$\phi_{X_t}(u) = \mathbb{E}[e^{iuX_t}], \quad t \geq 0, u \in \mathbb{R}. \tag{2.24}$$

Let $f(\mathrm{d}y)$ be the jump size distribution of Y_i. By following a similar derivation procedure as that for the mgf of a compound Poisson process [see (2.10a)], we obtain

$$E[e^{iuX_t}] = \exp\left(t\left[i\mu u - \frac{\sigma^2 u^2}{2} + \lambda \int_{-\infty}^{\infty} (e^{iux} - 1) f(\mathrm{d}y)\right]\right). \tag{2.25}$$

2.2.1 Itô's formula

We would like to derive the Itô formula for the jump-diffusion process defined in (2.23). Let $V_t = f(X_t)$ be twice differentiable function. Between two jump times T_i and T_{i+1}, X_t evolves continuously according to

$$\mathrm{d}X_{t^-} = \mathrm{d}X_t^c = \mu\ \mathrm{d}t + \sigma\ \mathrm{d}W_t.$$

The Itô formula for the diffusion part of the jump-diffusion process gives

$$V_{T_{i+1}^-} - V_{T_i} = \int_{T_i}^{T_{i+1}} f'(X_t)\ \mathrm{d}X_t^c + \int_{T_i}^{T_{i+1}} \frac{\sigma^2}{2} f''(X_t)\ \mathrm{d}t. \tag{2.26}$$

If a jump ΔX_t occurs at t, then V_t changes by $f(X_{t^-} + \Delta X_t) - f(X_{t^-})$. The total change in $V_t = f(X_t)$ over $[0,t]$ is given by the sum of changes due to diffusion and jumps, so we obtain

$$\begin{aligned} f(X_t) - f(X_0) = & \int_0^t f'(X_s)\ \mathrm{d}X_s^c + \int_0^t \frac{\sigma^2}{2} f''(X_s)\ \mathrm{d}s \\ & + \sum_{\substack{0 \leq s \leq t \\ \Delta X_s \neq 0}} [f(X_{s^-} + \Delta X_s) - f(X_{s^-})]. \end{aligned} \tag{2.27}$$

By observing $dX_s^c = dX_s - \Delta X_s$, we can deduce an alternative Ito's formula in terms of the jump measure J_X:

$$\begin{aligned} f(X_t) - f(X_0) &= \int_0^t f'(X_{s^-})\, dX_s + \int_0^t \frac{\sigma^2}{2} f''(X_s)\, ds \\ &\quad + \sum_{\substack{0 \le s \le t \\ \Delta X_s \neq 0}} [f(X_{s^-} + \Delta X_s) - f(X_{s^-}) - \Delta X_s f'(X_{s^-})] \\ &= \int_0^t f'(X_{s^-})\, dX_s + \int_0^t \frac{\sigma^2}{2} f''(X_s)\, ds \\ &\quad + \int_0^t \int_{\mathbb{R}} [f(X_{s^-} + x) - f(X_{s^-}) - x f'(X_{s^-})]\, J_X(dx, ds). \end{aligned} \tag{2.28}$$

2.2.2 Asset price process: Geometric Brownian motion with compound Poisson jumps

As an extension to the assumption of Geometric Brownian motion in asset price dynamics (Black and Scholes, 1973; Merton, 1973), we consider the inclusion of Poisson type proportional jump Y_t on the asset price S_t at jump time t, where $S_t = Y_t S_{t^-}$. The contribution to dS_t when there is an occurrence of proportional jump Y_t is given by $dS_t = (Y_t - 1)S_{t^-}$. Let Y_t be a compound Poisson process with intensity λ and iid increments Y_i, $i = 1, 2, \ldots$. We write $\mu_J = E[Y_i]$ for all i. Let μ be the drift and σ be the volatility of the continuous component modeled by a Geometric Brownian motion. The compound Poisson process and Geometric Brownian motion are assumed to be independent.

The stochastic differential equation for S_t with the inclusion of compound Poisson jumps under the statistical measure is given by

$$dS_t = \mu S_t\, dt + \sigma S_t\, dW_t + (Y_t - 1)S_{t^-}\, dN_t. \tag{2.29}$$

Taking $f(S_t) = \ln S_t$ in the Itô formula (2.27), we obtain

$$\ln S_t - \ln S_0 = \left(\mu - \frac{\sigma^2}{2}\right) t + \sigma W_t + \sum_{i=1}^{N_t} Y_i. \tag{2.30}$$

Note that $\ln S_t$ assumes the same form of jump-diffusion dynamics as depicted in (2.23), except with the adjustment of the drift from μ to $\mu - \frac{\sigma^2}{2}$. Between two consecutive jumps, the asset price process S_t behaves like the usual Geometric Brownian motion.

Martingale condition

Under a risk neutral measure Q, the model parameters μ, λ and μ_J must be constrained by the martingale condition such that the expected rate of return of S_t

under Q is $r-q$, where r is the riskless interest rate and q is the dividend yield. The martingale condition dictates

$$E_Q\left[\frac{S_t}{S_0}\right] = e^{(r-q)t}. \tag{2.31}$$

We rewrite (2.30) in the form:

$$S_t = S_0 e^{\left(\mu-\frac{\sigma^2}{2}\right)t+\sigma W_t} \prod_{i=1}^{N_t} Y_i, \tag{2.32a}$$

and note that

$$E_Q\left[e^{\left(\mu-\frac{\sigma^2}{2}\right)t+\sigma W_t}\right] = e^{\mu t}. \tag{2.32b}$$

By virtue of independence of N_t and W_t, and observing that Y_i are iid, we obtain

$$\begin{aligned} E_Q[S_t] &= S_0 e^{\mu t} E_Q\left[\prod_{i=1}^{N_t} Y_i\right] = S_0 e^{\mu t} \sum_{n=0}^{\infty} E_Q\left[\prod_{i=1}^{N_t} Y_i \middle| N_t = n\right] P[N_t = n] \\ &= S_0 e^{\mu t} \sum_{n=0}^{\infty} e^{-\lambda t} \frac{(\lambda t)^n}{n!} \mu_J^n = S_0 e^{\mu t + \lambda t(\mu_J - 1)}. \end{aligned} \tag{2.33}$$

The martingale condition (2.31) dictates that the drift and jump parameters must be constrained by

$$\mu + \lambda(\mu_J - 1) = \mu + \lambda E[Y-1] = r - q. \tag{2.34}$$

Substituting the constrained condition on drift into (2.29), the dynamic equation of S_t under Q is then given by

$$\mathrm{d}S_t = \{(r-q) - \lambda E[Y_t - 1]\} S_t\, \mathrm{d}t + \sigma S_t\, \mathrm{d}W_t + (Y_t - 1) S_{t^-}\, \mathrm{d}N_t. \tag{2.35}$$

With the inclusion of jumps, the risk neutral drift is reduced by the compensated amount $\lambda E[Y_t - 1] = \lambda(\mu_J - 1)$.

Market incompleteness due to jumps

In the dynamic equation for $\mathrm{d}S_t$ in (2.29), there are two stochastic differential terms associated with $\mathrm{d}W_t$ and $\mathrm{d}N_t$, corresponding to the risks from diffusion and jumps. If we use the underlying asset as the only hedging instrument, we cannot hedge both types of risks. In this sense, the market is said to be incomplete. To illustrate the argument, we consider the standard hedging procedure of short holding of an option and long holding of Δ units of the underlying asset. Let $V(S,t)$ denote the time-t option price with asset price level S, so the value of the hedging portfolio is

$$\Pi(S,t) = -V(S,t) + \Delta S.$$

The differential change of $\Pi(S,t)$ over differential time interval dt is given by

$$\begin{aligned}
d\Pi(S,t) &= -dV(S,t) + \Delta\, dS \\
&= -\frac{\partial V}{\partial t}\, dt - (\mu S\, dt + \sigma S\, dW_t)\frac{\partial V}{\partial S} - \frac{\sigma^2}{2}S^2\frac{\partial^2 V}{\partial S^2}\, dt \\
&\quad - [V(YS,t) - V(S,t)]\, dN_t + \Delta[\mu S\, dt + \sigma S\, dW_t + (Y-1)S\, dN_t]. \qquad (2.36)
\end{aligned}$$

We observe that there is no choice of Δ that makes this portfolio risk-free. With the usual choice of the hedge ratio $\Delta = \frac{\partial V}{\partial S}$, we can eliminate the randomness due to the Brownian motion but not jump risk. Due to loss of market completeness, risk neutral measures would not be unique.

There are numerous methods to choose a martingale measure that is equivalent to the statistical measure. The common methods include the Esscher transform, minimum entropy approach and rational expectations equilibrium. When the drift is nonzero, one may obtain an equivalent martingale measure by changing the drift. In an asset price model with pure jumps, there is no drift term so we cannot change the drift. Cont and Tankov (2003) illustrate a systematic procedure to use the Esscher transform to find a martingale measure. For the general cases, they show how to perform calibration of market prices of risks using traded option prices to find a martingale measure that is both equivalent to the statistical measure and reproducing traded option prices as well.

In his pioneering paper on option pricing under the jump-diffusion framework, Merton (1976) proposes a specific approach of assuming zero risk premia for jumps. He assumes that the jump risks are diversifiable so that these risks should not be priced into the option. Under this assumption, the beta in the Capital Asset Pricing Model is zero. Since the expected return on any zero beta securities is equal to the riskless interest rate r, so we have

$$E[d\Pi_t] = r\Pi_t\, dt. \qquad (2.37)$$

With $\Delta = \frac{\partial V}{\partial S}$ in (2.36) and $E[(Y-1)S\, dN_t] = \lambda E_Y[(Y-1)]S\, dt$, where E_Y is expectation taken with respect to the distribution of the jump ratio Y, we obtain

$$\begin{aligned}
&-\left(\frac{\partial V}{\partial t} + \frac{\sigma^2}{2}S^2\frac{\partial^2 V}{\partial S^2}\right) dt - \lambda E_Y[V(YS,t) - V(S,t)]\, dt + \lambda E_Y[(Y-1)]S\frac{\partial V}{\partial S}\, dt \\
&= r\left(S\frac{\partial V}{\partial S} - V\right) dt.
\end{aligned}$$

Rearranging the terms, we obtain the following integro-differential equation for option pricing:

$$\frac{\partial V}{\partial t} + (r - \lambda E_Y[Y-1])\, S\frac{\partial V}{\partial S} + \frac{\sigma^2}{2}S^2\frac{\partial^2 V}{\partial S^2} - rV + \lambda E_Y[V(YS,t) - V(S,t)] = 0, \qquad (2.38)$$

where the $E_Y[\cdot]$ terms can be computed by performing integration over the distribution of the jump ratio Y.

The specific jump-diffusion model is specified by the distribution of jump sizes. The two popular choices are the Gaussian jumps: $Y_i \sim N(\mu_J, \sigma_J^2)$ and exponential jumps. These two models are coined as Merton's model (1976) and Kou's model (2002), respectively.

2.2.3 Merton's model with Gaussian jumps

Merton (1976) assumes that the jump size of the logarithm of asset price, $X_t = \ln S_t$, follows the Gaussian jump distribution with mean μ_J and variance σ_J^2. The absolute jump in $\ln S_t$ is $y_i = \ln Y_i$, where Y_i is the proportional jump in S_t. The density function $f_J(y)$ of the Gaussian jump distribution is given by

$$f_J(y) = \frac{1}{\sqrt{2\pi\sigma_J^2}} \exp\left(-\frac{(y-\mu_J)^2}{2\sigma_J^2}\right), \quad -\infty < y < \infty. \tag{2.39}$$

This is equivalent to say that the relative price jump size $Y_t - 1$ is lognormally distributed with mean

$$E[Y_t - 1] = e^{\mu_J + \frac{\sigma_J^2}{2}} - 1$$

and variance

$$\mathrm{var}(Y_t - 1) = e^{2\mu_J + \sigma_J^2}\left(e^{\sigma_J^2} - 1\right).$$

According to (2.34), the risk neutral drift μ_{M} under Merton's jump-diffusion model is found to be

$$\mu_{\mathrm{M}} = r - q - \lambda E[Y_t - 1] = r - q - \lambda\left(e^{\mu_J + \frac{\sigma_J^2}{2}} - 1\right). \tag{2.40}$$

The density function f_{M} of the Merton jump-diffusion process can be expressed in an infinite series expansion. First, we observe that

$$\begin{aligned} f_{\mathrm{M}}(x)\,\mathrm{d}x &= P[X_t \in (x, x+\mathrm{d}x)] \\ &= \sum_{n=0}^{\infty} P[X_t \in (x, x+\mathrm{d}x) | N_t = n] P[N_t = n], \end{aligned} \tag{2.41}$$

where

$$P[N_t = n] = e^{-\lambda t}\frac{(\lambda t)^n}{n!}.$$

When $N_t = n$, by virtue of independence of N_t and W_t, we have

$$\mu_{\mathrm{M}} t + \sigma W_t + \sum_{i=1}^{n} y_i \sim N(\mu_{\mathrm{M}} t + n\mu_J, \sigma^2 t + n\sigma_J^2),$$

where $N(\mu, \sigma^2)$ denotes the normal distribution with mean μ and variance σ^2. Combining the results together, the density function is found to be

$$f_{\mathrm{M}}(x) = e^{-\lambda t} \sum_{n=0}^{\infty} \frac{(\lambda t)^n}{n!} \frac{\exp\left(-\frac{(x-\mu_{\mathrm{M}}t-n\mu_J)^2}{2(\sigma^2 t+n\sigma_J^2)}\right)}{\sqrt{2\pi(\sigma^2 t + n\sigma_J^2)}}. \tag{2.42}$$

Once we have found $f_{\mathrm{M}}(x)$, the price of the European call option $c_{\mathrm{M}}(S, \tau)$ with strike K and time to maturity τ under Merton's jump-diffusion model can be obtained as an infinite series in terms of the Black-Scholes call option price $c_{BS}(S, \tau; \sigma)$, where

$$c_{\mathrm{M}}(S, \tau) = e^{-\lambda\tau} \sum_{n=0}^{\infty} \frac{(\lambda\tau)^n}{n!} c_{BS}(S_n, \tau; \sigma_n), \tag{2.43}$$

where

$$S_n = S\exp\left(n\left(\mu_J + \frac{\sigma_J^2}{2}\right) - \lambda\tau\left(e^{\mu_J + \frac{\sigma_J^2}{2}} - 1\right)\right),$$

$$\sigma_n^2 = \sigma^2 + \frac{n\sigma_J^2}{\tau}.$$

The analytic forms of S_n and σ_n reflect the adjustment of the mean and variance in the density function $f_{\mathrm{M}}(x)$ for each term indexed with n in (2.42). In particular, note that μ_{M} is risk neutral adjusted due to jumps by subtracting $\lambda\left(e^{\mu_J + \frac{\sigma_J^2}{2}} - 1\right)$ [see (2.40)]. The mgf of Merton's jump-diffusion model is found to be

$$M_{\mathrm{M}}(\alpha) = \exp\left(\left[\mu_{\mathrm{M}}\alpha + \frac{\sigma^2\alpha^2}{2} + \lambda\left(e^{\mu_J\alpha + \frac{\sigma_J^2\alpha^2}{2}} - 1\right)\right]t\right). \tag{2.44a}$$

The corresponding cumulant generating function (cgf) is defined by

$$C_{\mathrm{M}}(\alpha) = \ln M_{\mathrm{M}}(\alpha) = \left[\mu_{\mathrm{M}}\alpha + \frac{\sigma^2\alpha^2}{2} + \lambda\left(e^{\mu_J\alpha + \frac{\sigma_J^2\alpha^2}{2}} - 1\right)\right]t. \tag{2.44b}$$

Using the cgf $C_{\mathrm{M}}(\alpha)$, the mean and variance of the Merton jump-diffusion process X_t are found to be

$$\left.\frac{\mathrm{d}C_{\mathrm{M}}(\alpha)}{\mathrm{d}\alpha}\right|_{\alpha=0} = E[X_t] = (\mu_{\mathrm{M}} + \lambda\mu_J)t, \tag{2.45a}$$

$$\left.\frac{\mathrm{d}^2C_{\mathrm{M}}(\alpha)}{\mathrm{d}\alpha^2}\right|_{\alpha=0} = \mathrm{var}(X_t) = [(\sigma_J^2 + \mu_J^2)\lambda + \sigma^2]t. \tag{2.45b}$$

2.2.4 Kou's model with exponential jumps

In the Kou jump-diffusion model, the distribution of jumps in the compound Poisson process is specified in terms of an asymmetric double exponential density, where

$$f_J(x) = p\eta_1 e^{-\eta_1 x}\mathbf{1}_{\{x>0\}} + (1-p)\eta_2 e^{\eta_2 x}\mathbf{1}_{\{x<0\}}, \tag{2.46}$$

where p is the probability of an upward jump, $1-p$ is the probability of downward jump, $\eta_1 > 0$ and $\eta_2 > 0$. We consider the combination of Brownian motion with drift μ and volatility σ together with the Kou double exponential jumps [see (2.23)]. The mgf and cgf of the Kou jump-diffusion model are found to be

$$\begin{aligned} M_{\text{Kou}}(\alpha) &= \exp\left(\left\{\mu\alpha + \frac{\sigma^2\alpha^2}{2} + \lambda\left[\int_0^\infty (e^{\alpha x}-1)p\eta_1 e^{-\eta_1 x}\,\mathrm{d}x \right.\right.\right. \\ &\qquad \left.\left.\left. + \int_{-\infty}^0 (e^{\alpha x}-1)(1-p)\eta_2 e^{\eta_2 x}\,\mathrm{d}x\right]\right\}t\right) \\ &= \exp\left(\left\{\mu\alpha + \frac{\sigma^2\alpha^2}{2} + \lambda\left[\frac{p\eta_1}{\eta_1-\alpha} + \frac{(1-p)\eta_2}{\eta_2+\alpha} - 1\right]\right\}t\right), \end{aligned} \tag{2.47a}$$

$$C_{\text{Kou}}(u) = \ln M_{\text{Kou}}(u) = \left\{\mu\alpha + \frac{\sigma^2\alpha^2}{2} + \lambda\left[\frac{p\eta_1}{\eta_1-\alpha} + \frac{(1-p)\eta_2}{\eta_2+\alpha} - 1\right]\right\}t, \tag{2.47b}$$

respectively. Using the cgf $C_{\text{Kou}}(\alpha)$, the mean and variance of the Kou jump-diffusion model are found to be

$$\left.\frac{\mathrm{d}C_{\text{Kou}}(\alpha)}{\mathrm{d}\alpha}\right|_{\alpha=0} = E[X_t] = \left[\mu + \frac{p\lambda}{\eta_1} - \frac{(1-p)\lambda}{\eta_2}\right]t, \tag{2.48a}$$

$$\left.\frac{\mathrm{d}^2C_{\text{Kou}}(\alpha)}{\mathrm{d}\alpha^2}\right|_{\alpha=0} = \operatorname{var}(X_t) = \left[\frac{2p\lambda}{\eta_1^2} + \frac{2(1-p)\lambda}{\eta_2^2} + \sigma^2\right]t. \tag{2.48b}$$

Kou (2008) presents a comprehensive survey on the merits of the double exponential jump-diffusion model. First, empirical tests reveal that the double exponential jump-diffusion model fits stock data better than the Merton jump-diffusion model, and both of them outperform the classical Geometric Brownian motion. Second, the exponential jump model enjoys nice analytic tractability in pricing various path-dependent options, like lookback options, barrier options, and interest rate derivatives. Third, the exponential jumps agree better with market behaviors that market participants tend to have overreaction and underreaction to good or bad news. The exponential jump distribution captures these behaviors more appropriately.

2.3 Lévy processes

Jumps in the asset price processes have been commonly observed in the financial markets (Geman *et al.*, 2001). Within a small time interval, there may be a lot of small jumps in the stock price movement. It may be difficult to distinguish if the price movement comes from the diffusion or jump component [see Aït-Sahalia (2004) for the methods of detecting jumps]. The financial models with compound Poisson jumps exhibit finite activity of the jump component. However, the Poisson process is incapable of capturing the market feature of high frequency of small jumps.

To provide more modeling tools on jump types, the Lévy process is a generalization of the jump-diffusion processes by allowing infinite jump activity. The standard form of the Lévy process assumes stationary increments, so resulting in nice analytic tractability. This makes it a good choice for pricing equity derivatives. With provision of randomizing the time clock on which the Lévy process is run, the time-changed Lévy processes exhibit stochastic volatility. They become capable of capturing volatility smile, smile skew and term structure of the smile.

This section and the next section provide a brief exposition on the key properties of Lévy processes. Readers may consult various comprehensive books for fuller discussion of Lévy processes and their applications in finance (Sato, 1999; Schoutens, 2003; Cont and Tankov, 2003; Kyprianou, 2006; Applebaum, 2009).

2.3.1 Definition

A càdlàg real-valued stochastic process X_t defined on a filtered probability space $(\Omega, \mathfrak{F}, P)$, endowed with a standard complete filtration $\mathscr{F} = \{\mathfrak{F}_t | t \geq 0\}$ and $X_0 = 0$, is a Lévy process if it possesses the following properties:

1. Independent increments: For any increasing sequence of times $0 \leq t_0 < t_1 < \cdots < t_n$, the random variables $X_{t_0}, X_{t_1} - X_{t_0}, \ldots, X_{t_n} - X_{t_{n-1}}$ are all independent.

2. Stationary increments: $X_{t+h} - X_t$ has the same distribution as X_h for all $h, t \geq 0$.

3. Stochastic continuity: For any $t \geq 0$ and $\varepsilon > 0$, $\lim_{s \to t} P[|X_s - X_t| > \varepsilon] = 0$.

The last property does not imply continuous sample paths of X_t. Rather, the property of seeing a jump at a given t is zero. Discontinuities occur at random times, so processes with jumps at deterministic times are excluded.

2.3.2 Infinite divisibility

A Lévy process is closely related to the property of infinite divisibility. A random variable Y is said to be infinite divisible if there exists iid random variables $Y_1^{(n)}, Y_2^{(n)}, \ldots, Y_n^{(n)}$, $n \geq 2$, such that

$$Y \stackrel{\mathrm{d}}{=} Y_1^{(n)} + Y_2^{(n)} + \cdots + Y_n^{(n)}. \tag{2.49}$$

A typical example of an infinitely divisible distribution is the normal distribution. Suppose $Y \sim N(\mu, \sigma^2)$, where μ and σ^2 are the mean and variance, respectively, then

$$Y \stackrel{\mathrm{d}}{=} \sum_{k=1}^{n} Y_k^{(n)},$$

where $Y_k^{(n)}$ are iid with law $N\left(\frac{\mu}{n}, \frac{\sigma^2}{n}\right)$ (see the proof in Sec. 2.3.3). Also, it can be shown that the Poisson distribution and Gamma distribution are infinitely divisible distributions. However, the discrete Bernuolli distribution is not infinitely divisible.

If X_t is a Lévy process, then X_t is infinitely divisible for each $t > 0$. To show the claim, for any $n \geq 2$, we define

$$Y_k^{(n)} = X_{\frac{k}{n}t} - X_{\frac{k-1}{n}t}, \quad k = 1, 2, \ldots, n.$$

By the property of independent and stationary increments of X_t, $Y_k^{(n)} \stackrel{\mathrm{d}}{=} X_{\frac{t}{n}}$. It is seen that

$$X_t = Y_1^{(n)} + Y_2^{(n)} + \cdots + Y_n^{(n)}$$

so that X_t is infinitely divisible for each $t > 0$.

The converse statement also holds. If F is an infinitely divisible distribution, then there exists a Lévy process X_t such that the distribution of X_1 is given by F. Readers may refer to Corollary 11.6 in Sato (1999) for its proof.

2.3.3 Characteristic exponent and Lévy-Khintchine representation

By virtue of infinite divisibility property (2.49) and independence of the random variables $Y_k^{(n)}$, the characteristic function ϕ_{X_t} of the infinitely divisible distribution X_t observes the following property:

$$\phi_{X_t}(u) = \left[\phi_{X_{\frac{t}{n}}}(u)\right]^n, \quad n \geq 2. \tag{2.50}$$

We would like to verify that the Brownian motion with constant drift μ and volatility σ is infinitely divisible for each $t > 0$. For $X_t = \mu t + \sigma W_t$, where W_t is the standard Brownian motion and n is any positive integer, we observe that

$$\begin{aligned} &\phi_{X_t}(u) \\ &= \int_{-\infty}^{\infty} e^{iux} \frac{1}{\sqrt{2\pi\sigma^2 t}} \exp\left(-\frac{(x-\mu t)^2}{2\sigma^2 t}\right) \mathrm{d}x = \exp\left(iu\mu t - \frac{u^2\sigma^2 t}{2}\right) \\ &= \exp\left(n\left(iu\frac{\mu t}{n} - \frac{u^2\sigma^2 t}{n}\right)\right) = \left[\exp\left(iu\frac{\mu t}{n} - \frac{u^2\sigma^2 t}{n}\right)\right]^n = \left[\phi_{X_{\frac{t}{n}}}(u)\right]^n. \end{aligned} \tag{2.51a}$$

Similarly, to check infinite divisibility for the Poisson process N_t with parameter λ, we replace α by iu and set $f(\mathrm{d}y) = 1$ in (2.10a) to obtain

$$\phi_{N_t}(u) = \exp\left(\lambda t\left(e^{iu} - 1\right)\right) = \left[\exp\left(\frac{\lambda t}{n}\left(e^{iu} - 1\right)\right)\right]^n = \left[\phi_{N_{\frac{t}{n}}}(u)\right]^n, \tag{2.51b}$$

where n is any positive integer.

Since a Lévy process is an infinitely divisible distribution, the Lévy-Khintchine Theorem [see Theorem 8.1 in Sato (1999)] for an infinitely divisible distribution can be applied to establish that the characteristic function of a Lévy process X_t admits the Lévy-Khintchine representation. Let the characteristic exponent $\psi_X(u)$ be defined by

$$\psi_X(u) = -\frac{1}{t}\ln\phi_{X_t}(u), \quad t \geq 0, \tag{2.52}$$

then $\psi_X(u)$ admits the representation

$$\psi_X(u) = -i\mu u + \frac{\sigma^2 u^2}{2} + \int_{\mathbb{R}\setminus\{0\}} (1 - e^{iux} + iux\mathbf{1}_{\{|x|<1\}})\,\Pi(\mathrm{d}x). \tag{2.53}$$

The Lévy measure $\Pi(\mathrm{d}x)$ is defined on the real domain excluding zero, with $\Pi(\{0\}) = 0$. The triplet (μ, σ^2, Π) is called the Lévy characteristic of X_t, where $\mu \in \mathbb{R}$ is the constant drift, $\sigma \geq 0$ is the constant volatility of the continuous component and Π is the Lévy measure that represents the expected number of jumps per unit time. The distribution of jump sizes has to satisfy the technical condition of finite quadratic variation:

$$\int_{\mathbb{R}\setminus\{0\}} \min(1, x^2)\,\Pi(\mathrm{d}x) < \infty, \tag{2.54}$$

a necessary condition that the jump process is a semimartingale. A Lévy process is completely characterized by its triplet (μ, σ^2, Π). Lévy processes are within a subclass of semimartingale.

From (2.52), it is seen that

$$\phi_{X_t}(u) = \left[e^{-\frac{t}{n}\psi_X(u)}\right]^n = \left[\phi_{X_{\frac{t}{n}}}(u)\right]^n, \tag{2.55a}$$

so the property of infinite divisibility is verified. Also, $\phi_{X_t}(u)$ can be decomposed into the form:

$$\begin{aligned}\phi_{X_t}(u) &= \exp\left(iu\mu t - \frac{u^2\sigma^2 t}{2}\right)\exp\left(t\int_{\mathbb{R}\setminus\{0\}} e^{iux} - 1 - iux\mathbf{1}_{\{|x|<1\}}\,\Pi(\mathrm{d}x)\right)\\ &= \phi_{B_t}(u)\phi_{\text{jump}}(u),\end{aligned} \tag{2.55b}$$

which is the product of the characteristic function $\phi_{B_t}(u)$ of the Brownian motion (with drift μ and volatility σ) and characteristic function $\phi_{\text{jump}}(u)$ of the jump component [see (2.51a,b) for reference].

Consider a pure jump Lévy process without the diffusion term, if the integral of the Lévy measure is finite, where

$$\int_{\mathbb{R}\setminus\{0\}} \Pi(\mathrm{d}x) < \infty, \tag{2.56}$$

then the sample paths exhibit finite activity, meaning that only a finite number of jumps occur within any finite interval of time. Otherwise, for infinite activity process, the number of jumps is infinite in any fixed time interval. Numerous examples of infinite activity pure jump Lévy processes, like the CGMY model, Variance Gamma model and Normal Inverse Gaussian model will be discussed in later sections.

A pure jump Lévy process is said to exhibit infinite variation if

$$\int_{\mathbb{R}\setminus\{0\}} \min(1,|x|)\,\Pi(\mathrm{d}x) = \infty. \tag{2.57}$$

Otherwise, the Lévy process shows finite variation.

For mathematical purpose, the truncation function $h(x) = x\mathbf{1}_{\{|x|<1\}}$ in the jump integral in $\psi_X(u)$ [see (2.53)] is needed in analyzing the jump properties around the singular point of zero jump size. The choice of $h(x)$ is not unique, provided that $h(x) \approx x$ in the neighborhood of zero. Some other common choices of the truncation function include $h(x) = \dfrac{x}{1+x^2}$ and $h(x) = \min(1,|x|)$.

2.3.4 Lévy-Itô decomposition theorem

The Lévy-Itô decomposition theorem entails that a Lévy process is made of drift and diffusion, together with a set of finite jumps of large size and a set of infinite jumps of infinitesimal size. The details of the decomposition theorem are summarized in Theorem 2.1.

Theorem 2.1
Let X_t be a Lévy process and $\Pi(\mathrm{d}x)$ denote its Lévy measure on $\mathbb{R}\setminus\{0\}$ that satisfies finiteness of quadratic variation:

$$\int_{\mathbb{R}\setminus\{0\}} \min(1,x^2)\,\Pi(\mathrm{d}x) < \infty.$$

The jump measure J_X of X is a Poisson random measure with intensity measure $\Pi(\mathrm{d}x)\,\mathrm{d}t$. The corresponding compensated jump measure $\widetilde{J}_X$ is defined by (2.22). The Lévy-Itô decomposition theorem states that there exists a Brownian motion B_t and drift μ such that

$$X_t = \mu t + B_t + C_t + M_t, \tag{2.58}$$

where

$$C_t = \int_0^t \int_{|x|\geq 1} x\,J_X(\mathrm{d}x,\mathrm{d}s), \tag{2.59a}$$

$$\begin{aligned} M_t &= \lim_{\varepsilon\to 0^+} \int_0^t \int_{\varepsilon\leq|x|<1} x\,\widetilde{J}_X(\mathrm{d}x,\mathrm{d}s) \\ &= \lim_{\varepsilon\to 0^+} \int_0^t \int_{\varepsilon\leq|x|<1} x\,[J_X(\mathrm{d}x,\mathrm{d}s) - \Pi(\mathrm{d}x)\,\mathrm{d}s]. \end{aligned} \tag{2.59b}$$

The Lévy process is specified by the Lévy-Khintchine triplet (μ,σ^2,Π). The proof of the Lévy-Itô decomposition theorem can be found in Cont and Tankov (2003).

The first two terms in (2.58) represent a continuous Gaussian Lévy process, which corresponds to a Brownian motion with drift. Indeed, every Gaussian Lévy process is continuous and can be represented by this form. The other two terms C_t

and M_t represent the respective large jumps and small jumps of X that are dependent only on the jump measure J_X. We may write

$$C_t = \sum_{0<s\leq t} \Delta X_s \mathbf{1}_{\{|\Delta X_s|\geq 1\}}, \tag{2.60a}$$

which represents a compound Poisson process with finite number of large jumps of X within $[0,t]$ that are greater than or equal to the threshold level of one. The threshold level can be any finite positive real number, not limited to one. The second small jump term M_t can be written as

$$M_t = \sum_{\substack{0<s\leq t \\ \varepsilon \leq |\Delta X_s| < 1}} \Delta X_s - tE\left[\Delta X_1 \mathbf{1}_{\{\varepsilon\leq|\Delta X_1|<1\}}\right], \tag{2.60b}$$

which represents the compensated compound Poisson process of the jumps of X with size within $[\varepsilon, 1)$. The compensated version is used since Π may have a singularity at zero; that is, there can be infinitely many small jumps and their sum does not necessarily converge.

Cumulant exponent

Note that $\phi_{X_t}(u) = [\phi_{X_1}(u)]^t$, so the law of X_t is determined by the knowledge of the law of X_1. To specify a Lévy process, it suffices to specify the distribution of X_t for a single time. The cumulant exponent $\kappa_X(u)$ is defined to be

$$\kappa_X(u) = \frac{1}{t}\ln E\left[e^{uX_t}\right] = -\psi_X(-iu). \tag{2.61}$$

The Lévy-Itô decomposition theorem shows that one can define a Lévy process by specifying the Lévy measure Π. The law of X_t is characterized by the characteristic exponent ψ_X, which can be computed using the Lévy-Khintchine representation formula (2.53). We have seen some simple choices of $\Pi(\mathrm{d}x)$, like the Gaussian jump distribution and exponential jump distribution, corresponding to Merton's jump-diffusion model (1976) and Kou's exponential jump model (2002), respectively (see Sec. 2.2). Next, we illustrate how the CGMY Lévy process (Carr *et al.*, 2002) is defined by choosing $\Pi(\mathrm{d}x)$ to be the dampened power law.

2.3.5 CGMY model: Dampened power law as Lévy measure

The Lévy measure $\Pi(\mathrm{d}x)$ of the CGMY (Carr-Geman-Madan-Yor) model (Carr *et al.*, 2002) is specified by the following four-parameter function:

$$\Pi_{\mathrm{CGMY}}(\mathrm{d}x) = \begin{cases} \dfrac{Ce^{-G|x|}}{|x|^{1+Y}}\,\mathrm{d}x, & x<0 \\ \dfrac{Ce^{-MY}}{x^{1+Y}}\,\mathrm{d}x, & x>0 \end{cases}, \tag{2.62}$$

where the four parameters (denoted by the names of the four authors) observe the following constraints:

$$C > 0, \quad M \geq 0, \quad G \geq 0 \quad \text{and} \quad Y < 2.$$

The continuous component in the Lévy triplet is taken to be zero, so the CGMY process is a pure jump Lévy process. The Lévy measure of the CGMY model can be seen to be a generalization of the Kou double exponential model with the additional dampened power law: $1/|x|^{1+Y}$ on $x < 0$ and $1/x^{1+Y}$ on $x > 0$.

The power coefficient Y controls the arrival frequency of small jumps, playing a similar role as the Poisson intensity parameter λ. It can be shown that when $Y < 0$, we observe

$$\int_{\mathbb{R}\setminus\{0\}} \Pi_{\text{CGMY}}(\mathrm{d}x) < \infty,$$

so it has a finite activity. For $Y < 0$, the CGMY process corresponds to a compound Poisson process with large jumps and jump events are rare.

Recall that asset price models with infinite-activity jumps confirm better with empirical observations of asset returns. Therefore, we are more interested in the case with $Y \geq 0$, where the CGMY process has an infinite activity. We enforce $Y < 2$ due to the finite quadratic variation requirement [see (2.54)], where the Lévy measure integrates with weight x^2 near the neighborhood of 0. The CGMY process exhibits finite variation when $0 \leq Y < 1$ and infinite variation when $1 \leq Y < 2$. When $Y \to 2^-$, the CGMY model behaves like a diffusion process (see explanation below).

The two exponential parameters G and M control the rate of exponential decay on the left and right side of the Lévy measure, respectively. Unequal values of G and M lead to skewed distribution. Since the risk neutral distribution implied from option prices show heavier left tail of the distribution, so we typically choose $G < M$. The difference between G and M is dictated by the amount of price drop relative to price increase. The sum of G and M is related to the price of large move relative to a small one. The exponential damping factors in the Lévy measure give finiteness of all moments of the CGMY process, which is a desirable property from modeling perspective. The parameter C is related to the overall level of activity. When $G = M$, the Lévy jumps are symmetric. In this case, the parameter C provides control over the kurtosis of the distribution.

Characteristic function

Recall that $\mu = 0$ and $\sigma = 0$ in the triplet for a pure jump Lévy process. We employ the Lévy-Khintchine representation to find the characteristic function of the CGMY process. The characteristic function $\phi_{\text{CGMY}}(u)$ is given by

$$\phi_{\text{CGMY}}(u) = \exp\left(Ct\int_0^{\infty} (e^{iux} - 1)\frac{e^{-Mx}}{x^{1+Y}}\,\mathrm{d}x + Ct\int_{-\infty}^{0} (e^{iux} - 1)\frac{e^{Gx}}{(-x)^{1+Y}}\,\mathrm{d}x\right).$$

Consider the first integral, assuming $Y \neq 1$ and $Y \neq 0$, we have

$$\begin{aligned} Ct \int_0^\infty \left(e^{iux} - 1\right) \frac{e^{-Mx}}{x^{1+Y}} \, \mathrm{d}x &= Ct \int_0^\infty \frac{e^{-(M-iu)x} - e^{-Mx}}{x^{1+Y}} \, \mathrm{d}x \\ &= Ct\Gamma(-Y) \left[(M-iu)^Y - M^Y\right], \end{aligned}$$

where the Gamma function $\Gamma(\alpha+1)$ is defined by

$$\Gamma(\alpha+1) = \int_0^\infty e^{-y} y^\alpha \, \mathrm{d}y. \tag{2.63}$$

The second integral can be evaluated in a similar manner. We obtain the analytic formula for $\phi_{\mathrm{CGMY}}(u)$ as follows:

$$\phi_{\mathrm{CGMY}}(u) = \exp\left(Ct\Gamma(-Y)[(M-iu)^Y - M^Y + (G+iu)^Y - G^Y]\right), \; Y \neq 1 \text{ and } Y \neq 0. \tag{2.64}$$

When $Y \to 2^-$, $\phi_{\mathrm{CGMY}}(u)$ approaches a quadratic function in u, exhibiting close resemblance to the characteristic function of a diffusion process. This explains why the CGMY behaves like a diffusion process as $Y \to 2^-$. Under the special cases $Y = 1$ or $Y = 0$, a simplified expression for $\phi_{\mathrm{CGMY}}(u)$ can be obtained in a more direct manner. Indeed, the CGMY model reduces to the renowned Variance Gamma model when $Y = 0$ (see Sec. 2.3.2).

Though the characteristic function of the CGMY model admits nice analytic form, the corresponding density function is not available in an analytic form.

2.3.6 Generalized Hyperbolic model

Besides constructing a Lévy process via the specification of its Lévy measure, we may construct a Lévy process by specifying its probability density. One example is the Generalized Hyperbolic (GH) model (Eberlein *et al.*, 1998), whose density function f_{GH} is defined by the following five-parameter function:

$$\begin{aligned} &f_{\mathrm{GH}}(x;\alpha,\beta,\delta,\mu,\nu) \\ &= \frac{(\alpha^2-\beta^2)^{\nu/2}\left[\delta^2+(x-\mu)^2\right]^{\frac{\nu}{2}-\frac{1}{4}}}{\sqrt{2\pi}\alpha^{\nu-\frac{1}{2}}\delta^\nu \kappa_\nu\left(\delta\sqrt{\alpha^2-\beta^2}\right)} e^{\beta(x-\mu)} \kappa_{\nu-\frac{1}{2}}\left(\alpha\sqrt{\delta^2+(x-\mu)^2}\right), \end{aligned} \tag{2.65}$$

where $\kappa_p(\cdot)$ is the pth-order modified Bessel function of the second kind, $\mu \in \mathbb{R}$ and the other three parameters satisfy the following constraints:

$$\begin{aligned} &\delta \geq 0, \; |\beta| < \alpha \quad \text{if} \quad \nu > 0, \\ &\delta > 0, \; |\beta| < \alpha \quad \text{if} \quad \nu = 0, \\ &\delta > 0, \; |\beta| \leq \alpha \quad \text{if} \quad \nu < 0. \end{aligned}$$

The GH distribution embeds various distributions under different special choices of the parameter values. When $\nu = 1$, the GH distribution reduces to the hyperbolic distribution, whose logarithm of its density is a hyperbolic (this is how the name "hyperbolic" is derived). In addition, when $\delta \to \infty$ and $\frac{\delta}{\alpha} \to \sigma^2$, the GH distribution reduces to the normal distribution. Furthermore, when $\delta = 0$ and $\mu = 0$, the GH distribution becomes the Variance Gamma distribution (see Sec. 2.4.2); when $\nu = -\frac{1}{2}$, it becomes the Normal Inverse Gaussian distribution (see Sec. 2.4.3).

One manages to derive the characteristic function of the GH model, which assumes the following analytic form:

$$\phi_{\mathrm{GH}}(x;\alpha,\beta,\delta,\mu,\nu) = e^{i\mu u}\left[\frac{\alpha^2-\beta^2}{\alpha^2-(\beta+iu)^2}\right]^{\nu/2} \frac{\kappa_\nu\left(\delta\sqrt{\alpha^2-(\beta+iu)^2}\right)}{\kappa_\nu(\delta\sqrt{\alpha^2-\beta^2})}. \quad (2.66)$$

Unfortunately, the Lévy measure of the GH model has no explicit analytic form. It can be expressed only in terms of integrals of the modified Bessel functions [see Sec. 5.3.11 in Schoutens (2003) for the Lévy measure formula].

2.3.7 Martingale condition on drift under risk neutral measure

Let the asset price process S_t be defined as an exponential Lévy process X_t, where

$$S_t = S_0 e^{X_t} \quad (2.67)$$

under the risk neutral valuation framework in option pricing. Since the Lévy process X_t exhibits jumps, the market is not complete and uniqueness of equivalent martingale measure fails.

Under a risk neutral measure Q, $S_t e^{-(r-q)t}$ is a martingale, where r and q are the constant riskless interest rate and dividend yield, respectively. Let X_t have the Lévy triplet (μ_Q, σ^2, Π), so X_t admits the Lévy-Khintchine representation [see (2.53)]:

$$\begin{aligned} E\left[e^{iuX_t}\right] &= \phi_{X_t}(u) \\ &= \exp\left(t\left[i\mu_Q u - \frac{\sigma^2 u^2}{2} + \int_{\mathbb{R}\setminus\{0\}} \left(e^{iux} - 1 - iux\mathbf{1}_{\{|x|<1\}}\right)\,\Pi(\mathrm{d}x)\right]\right). \end{aligned} \quad (2.68)$$

Under the martingale condition, we must observe

$$\begin{aligned} 1 = E_Q\left[\frac{S_t e^{-(r-q)t}}{S_0}\right] &= e^{-(r-q)t}E_Q\left[e^{X_t}\right] = e^{-(r-q)t}\phi_{X_t}(-i) \\ &= \exp\left(-(r-q)t + t\left[\mu_Q + \frac{\sigma^2}{2} + \int_{\mathbb{R}\setminus\{0\}} \left(e^x - 1 - x\mathbf{1}_{\{|x|<1\}}\right)\,\Pi(\mathrm{d}x)\right]\right). \end{aligned}$$

The martingale condition dictates that the drift μ_Q under the risk neutral measure Q has to satisfy

$$\mu_Q = (r-q) - \frac{\sigma^2}{2} - \int_{\mathbb{R}\setminus\{0\}} \left(e^x - 1 - x\mathbf{1}_{\{|x|<1\}}\right)\,\Pi(\mathrm{d}x). \quad (2.69)$$

As an example, we use (2.69) to determine μ_Q of the CGMY model based on $\phi_{\text{CGMY}}(u)$ in (2.64). Recall that σ^2 is taken to be zero in the CGMY model. Under a risk neutral measure Q, the drift μ_Q of the CGMY model is given by

$$\mu_Q = r - q + C\Gamma(-Y)\left[G^Y - (1+G)^Y + M^Y - (M-1)^Y + Y(G^{Y-1} - M^{Y-1})\right]. \tag{2.70}$$

2.4 Time-changed Lévy processes

Empirical data of asset returns reveal phenomena of volatility clustering. Asset return volatilities are well documented to be stochastic. Higher order moments, like skewness and kurtosis, also vary significant over time. In this section, we show how to adopt the time-change technique to capture stochastic volatility. The concept of random time change amounts to changing the time clock on which the Lévy process is run. We change the calendar time t to a nonnegative process with non-decreasing sample paths T_t, which may be visualized financially as the business activity time. The most desirable merit of time-changed Lévy processes is that they inherit nice analytic tractability of their base Lévy processes.

The more effective and versatile method of constructing Lévy processes is applying time change to some basic Lévy processes. This starts with choosing an almost surely increasing Lévy process as the subordinator. The time variable of another independent Lévy process is made stochastic by choosing the process of the subordinator as the stochastic clock. For example, the Variance Gamma process is obtained by subordinating a Brownian motion by a Gamma process as the subordinator and the Normal Inverse Gaussian process is a Brownian motion subordinated by an Inverse Gaussian process. By applying stochastic activity rate dynamics underlying the time change to different Lévy components, one can generate both stochastic volatility and higher order moments of the resulting time-changed Lévy processes (Carr *et al.*, 2003; Carr and Wu, 2004; Huang and Wu, 2004; Chourdakis, 2005; Wu, 2008).

2.4.1 Time-change techniques: Subordinators and activity rates

We formalize the change of time frame from the calendar time t to random clock T_t, which symbolizes random business time. The random clock is formulated as a subordinator, which is an almost surely increasing Lévy process with finite variation and some additional regularity properties. We discuss how a time-changed Lévy process is generated by the subordination procedure via composition of a base Lévy process and another subordinator process. The characteristic function of the time-changed Lévy process can be shown to be composition of the characteristic function of the base Lévy process and Laplace exponent of the surbordinator. Lastly, we discuss the use of stochastic activity rate to define the random clock T_t.

Let X_t be a stochastic process and $T_t(\omega)$ be an almost surely increasing sample path with $T_0 = 0$ and tending to infinity as $t \to \infty$; that is, for $t \geq s$, we have almost surely $X_t \geq X_s$. The time-changed process:

$$Z_t = X_{T_t} \tag{2.71}$$

assumes the same values as X, in the same sequential order but at different clock times. Let the base process be a Lévy process. What is the class of processes for T_t such that the time-changed process is also a Lévy process? We state the following results [for proofs, see Theorems 3.2.2 and 3.2.4 in Cherubini *et al.* (2010)]:

1. An almost surely increasing Lévy process T_t is a subordinator if and only if it has finite variation of the form:

$$T_t = \mu t + \sum_{s \in [0,t]} \Delta X_s, \tag{2.72}$$

   with $\mu \geq 0$, $\Pi((-\infty, 0]) = 0$ and $\int_0^\infty \min(1, x)\, \Pi(\mathrm{d}x) < \infty$.

2. Suppose X_t is a Lévy process and T_t is a subordinator, then $Z_t = X_{T_t}$ is also a Lévy process.

The Lévy process Z_t generated by time change of X_t with subordinator T_t exhibits nice analytic tractability due to the Bochner formula for the subordinated Lévy process $Z_t = X_{T_t}$.

Bochner formula

Let $E[e^{iuX_t}] = \phi_{X_t}(u) = e^{-t\psi_{X_t}(u)}$ denote the characteristic function of X_t and

$$E\left[e^{-uT_t}\right] = e^{-t\Phi_{T_t}(u)}$$

defines the Laplace exponent Φ_{T_t} of the subordinator T_t. Assume that X_t and T_t are independent. The characteristic function of the time-changed Lévy process $Z_t = X_{T_t}$ is given by the composite of ψ_{X_t} and Φ_{T_t}, known as the Bochner formula:

$$\phi_{Z_t}(u) = e^{-t\Phi_{T_t}\left(\psi_{X_t}(u)\right)}. \tag{2.73}$$

It is quite straightforward to derive the formula. Let $f_{T_t}(\mathrm{d}s)$ denote the distribution of T_t. By virtue of the time change procedure, together with independence of X_t and T_t, we have

$$\begin{aligned}
\phi_{Z_t}(u) = E\left[e^{iuX_{T_t}}\right] &= \int_0^\infty E\left[e^{iuX_s}\right] f_{T_t}(\mathrm{d}s) \\
&= \int_0^\infty \phi_{X_s}(u) f_{T_t}(\mathrm{d}s) = \int_0^\infty e^{-s\psi_{X_s}(u)} f_{T_t}(\mathrm{d}s) \\
&= E\left[e^{-\psi_{X_t}(u)T_t}\right] = e^{-t\Phi_{T_t}(\psi_{X_t}(u))}.
\end{aligned}$$

Stochastic activity rates

We may set the random time T_t to be the integrated non-negative instantaneous process v_t as defined by

$$T_t = \int_0^t v_s \, ds. \tag{2.74}$$

A more active business day corresponds to a higher instantaneous activity rate v_t and so the corresponding time-changed process generates a higher volatility. When the time-changed Lévy process Z_t is a Brownian motion, v_t corresponds to the instantaneous variance of the Brownian motion. When Z_t is a pure jump Lévy process, like the compound Poisson jump process, then v_t is proportional to the jump arrival rate.

Note that the stochastic time change can be applied separately to the jump component or diffusion component or both so as to generate stochastic volatility arising from different scenarios. The correlation between the time-changed process and the underlying Lévy process can be embedded into the model to represent the leverage effect (Huang and Wu, 2004; Carr and Wu, 2004).

The role played by the activity rate in the time-changed Lévy model is analogous to that of the stochastic interest rate in the term structure model for pricing zero coupon bonds. In the literature of interest rate models, the two popular choices are the Cox-Ingersoll-Ross (CIR) process:

$$dv_t = \kappa(\theta - v_t)dt + \sigma_v \sqrt{v_t} \, dW_t^v, \tag{2.75a}$$

and Ornstein-Uhlenbeck (OU) process:

$$dv_t = \kappa(\theta - v_t)dt + \sigma_v \, dW_t^v. \tag{2.75b}$$

Here, κ is the reversion speed, θ is the mean reversion level and σ_v is the volatility of the activity rate. To capture the term structure of at-the-money implied volatilities across a wide range of maturities, a multi-factor activity rate dynamics gives better performance. For instance, one may include an additional stochastic mean level process θ_t as follows:

$$\begin{aligned} dv_t &= \kappa(\theta_t - v_t)dt + \sigma_v \sqrt{v_t} \, dW_t^v, \\ d\theta_t &= \kappa_\theta(m - \theta_t)dt + \sigma_\theta \sqrt{\theta_t} \, dW_t^\theta. \end{aligned} \tag{2.75c}$$

Intuitively, the activity rate v_t affects the short-term implied volatilities while the mean level factor θ_t controls the long-term implied volatilities (Wu, 2008).

In most cases, the affine activity rate models are adopted for the purpose of analytic tractability. However, there exist other non-affine model specifications, like the 3/2 model:

$$dv_t = \kappa v_t(\theta - v_t)dt + \sigma_v v_t^{3/2} \, dW_t^v. \tag{2.75d}$$

The 3/2-model becomes more popular recently since the resulting time-changed Lévy price process confirms better with empirical studies, though it is less tractable when compared with the affine models.

For numerical pricing calculations under the time-changed Lévy process, it is necessary to have an explicit formula for the Laplace transform of the transition density in $\int_0^t v_s\, ds$. For the CIR process and the 3/2-process, their corresponding explicit Laplace transformed formulas are readily available, though the confluent hypergeometric functions are involved.

Asset price process

The asset price process S_t can be modeled by a multi-dimensional time-changed Lévy process. Suppose there exists a risk neutral measure Q under which we obtain the K-dimensional Lévy process $\mathbf{L}_t = (L_t^1, \cdots, L_t^K)^T$ and K-dimensional random time process $\mathbf{Z}_t = (Z_t^1, \cdots, Z_t^K)^T$. Note that in general we may allow correlation between L_t^i and Z_t^j for $i, j = 1, \cdots, K$. We define the log of asset price process by

$$\ln \frac{S_t}{S_0} = (r-q)t + \sum_{k=1}^{K} [b^k Y_t^k - \kappa_{L^k}(b^k) Z_t^k], \tag{2.76}$$

where r and q are the constant risk-free rate and dividend yield of the underlying asset, respectively, b^k denotes constant loading coefficient in the kth component, $k = 1, \cdots, K$, and $Y_t^k = L_{Z_t^k}^k$. Here, κ_{L^k} is the cumulant exponent of L^k satisfying

$$E_0^Q \left[e^{b^k L_t^k} \right] = e^{\kappa_{L_t^k}(b^k) t}.$$

Recall that the relation between the characteristic exponent ψ_X and cumulant exponent κ_X is stated in (2.61). Note that the term $-\kappa_{L_t^k}(b^k) Z_t^k$ is appended to ensure that the discounted asset price process $S_t e^{-(r-q)t}$ is a Q-martingale. To see this, we let $t > s \geq 0$, and observe that

$$\begin{aligned}
& E^Q\left[\exp\left(\sum_{k=1}^{K} [b^k L_t^k - \kappa_{L_t^k}(b^k) t]\right) \Big| \mathcal{F}_s\right] \\
= \; & E^Q\left[\exp\left(\sum_{k=1}^{K} [b^k (L_t^k - L_s^k) - \kappa_{L_t^k}(b^k)(t-s)]\right) \Big| \mathcal{F}_s\right] \\
& + \exp\left(\sum_{k=1}^{K} [b^k L_s^k - \kappa_{L_t^k}(b^k) s]\right) \\
= \; & \exp\left(\sum_{k=1}^{K} [b^k L_s^k - \kappa_{L_t^k}(b^k) s]\right),
\end{aligned}$$

which implies that $\exp\left(\sum_{k=1}^{K} [b^k L_t^k - \kappa_{L_t^k}(b^k) t]\right)$ is a Q-martingale. We deduce that the stopped process $\exp\left(\sum_{k=1}^{K} [b^k Y_t^k - \kappa_{L_t^k}(b^k) Z_t^k]\right)$ is a Q-martingale as well. Therefore, the discounted asset price process define in (2.76) is a Q-martingale.

To give an example of a time-changed Lévy process, we start with the simple case where the return innovation is driven by a Brownian motion with drift. We choose the base model to be the Black-Scholes model, where

$$\ln \frac{S_t}{S_0} = (r-q)t + \sigma W_t^s - \frac{\sigma^2 t}{2}.$$

Here, W_t^s is the standard Brownian motion. Let the activity rate follows the CIR process as specified by

$$dv_t = \kappa(\theta - v_t)dt + \sigma_v \sqrt{v_t}\, dW_t^v. \tag{2.77a}$$

Applying the random time change $Z_t = \int_0^t v_s\, ds$ to the diffusion component, we obtain a time-changed Lévy process as follows

$$\ln \frac{S_t}{S_0} = (r-q)t + \sigma W^s_{\int_0^t v_s\, ds} - \frac{\sigma^2}{2}\int_0^t v_s\, ds, \tag{2.77b}$$

As a remark, the resulting asset price process can be identified as a variation of the Heston stochastic volatility model. To see this, we note that

$$W^s_{\int_0^t v_s\, ds} \overset{d}{=} \int_0^t \sqrt{v_s}\, d\tilde{W}_t,$$

where $\tilde{W}_t$ is another standard Brownian motion different from W_t^s. Combining (2.77a) and (2.77b), we obtain

$$\begin{aligned} d\ln S_t &= (r-q)dt + \sigma\sqrt{v_t}\, d\tilde{W}_t - \frac{\sigma^2}{2} v_t\, dt, \\ dv_t &= \kappa(1-v_t)dt + \sigma_v\sqrt{v_t}\, dW_t^v. \end{aligned} \tag{2.78}$$

The long-term mean of the activity rate v_t is normalized to one for convenience since there exists σ already that captures the mean level of volatility. More examples of time-changed Lévy processes can be found in Wu (2008).

2.4.2 Variance Gamma model

The Variance Gamma (VG) model is first initiated by Madan and his coauthors in a series of papers (Madan and Seneta, 1990; Madan and Milne, 1991; Madan *et al.*, 1998). Seneta (2007) presents an interesting historical account on the development of the VG model. The VG process for the logarithm of asset price has no continuous martingale component. It is a pure jump process with finite variation. The VG model has been applied to price a wide variety of path-dependent options, like the American options (Hirsa and Madan, 2004; Almendral and Oosterlee, 2007). The VG model is popular since it can capture volatility skewness and kurtosis in the risk neutral asset price distribution with relatively simple procedure. The model formulation involves only 3 parameters. Most pricing algorithms take advantage of the simple analytic form of the characteristic function of the VG process. Efficient Monte Carlo simulation (Avramidis and L'Ecuyer, 2006) and finite difference scheme (Hirsa and Madan, 2004) can be conveniently constructed. Also, analytic approximation formulas for pricing options can be derived (Aguilar, 2020) using the Mellin transform (closely related to the bilateral Laplace transform).

There are three aspects of the VG model. It can be visualized as a special case of the CGMY model as well as the GH model. Also, it can be expressed as the difference of two Gamma processes. Lastly, it can be defined as a time-changed Brownian motion subordinated by a Gamma process.

Recall that the VG model is seen as a special case of the CGMY model with $Y = 0$ in $\Pi_{\mathrm{CGMY}}(\mathrm{d}x)$ [see (2.62)]. The Lévy measure of the VG model is given by

$$\Pi_{\mathrm{VG}}(\mathrm{d}x) = \begin{cases} \dfrac{C\exp(-G|x|)}{|x|}\,\mathrm{d}x, & x < 0 \\ \dfrac{C\exp(-Mx)}{x}\,\mathrm{d}x, & x > 0 \end{cases}, \tag{2.79}$$

where C, G, and M are positive real parameters.

We would like to explore the more interesting aspect of the VG model, where it can be obtained by subordinating a Brownian motion by an independent Gamma process. The Gamma process $\gamma(t;\mu,\nu)$ is the random process of independent Gamma increment over non-overlapping time intervals with mean rate μ and variance rate ν. The characteristic function of $\gamma(t;\mu,\nu)$ is given by

$$\phi_{\gamma(t)}(u) = \left(\frac{1}{1 - iu\frac{\nu}{\mu}}\right)^{\frac{\mu^2 t}{\nu}}. \tag{2.80}$$

Let $W(t;\theta,\sigma)$ denote a Brownian motion with drift θ and volatility σ, where

$$U(t;\theta,\sigma) = \theta t + \sigma W_t.$$

Here, W_t is a standard Brownian motion. The VG process $X(t;\sigma,\nu,\theta)$ is defined to be the Brownian motion $U(t;\theta,\sigma)$ with drift subordinated by the Gamma process with unit mean rate $\gamma(t;1,\nu)$, where

$$X(t;\sigma,\nu,\theta) = U(\gamma(t;1,\nu);\theta,\sigma). \tag{2.81}$$

The drift parameter θ generates skew and variance parameter ν generates kurtosis over the volatility σ of the underlying Brownian motion. The subordinating Gamma process is chosen to have mean rate of one to ensure that the time-changed Brownian motion has an expected time length of t. The Gamma process is an increasing and positive random process, so $U(\gamma(t;1,\nu);\theta,\sigma)$ is well defined.

Conditional on the Gamma time change value y, $X(t)$ has a normal density. The unconditional density $f_{X(t)}$ of the VG process is obtained by integrating out y using the normal density for y. This gives

$$f_{X(t)}(x) = \int_0^\infty \frac{1}{\sigma\sqrt{2\pi y}} e^{-\frac{(x-\theta y)^2}{2\sigma^2 y}} \frac{y^{\frac{t}{v}-1}}{v^{\frac{t}{v}}\Gamma\left(\frac{t}{v}\right)} e^{-\frac{y}{v}}\,\mathrm{d}y$$

$$= \frac{2e^{\frac{\theta x}{\sigma^2}}}{v^{\frac{t}{v}}\sqrt{2\pi}\sigma\Gamma\left(\frac{t}{v}\right)} \left(\frac{x^2}{\frac{2\sigma^2}{v}+\theta^2}\right)^{\frac{t}{2v}-\frac{1}{4}} \kappa_{\frac{t}{v}-\frac{1}{2}}\left(\frac{\sqrt{\frac{2\sigma^2}{v}+\theta^2}\,|x|}{\sigma^2}\right), \tag{2.82}$$

where $\Gamma(\cdot)$ is the Gamma function and $\kappa_p(\cdot)$ is the pth-order modified Bessel function of the second kind. The characteristic function $\phi_{\mathrm{VG}}(u)$ admits a simple analytic form:

$$\phi_{\mathrm{VG}}(u) = E\left[e^{iuX(t)}\right] = \left(\frac{1}{1-i\theta vu+\frac{\sigma^2 v}{2}u^2}\right)^{\frac{t}{v}}. \tag{2.83a}$$

Correspondingly, the characteristic exponent $\psi_{\mathrm{VG}}(u)$ is found to be

$$\psi_{\mathrm{VG}}(u) = -\frac{1}{t}\ln E\left[e^{iuX(t)}\right] = \frac{1}{v}\ln\left(1-i\theta vu+\frac{\sigma^2 v}{2}u^2\right). \tag{2.83b}$$

One may use the Bochner formula (2.73) to derive $\psi_{\mathrm{VG}}(u)$, where

$$\psi_{\mathrm{VG}}(u) = \Phi_{\gamma(t;1,v)}\left(\psi_{U(t;\theta,\sigma)}(u)\right). \tag{2.84}$$

The characteristic exponent of $U(t;\theta,\sigma)$ is

$$\psi_{U(t;\theta,\sigma)} = -iu\theta + \frac{u^2\sigma^2}{2}.$$

The Laplace exponent of $\gamma(t;1,v)$ is

$$\Phi_{\gamma(t;1,v)} = -\frac{1}{t}\ln\left(\frac{1}{1+uv}\right)^{t/v} = \frac{1}{v}\ln(1+uv),$$

which is obtained by replacing iu by $-u$ in (2.81). Substituting these results into the above Bochner formula, we obtain the formula for $\psi_{\mathrm{VG}}(u)$ in (2.84).

Another important property of the VG process is that it can be decomposed into the difference of two independent Gamma processes. We observe that

$$\frac{1}{1-i\theta vu+\frac{\sigma^2 v}{2}u^2} = \frac{1}{1-i\eta_p u}\frac{1}{1+i\eta_n u},$$

where η_p and η_n satisfy

$$\eta_p - \eta_n = \theta v \quad \text{and} \quad \eta_p\eta_n = \frac{\sigma^2 v}{2}.$$

The solution of the above two simultaneous equations gives

$$\eta_p = \sqrt{\frac{\theta^2 v^2}{4} + \frac{\theta^2 v}{2}} + \frac{\theta v}{2},$$
$$\eta_n = \sqrt{\frac{\theta^2 v^2}{4} + \frac{\theta^2 v}{2}} - \frac{\theta v}{2}.$$

Let $G(t;\mu_p, v_p)$ and $G(t;\mu_n, v_n)$ denote the two Gamma processes with respective mean rate: μ_p and μ_n and variance rate: v_p and v_n. Their mean rates and variance rates are found to be

$$\mu_p = \frac{\eta_p}{v}, \; \mu_n = \frac{\eta_n}{v}, \; v_p = \mu_p^2 v \; \text{and} \; v_n = \mu_n^2 v.$$

We then obtain

$$X(t;\sigma, v, \theta) = G(t;\mu_p, v_p) - G(t;\mu_n, v_n). \tag{2.85}$$

The decomposition is possible by virtue of the finite variation property while the VG process shares the infinite activity property as a Gamma process. Based on the above decomposition and the known analytic form of the Lévy measure of a Gamma process, we obtain the Lévy measure of the VG process as follows:

$$\Pi_{\mathrm{VG}}(\mathrm{d}x) = \begin{cases} \dfrac{\mu_n^2}{v_n} \dfrac{e^{-\frac{\mu_n}{v_n}|x|}}{|x|} \, \mathrm{d}x, & x < 0 \\ \dfrac{\mu_p^2}{v_p} \dfrac{e^{-\frac{\mu_p}{v_p}x}}{x} \, \mathrm{d}x, & x > 0 \end{cases}. \tag{2.86}$$

By comparing with $\Pi_{\mathrm{VG}}(\mathrm{d}x)$ defined in (2.79), we deduce that

$$C = \frac{1}{v}, \; G = \frac{1}{\eta_n} \; \text{and} \; M = \frac{1}{\eta_p}. \tag{2.87}$$

When $\theta = 0$, the Lévy measure becomes symmetric around the origin.

Using the relations in (2.87) and $\psi_{\mathrm{VG}}(u)$ in (2.83b), we obtain the characteristic exponent of the VG model in terms of C, G and M as follows:

$$\psi_{\mathrm{VG}}(u) = C \ln\left(\frac{GM + (M - G)iu + u^2}{GM}\right). \tag{2.88}$$

As a remark, to generate stochastic volatility for the VG process, one may apply the time change T_t using the stochastic activity rate dynamics v_t [see (2.74)]. The details of applying the second time change for stochastic volatility can be found in Carr *et al.* (2003).

Given that the CGMY process nests the VG process and the VG process is a time-changed Brownian motion subordinated by a Gamma process, can the CGMY process be expressed as a time-changed Brownian motion? This question is well answered by Madan and Yor (2008), who manage to show that the CGMY process is a time-changed Brownian motion with drift. The law of the time change is absolutely continuous over finite time intervals with respect to that of the one-sided stable $Y/2$ subordinator.

The three parameters, σ, ν and θ, control the skewness and kurtosis of the VG process. Madan *et al.* (1998) obtain the analytic formulas for the first four order central moments of the return distribution of $X(t)$ as follows:

$$
\begin{aligned}
& m = E[X(t)] = \theta t, \quad E[(X(t)-m)^2] = (\theta^2\nu + \sigma^2)t, \\
& E[(X(t)-m)^3] = (2\theta^3\nu^2 + 3\sigma^2\theta\nu)t, \\
& E[(X(t)-m)^4] = (3\sigma^4\nu + 12\sigma^2\theta^2\nu^2 + 6\theta^4\nu^3)t + (3\sigma^4 + 6\sigma^2\theta^2\nu + 3\theta^4\nu^2)t^2.
\end{aligned} \tag{2.89}
$$

Martingale adjustment

We consider the compensator or martingale adjustment ω_{VG} required such that the discounted asset price S_t modeled as an exponential VG process is a martingale under a risk neutral measure Q. Let r be the constant riskless interest rate. The risk neutral exponential VG process for the asset price is given by

$$S_t = S_0 e^{(r+\omega_{\mathrm{VG}})t + X(t;\sigma,\nu,\theta)}.$$

To satisfy the martingale condition: $E_Q\left[e^{-rt}S_t\right] = S_0$, we need to choose ω_{VG} such that

$$E_Q\left[e^{\omega_{\mathrm{VG}}t + X(t;\sigma,\nu,\theta)}\right] = 1$$

or equivalently,

$$\omega_{\mathrm{VG}} = \psi_{\mathrm{VG}}(-i).$$

Using (2.83b), the compensator for the exponential VG process is found to be

$$\omega_{\mathrm{VG}} = \frac{1}{\nu}\ln\left(1 - \theta\nu - \frac{\sigma^2\nu}{2}\right). \tag{2.90}$$

2.4.3 Normal Inverse Gaussian model

The Normal Inverse Gaussian (NIG) model is initiated in the two papers by Barndorff-Nielsen (1997, 1998). Similar to the VG model, it is generated by subordinating a Brownian motion through an inverse Gaussian process $I(t;a,b)$. Both the inverse Gaussian process and Gamma process belong to the family of tempered stable subordinators, where the choice of the index of stability α is $\alpha = \frac{1}{2}$ for $I(t;a,b)$ and $\alpha = 0$ for $\gamma(t;\mu,\nu)$ [see (4.19) in Cont and Tankov (2003)]. The Lévy measure of the inverse Gaussian process $I(t;a,b)$ is given by

$$\Pi_I(\mathrm{d}x) = \frac{ae^{-bx}}{x^{3/2}}\mathbf{1}_{\{x>0\}}\,\mathrm{d}x, \quad a>0 \text{ and } b>0. \tag{2.91}$$

The characteristic function and density function of $I(t;a,b)$ can be found to be

$$\phi_{I(t)}(u) = e^{2at\sqrt{\pi}(\sqrt{b}-\sqrt{b-iu})}, \tag{2.92a}$$

$$f_{I(t)}(x) = \frac{at}{x^{3/2}} \exp\left(-\frac{(a\sqrt{\pi}t - \sqrt{b}x)^2}{x}\right), \quad x > 0, \tag{2.92b}$$

respectively. The density function of the inverse Gaussian process $I(t;a,b)$ can be related to the density function of the first passage time of a Brownian motion with drift to a barrier threshold. Since the first passage time can be interpreted as the right inverse of the Brownian path, it is called the inverse Gaussian process. The term "Normal" in the NIG model refers to the subordination of a Brownian motion (Normal) by an inverse Gaussian process.

Similar to the VG model, we impose the condition that $I(t;a,b)$ has unit mean rate: $E[I(1;a,b)] = 1$ and write $\text{var}(I(1;a,b)) = \nu$. These conditions give

$$a = \frac{1}{\sqrt{2\pi\nu}} \text{ and } b = \frac{1}{2\nu}. \tag{2.93}$$

The NIG process is obtained by subordinating a Brownian motion W_t with volatility σ and drift μ through the inverse Gaussian process $I\left(1; \frac{1}{\sqrt{2\pi\nu}}, \frac{1}{2\nu}\right)$ with unit mean rate and variance rate ν, where

$$X(t;\mu,\sigma,\nu) = \mu t + \sigma W_{I(t)}. \tag{2.94}$$

Using the Bochner formula (2.84), the characteristic exponent of the NIG model is found to be

$$\psi_{\text{NIG}}(u) = \delta\left(\sqrt{\alpha^2 - (\beta + iu)^2} - \sqrt{\alpha^2 - \beta^2}\right), \tag{2.95}$$

where

$$\beta = \frac{\mu}{\sigma^2}, \ \delta = \frac{\sigma}{\sqrt{\nu}} \text{ and } \alpha^2 = \beta^2 + \frac{1}{\sigma^2\nu}.$$

The parameters observe the properties: $\alpha > 0$, $\delta > 0$ and $-\alpha < \beta < \alpha$. The first four cumulants of the NIG model are found to be

$$\begin{aligned}
c_1(X(1)) &= E[X_1] = \mu, \\
c_2(X(1)) &= \text{var}(X(1)) = \sigma^2 + \nu\mu^2, \\
c_3(X(1)) &= 3\nu\mu(\sigma^2 + \nu\mu^2), \\
c_4(X(1)) &= 3\nu(\sigma^2 + \nu\mu^2)(\sigma^2 + 5\nu\mu^2).
\end{aligned} \tag{2.96}$$

The Lévy measure of the NIG model is given by

$$\Pi_{\text{NIG}}(\mathrm{d}x) = \frac{\sqrt{\mu^2 + \frac{\sigma^2}{\nu}}}{2\pi\sigma\sqrt{\nu}} \frac{e^{\frac{\mu}{\sigma^2}x}}{|x|} \kappa_1\left(\frac{\sqrt{\mu^2 + \frac{\sigma^2}{\nu}}}{\sigma^2}|x|\right) \mathrm{d}x, \tag{2.97}$$

where $\kappa_1(\cdot)$ is the first-order modified Bessel function of the second kind.

2.4.4 Barndorff-Nielsen and Shephard model

The Barndorff-Nielsen and Shephard (BNS) model (2001) is a stochastic volatility model of the asset return with the inclusion of jump in variance. The proxy variance process is modeled as a non-Gaussian Ornstein-Uhlenbeck process with Lévy jumps, which may follow the Gamma process or inverse Gaussian process. Under the statistical measure P, the asset price S_t evolves in time as

$$\frac{\mathrm{d}S_t}{S_t} = (\mu + \beta V_t)\mathrm{d}t + \sigma_t \, \mathrm{d}W_t + \rho \, \mathrm{d}Z_{\lambda t}, \tag{2.98a}$$

where W_t is a standard Brownian motion, ρ, μ, and β are constant parameters (note that ρ is not interpreted as the correlation coefficient), and σ_t^2 follows a non-Gaussian Ornstein-Uhlenbeck process as depicted by

$$\mathrm{d}\sigma_t^2 = -\lambda \sigma_t^2 \, \mathrm{d}t + \mathrm{d}Z_{\lambda t}, \tag{2.98b}$$

where $\lambda > 0$. Here, $Z_{\lambda t}$ is a subordinator referred as the background driving Lévy process. We assume W_t and $Z_{\lambda t}$ to be independent. The negative multiplier $-\lambda$ gives the mean reversion feature of σ_t^2. Due to $Z_{\lambda t}$, σ_t^2 is non-Gaussian. The popular choices for $Z_{\lambda t}$ in the BNS model are the Gamma process or inverse Gaussian process. Both choices lead to nice analytic tractability of the characteristic functions of integrated variance and returns. For the dynamics of S_t in (2.98a), the term βV_t corresponds to volatility risk premium and the last term $\rho \, \mathrm{d}Z_{\lambda t}$ accounts for the leverage effect, typically with $\rho < 0$. According to (2.98b), σ_t^2 increases by positive jumps of $Z_{\lambda t}$ and decays exponentially between two consecutive jumps.

2.5 Stochastic volatility models with jumps

Volatility is interpreted as the standard deviation of asset returns. It is a hidden stochastic process on its own, which is not directly observable. In stochastic volatility models, we model the processes of asset return and instantaneous variance as correlated stochastic processes. In this section, we present an overview of various formulations of stochastic volatility models. We also present some useful formulas for the density function and characteristic function of the Cox-Ingersoll-Ross (CIR) process for the instantaneous variance. Lastly, we derive the fair strike formula for a swap on the continuous realized variance under the affine stochastic volatility model with simultaneous jumps in the asset price and variance processes.

Let S_t and V_t denote the time-t asset price and instantaneous variance, respectively. We consider the following stochastic volatility model with simultaneous jump as modeled by the common Poisson jump N_t in the asset return and variance under a martingale pricing measure Q:

$$\frac{dS_t}{S_t} = (r-q-\lambda m)\,dt + \sqrt{V_t}\left(\rho\, dW_t^1 + \sqrt{1-\rho^2}\, dW_t^2\right) + (e^{J^S}-1)\,dN_t,$$
$$dV_t = \alpha(V_t)\,dt + \beta(V_t)\,dW_t^1 + J^V\,dN_t, \tag{2.99}$$

where r is the constant risk-free rate, q is the constant continuous dividend yield, and ρ is the correlation coefficients between S_t and V_t. The Brownian motion W_t^S associated with S_t is expressed by the Cholesky decomposition:

$$W_t^S \overset{d}{\sim} \rho\, W_t^1 + \sqrt{1-\rho^2}\, W_t^2.$$

Also, λ is the intensity parameter of the Poisson process N_t and m is the compensator. It is reasonable to assume contemporaneous arrivals of jumps in asset return and variance, where both jumps arise from the same financial shock. Let J^S and J^V denote the random jump magnitude of $\ln S_t$ and V_t, respectively. A wide set of distributions can be used to model J^S and J^V. The following canonical jump size distributions are usually adopted:

$$J^V \sim \text{Exp}(1/\eta) \quad \text{and} \quad J^S | J^V \sim N(\nu + \rho_J J^V, \delta^2). \tag{2.100}$$

These distributions correspond to the exponential distribution for J^V with mean η and the Gaussian distribution for J^S with mean $\nu + \rho_J J^V$ and variance δ^2 conditional on J^V. Here, ρ_J models the impact of J^V on the mean of J^S. These choices of jump distributions somehow reflect the nature of jumps upon arrival of shocks in the markets, where the asset return may react with either an increase or decrease in value while the volatility of returns normally increases with shocks.

The parameter m in the compensator is found to be

$$m = E[e^{J^S} - 1] = E\left[E\left[e^{J^S} | J_V\right]\right] - 1 = E\left[e^{\nu+\delta^2/2+\rho_J J_V}\right] - 1$$
$$= e^{\nu+\delta^2/2}\int_0^\infty \frac{1}{\eta} e^{-\frac{u}{\eta}} e^{\rho_J u}\, du - 1 = \frac{e^{\nu+\delta^2/2}}{1-\eta\rho_J} - 1, \tag{2.101}$$

provided that $\eta\rho_J < 1$. Also, note that

$$E[J^V\, dN_t] = \lambda\eta. \tag{2.102}$$

Leverage effect

The negative correlation coefficient ρ between the level of asset return and its instantaneous variance captures the leverage effect where decreases in asset return are associated with increases in variance. This is an important observation for equity index valuation since it increases the probability of a large loss and consequently the value of the out-of-the-money put options. The leverage effect induces negative skewness in asset returns and yields a volatility smirk.

The general specification of the drift term $\alpha(V_t)$ and variance term $\beta(V_t)$ in V_t includes the Heston model, Hull-White model, 3/2-model and others. The Heston model (1993) corresponds to the choice of

$$\alpha(V_t) = \kappa(\theta - V_t) \quad \text{and} \quad \beta(V_t) = \varepsilon\sqrt{V_t}. \tag{2.103}$$

The drift term $\alpha(V_t)$ exhibits the mean reversion phenomenon, where κ and θ represent the mean reversion speed and level, respectively. The dynamics of V_t in the Heston model takes the form as the CIR short rate model (Cox *et al.*, 1985). The Hull-White model (1987) chooses

$$\alpha(V_t) = \kappa(\theta - V_t) \quad \text{and} \quad \beta(V_t) = \varepsilon V_t, \tag{2.104}$$

while the 3/2-model (Lewis, 2016) chooses

$$\alpha(V_t) = V_t(\theta - \kappa V_t) \quad \text{and} \quad \beta(V_t) = \varepsilon V_t\sqrt{V_t}. \tag{2.105}$$

We assume zero jump in V_t in the 3/2-model in order to achieve analytic tractability of the model. The power of V_t in $\beta(V_t)$ in these three popular stochastic volatility models are taken to be 1/2, 1 and 3/2, respectively. The 1/2-model is also called the affine model and enjoys nice analytic tractability due to the affine structure of the dynamic equations of the model. The 3/2-model can be transformed into the affine type if we take $1/V_t$ as the state variable. In our later exposition, we focus on the 1/2-model and 3/2-model due to their nice analytic tractability.

There are other variants of the stochastic volatility models. For example, the α-hypergeometric stochastic volatility model assumes the variance dynamics to be governed by (Da Fonseca and Martini, 2016)

$$\mathrm{d}V_t = (u - be^{\alpha V_t})\mathrm{d}t + \varepsilon\,\mathrm{d}W_t^1, \tag{2.106}$$

for some constant parameters u, b, α and ε. The 4/2-model specifies the dynamics of V_t to be the CIR type as specified by (2.103) and that of S_t to be governed by (Grasselli, 2017)

$$\begin{aligned}\frac{\mathrm{d}S_t}{S_t} &= (r - q - \lambda m)\mathrm{d}t \\ &\quad + \left(a\sqrt{V_t} + \frac{b}{\sqrt{V_t}}\right)\left(\rho\,\mathrm{d}W_t^1 + \sqrt{1-\rho^2}\,\mathrm{d}W_t^2\right) + (e^{J^S} - 1)\,\mathrm{d}N_t,\end{aligned} \tag{2.107}$$

for some constant parameters a and b.

Instead of using the instantaneous variance V_t as the state variable, one may use the instantaneous volatility σ_t as the state variable. Replacing $\sqrt{V_t}$ by σ_t as the state variable and assuming the dynamic of σ_t to be driven by the Ornstein-Uhlenbeck process, the Schöbel-Zhu (1999) model is specified by

$$\mathrm{d}\sigma_t = \kappa(\theta - \sigma_t)\mathrm{d}t + \varepsilon\,\mathrm{d}W_t^1, \tag{2.108a}$$

The dynamic of S_t remains to be the same form as the first equation in (2.99). Taking $V_t = \sigma_t^2$, the use of Itô calculus yields the dynamics of V_t as governed by

$$dV_t = 2\kappa\left(\frac{\varepsilon^2}{2\kappa} + \theta\sqrt{V_t} - V_t\right)dt + 2\varepsilon\sqrt{V_t}\, dW_t^1. \quad (2.108b)$$

Another stochastic volatility model that uses σ_t as the state variable is the Inverse Gamma model, where the dynamic of σ_t is governed by

$$d\sigma_t = \kappa(\theta - \sigma_t)dt + \varepsilon\sigma_t\, dW_t^1. \quad (2.109a)$$

In terms of V_t, the above dynamic equation is transformed into

$$dV_t = \left[2\kappa\theta\sqrt{V_t} - (2\kappa - \varepsilon^2)V_t\right]dt + 2\varepsilon V_t\, dW_t^1. \quad (2.109b)$$

Langrené *et al.* (2016) manage to derive closed form infinite series expansion for the European vanilla option price functions under the Inverse Gamma model.

Feller condition and its generalization for instantaneous variance process

We are concerned with the behavior of the instantaneous variance V_t around the boundaries of the domain $[0, \infty)$. We consider the general constant-elasticity-of-variance (CEV)-type stochastic process for V_t as specified by

$$dV_t = \kappa(\theta - V_t)dt + \varepsilon V_t^p\, dW_t, \quad (2.110)$$

where all parameters are positive constants. The following properties hold.

1. For $0 < p < \frac{1}{2}$, V_t is reflected at the origin; that is, 0 is always accessible. It is necessary to impose a boundary condition at $V = 0$.

2. For $p = \frac{1}{2}$, the origin is an attainable boundary if $2\kappa\theta < \varepsilon^2$ and becomes unattainable if otherwise.

3. For $p > \frac{1}{2}$, the origin is an unattainable boundary.

4. For $p > 0$, ∞ is an unattainable boundary.

The second property is the well-known Feller condition for the variance process of the CIR type. A brief proof of the Feller condition can be found in Cheang *et al.* (2013). The proof of the generalized results can be found in Andersen and Piterbarg (2007).

Multifactor Heston models

Empirical studies on asset returns reveal that the slope and level of smirk fluctuate almost independently. Single-factor stochastic volatility models manage to capture the slope of the smirk. However, they fail to exhibit such largely independent fluctuations in the slope and level of smirk. This is because the correlation between asset returns and variance is constant in a single-factor stochastic volatility model, so this limits the model's ability to capture the time varying nature of the smirk.

Christoffersen *et al.* (2009) consider the extension of the one-factor Heston model to allow the stochastic asset price dynamics to have two independent Brownian shocks. The instantaneous variance is the sum of two stochastic factors that are uncorrelated but each factor is correlated to the respective Brownian shock in the asset price process. Their empirical studies on the two-factor models show that they allow more flexibility in modeling the volatility term structure. One of the stochastic factors has high mean reversion and determines the correlation between short term asset returns and variance, while the other stochastic factor has lower mean reversion and determines the correlation between long-term returns and variance. Da Fonseca *et al.* (2008) consider further extension of the multifactor Heston model by assuming the variance of the asset return to follow the matrix Wishart affine process, which is a multidimensional matrix Brownian motion. Stochastic leverage effect appears in their model and nature of stochastic skew effects can be modeled. These multifactor Heston model retains the affine structure, so analytic tractability similar to the usual Heston model is maintained.

2.5.1 Distribution formulas of instantaneous variance of CIR type

For the CIR process of the instantaneous variance V_t as specified by

$$dV_t = \kappa(\theta - V_t)dt + \varepsilon\sqrt{V_t}\, dW_t, \tag{2.111}$$

corresponding to the choice of $p = \frac{1}{2}$ in (2.110), its probability distribution possesses rich analytic tractability. One manages to obtain analytic formulas for the density function, together with various forms of characteristic functions of V_t and its integrated variance $\int_0^t V_u\, du$.

Cox *et al.* (1985) show that the distribution of V_t given V_0 is a noncentral chi-squared distribution, whose transition law is given by

$$V_t = \frac{\varepsilon^2(1-e^{-\kappa t})}{4\kappa}\chi_d'^2\left(\frac{4\kappa e^{-\kappa t}}{\varepsilon^2(1-e^{-\kappa t})}V_0\right), \quad t > 0, \tag{2.112}$$

where $\chi_d'^2(\lambda)$ denotes the noncentral chi-squared random variable with d degrees of freedom, where $d = \dfrac{4\theta\kappa}{\varepsilon^2}$ and λ is the noncentrality parameter. Therefore, one can sample the distribution of V_t provided that one can sample the noncentral chi-squared distribution. Broadie and Kaya (2006) present more details on the simulation of V_t based on various properties of the chi-squared distribution.

Based on the relation between V_t and $\chi_d'^2$, one can derive the density function of V_t conditional on V_0. Writing $y = V_t$, the conditional density $p_{V_t}(y|V_0)$ is given by (Cox *et al.*, 1985)

$$p_{V_t}(y|V_0) = ce^{-w-cy}\left(\frac{cy}{w}\right)^{q/2} I_q(2\sqrt{wcy}), \tag{2.113}$$

where I_q is the modified Bessel function of the first kind of order q,

$$I_q(z) = \left(\frac{z}{2}\right)^q \sum_{k=0}^{\infty} \frac{\left(\frac{z^2}{2}\right)^k}{k!\Gamma(q+k+1)},$$

$$c = \frac{2\kappa}{\varepsilon^2(1-e^{-\kappa t})}, \quad w = ce^{-\kappa t}V_0 \quad \text{and} \quad q = \frac{2\kappa\theta}{\varepsilon^2} - 1.$$

Characteristic function of integrated variance conditional on terminal variance

Broadie and Kaya (2006) show that the characteristic function of the integrated variance $\int_0^t V_s \, \mathrm{d}s$ conditional on V_0 and V_t is given by

$$E\left[\exp\left(iu\int_0^t V_s \, \mathrm{d}s \middle| V_0, V_t\right)\right]$$
$$= \frac{\gamma(u)e^{-\frac{\gamma(u)-\kappa}{2}t}1-e^{-\kappa t}}{\kappa[1-e^{-\gamma(u)t}]} \frac{I_{\frac{d}{2}-1}\left(\sqrt{V_0V_t}\,\frac{4\gamma(u)e^{-\gamma(u)t/2}}{\varepsilon^2[1-e^{-\gamma(u)t}]}\right)}{I_{\frac{d}{2}-1}\left(\sqrt{V_0V_t}\,\frac{4\kappa e^{-\kappa t/2}}{\varepsilon^2[1-e^{-\kappa t}]}\right)}$$
$$\exp\left(\frac{V_0+V_t}{\varepsilon^2}\frac{\kappa(1+e^{-\kappa t})}{1-e^{-\kappa t}} - \gamma(u)\frac{1+e^{-\gamma(u)t}}{1-e^{-\gamma(u)t}}\right), \tag{2.114}$$

where $\gamma(u) = \sqrt{\kappa^2 - 2\varepsilon^2 iu}$, $d = \dfrac{4\theta\kappa}{\varepsilon^2}$ and $I_\nu(\cdot)$ is the modified Bessel function of the first kind.

By performing the Fourier inversion of the characteristic function of $\int_0^t V_s \, \mathrm{d}s$, Broadie and Kaya (2006) manage to generate a sample distribution of $\int_0^t V_s \, \mathrm{d}s$. This is a crucial step in their exact simulation algorithm for pricing European options under the Heston model.

Laplace transform of joint density function of V_t and its integrated variance

Assuming that the model parameters satisfy the Feller condition $2k\theta \geq \varepsilon^2$, V_t can never reach zero and stays positive. Let $p_{\text{CIR}}(V,y,t;V_0)$ be the joint transition density of V_t and $\int_0^t V_s \, \mathrm{d}s$ from $(V_0, 0)$ at time 0 to (V, y) at time t. The following lemma gives the Laplace transform of the joint density $p_{\text{CIR}}(V,y,t;V_0)$ with respect to y.

Lemma 2.1 *The Laplace transform $G_{\text{CIR}}(V,t;V_0,\eta)$ of the joint density $p_{\text{CIR}}(V,y,t;V_0)$ with respect to y is given by*

$$G_{\text{CIR}}(V,t;V_0,\eta) = \int_0^\infty e^{-\eta y} p_{\text{CIR}}(V,y,t;V_0) \, \mathrm{d}y$$

$$= 2\exp\left(-\varepsilon^2\kappa(V-V_0)-\frac{1+e^{-\delta t}}{1-e^{-\delta t}}(V+V_0)\delta\varepsilon^2\right)\left[\frac{e^{-\delta t}(e^{\delta t}-1)^2V\varepsilon^4}{V_0\delta^2}\right]^{\frac{\beta-1}{2}}$$
$$\left[\frac{e^{\frac{1}{2}(k+\delta)t}\delta}{(e^{\delta t}-1)\varepsilon^2}\right]^{\beta} I_{\beta-1}\left(4\sqrt{\frac{e^{\delta t}V_0V\delta^2}{(e^{\delta t}-1)^2\varepsilon^4}}\right), \tag{2.115}$$

where I_q is the modified Bessel function of the first kind of order q,

$$\delta = \sqrt{\kappa^2+2\varepsilon^2\eta} \text{ and } \beta = \frac{2\kappa\theta}{\varepsilon^2}.$$

The proof of Lemma 2.1 can be found in Lewis (2016).

2.5.2 Pricing of swap on continuous realized variance

Based on the dynamic of V_t and the correlated jumps between S_t and V_t, we manage to derive the fair strike of a swap on continuous realized variance without the necessity of solving the joint dynamic model of S_t and V_t. Readers should be aware that variance swap contracts use discrete realized variance calculated based on the sum of squares of logarithm returns. The pricing procedures for variance derivatives on discrete realized variance are more complex, the details of which are relegated to later chapters.

Given the dynamic equation of V_t of the CIR type with jump as specified by

$$\mathrm{d}V_t = \kappa(\theta - V_t)\mathrm{d}t + \varepsilon\sqrt{V_t}\,\mathrm{d}W_t + J^V\,\mathrm{d}N_t \tag{2.116}$$

and the jump distribution $J^V \sim \mathrm{Exp}(1/\eta)$, we derive the closed-form formula for its first-order moment $E_0^Q[V_t]$. Taking expectation on the above dynamic equation for V_t, we obtain

$$\frac{\mathrm{d}E_0^Q[V_t]}{\mathrm{d}t} = \kappa\{\theta - E_0^Q[V_t]\} + \lambda\eta, \quad E[V_0] = V_0.$$

Solving this ordinary differential equation for $E_0^Q[V_t]$, we obtain

$$E_0^Q[V_t] = \theta + e^{-\kappa t}(V_0-\theta) + \frac{\lambda\eta}{\kappa}(1-e^{-\kappa t}).$$

From the dynamic equation of S_t in (2.99), we rewrite the dynamic equation in terms of $\ln S_t$ as follows:

$$\mathrm{d}\ln S_t = \left(r-q-\lambda m-\frac{V_t}{2}\right)\mathrm{d}t + \sqrt{V_t}(\rho\,\mathrm{d}W_t^1 + \sqrt{1-\rho^2}\,\mathrm{d}W_t^2) + J_S\,\mathrm{d}N_t. \tag{2.117}$$

The continuous realized variance is the limit of discrete realized variance when the number of monitoring instants tends to infinity. From the dynamic equation of $\ln S_t$ in (2.117), we deduce that the continuous realized variance $V_c[0,T]$ over $[0,T]$ comes from two sources: accumulated variance from the dynamics of the instantaneous variance V_t and contribution from the jumps of asset price. By normalization with the annualization factor, we obtain

$$V_c[0,T] = \lim_{N\to\infty} \frac{A}{N}\sum_{i=1}^{N}\left(\ln\frac{S_{t_i}}{S_{i-1}}\right)^2 = \frac{1}{T}\left[\int_0^T V_t\, dt + \sum_{k=1}^{N_T}(J_k^S)^2\right], \tag{2.118a}$$

where N_T is the total random number of jumps over $[0,T]$. Mathematically, the quadratic variation of $\ln S_t$ over $[0,T]$ is given by

$$[\ln S_t, \ln S_t]_0^T = \int_0^T V_t\, dt + \sum_{k=1}^{N_T}(J_k^S)^2 = TV_c[0,T]. \tag{2.118b}$$

The fair strike of the swap on continuous realized variance is given by

$$K_v = \frac{1}{T}\left\{\int_0^T E_0^Q[V_t]\, dt + E_0^Q\left[\sum_{k=1}^{N_T}(J_k^S)^2\right]\right\}. \tag{2.119}$$

Using $E_0^Q[V_t]$ derived in (2.5.2), we obtain

$$\frac{1}{T}\int_0^T E_0^Q[V_t]\, dt = \frac{1}{T}\left[\frac{\theta}{\kappa}(\kappa T - 1 + e^{-\kappa T}) + \frac{1-e^{-\kappa T}}{\kappa}V_0 - \frac{\lambda\eta}{\kappa^2}(1-e^{-\kappa T} - \kappa T)\right].$$

Given the jump distribution in (2.100), where

$$J^S|J^V \sim N(\nu + \rho_J J^V, \delta^2) \quad \text{and} \quad J^V \sim \text{Exp}(1/\eta),$$

we observe

$$E_0^Q[(J_k^S)^2] = \text{var}(J_k^S) + \left(E_0^Q\left[J_k^S\right]\right)^2 = \delta^2 + \rho_J^2\eta^2 + (\nu + \rho_J\eta)^2$$

while the expected number of Poisson jumps over $[0,T]$ is λT. We then obtain

$$\frac{1}{T}E_0^Q\left[\sum_{k=1}^{N_T}(J_k^S)^2\right] = \frac{1}{T}\{\lambda T\left[\delta^2 + \rho_J^2\eta^2 + (\nu+\rho_J\eta)^2\right]\}.$$

Combining the above results together, the fair strike of the variance swap on continuous realized variance is given by

$$K_v = \frac{1}{T}\left\{\frac{\theta}{\kappa}(\kappa T - 1 + e^{-\kappa T}) + \frac{1-e^{-\kappa T}}{\kappa}V_0 - \frac{\lambda\eta}{\kappa^2}(1-e^{-\kappa T}-\kappa T)\right.$$
$$\left. + \lambda[\delta^2 + \rho_J^2\eta^2 + (\nu+\rho_J\eta)^2]T\right\}. \tag{2.120}$$

Alternatively, one may compute the fair strike of the swap on discrete realized variance and take the limit of infinite number of monitoring instants to obtain the limit of continuous sampling. As expected, both procedures give the same fair strike formula for the swap on continuous realized variance [see (4.12) and Sec. 4.5].

To price swaps on continuous realized volatility, it may be more convenient to use stochastic volatility σ_t as the state variable instead of stochastic variance V_t. Pricing formulas of continuously sampled volatility swaps on various model assumptions on σ_t can be found in Howison *et al.* (2004).

Remarks

1. Besides the expectation approach, Javaheri *et al.* (2004) use the alternative approach of solving partial differential equation to derive the fair strike formula for continuously sampled variance under the GARCH (1,1) models.

2. One may follow similar procedures to price continuously sampled variance swap under various stochastic volatility models with regime switching, like the Heston model (Elliott *et al.*, 2007), non-Gaussian Ornstein-Uhlenbeck model (Benth *et al.*, 2007) and Schöbel-Zhu-Hull-White hybrid model (Shen and Siu, 2013).

2.6 Affine jump-diffusion stochastic volatility models

In the affine stochastic volatility model with simultaneous jumps in the asset return S_t and instantaneous variance V_t, commonly called the SVSJ model, the joint dynamics under a risk neutral measure Q are given by

$$\begin{aligned} \frac{\mathrm{d}S_t}{S_t} &= (r-q-\lambda m)\,\mathrm{d}t+\sqrt{V_t}\,\mathrm{d}W_t^S+(e^{J^S}-1)\,\mathrm{d}N_t, \\ \mathrm{d}V_t &= \kappa(\theta-V_t)\,\mathrm{d}t+\varepsilon\sqrt{V_t}\,\mathrm{d}W_t^V+J^V\,\mathrm{d}N_t, \end{aligned} \tag{2.121}$$

where r is the interest rate, q is the dividend yield, ε is the volatility of V_t, κ is the mean reversion speed and θ is the long-term mean of V_t. Here W_t^S and W_t^V are correlated standard Brownian motions that observes $\mathrm{d}W_t^S\mathrm{d}W_t^V=\rho\;\mathrm{d}t$, where ρ is the correlation coefficient. Also, N_t is a Poisson process with constant intensity λ that is independent of the two Brownian motions. We use J^S and J^V to denote the random jump sizes of the log asset price and its variance, respectively. These random jump sizes are assumed to be independent of W_t^S, W_t^V and N_t. We assume the jump size distributions to follow the same canonical forms as stated in (2.100). Accordingly, m is given by (2.101).

2.6.1 Joint moment generating function of the affine model

Let $X_t=\ln S_t$, and we define the joint moment generating function (mgf) of X_t and V_t to be

$$E_t^Q[\exp(\phi X_T+bV_T+\gamma)],$$

where ϕ, b, and γ are constant parameters. The inclusion of the constant term γ facilitates the later iterated expectation calculation required in pricing variance swaps (see Sec. 4.3). The joint mgf can be regarded as the time-t forward value of the contingent claim with the time-T terminal payoff: $\exp(\phi X_T + bV_T + \gamma)$.

Let $U(X_t, V_t, t)$ denote the non-discounted time-t value of a European contingent claim with the terminal payoff function: $U(X_T, V_T, T)$, where T is the maturity date. By adopting the temporal variable, $\tau = T - t$, it can be deduced from the Feynman-Kac theorem that $U(X, V, \tau)$ is governed by the following partial integro-differential equation (PIDE):

$$\begin{aligned}\frac{\partial U}{\partial \tau} &= \left(r - q - m\lambda - \frac{V}{2}\right)\frac{\partial U}{\partial X} + \kappa(\theta - V)\frac{\partial U}{\partial V} + \frac{V}{2}\frac{\partial^2 U}{\partial X^2} + \frac{\varepsilon^2 V}{2}\frac{\partial^2 U}{\partial V^2} \\ &\quad + \rho\varepsilon V\frac{\partial^2 U}{\partial X \partial V} + \lambda E^Q\left[U(X + J^s, V + J^V, \tau) - U(X, V, \tau)\right].\end{aligned} \tag{2.122}$$

The terminal payoff in the price function of the contingent claim becomes the initial condition at $\tau = 0$ of the PIDE. The mgf is seen to satisfy PIDE (2.122). Once the joint mgf is known, the respective marginal mgf can be obtained easily by setting the irrelevant parameters in the joint mgf to be zero. For example, the marginal mgf with respect to the state variable V can be obtained by setting $\phi = \gamma = 0$.

Assume that the jump distributions J^V and $J^S|J^V$ are exponential and conditional normal, respectively, which are governed by (2.100). Also, observing $\eta\rho_J < 1$, we solve for the joint mgf $U(X, V, \tau)$. Thanks to the affine structure in the SVSJ model, $U(X, V, \tau)$ admits an analytic solution of the following exponential affine form (Duffie *et al.*, 2000):

$$U(X, V, \tau) = \exp\left(\phi X + B(\tau; \Theta, \mathbf{q})V + \Gamma(\tau; \Theta, \mathbf{q}) + \Lambda(\tau; \Theta, \mathbf{q})\right), \tag{2.123}$$

where the parameter functions $B(\tau; \Theta, \mathbf{q})$, $\Gamma(\tau; \Theta, \mathbf{q})$ and $\Lambda(\tau; \Theta, \mathbf{q})$ satisfy the following Riccati system of ordinary differential equations:

$$\begin{aligned}\frac{dB}{d\tau} &= -\frac{1}{2}(\phi - \phi^2) - (\kappa - \rho\varepsilon\phi)B + \frac{\varepsilon^2}{2}B^2 \\ \frac{d\Gamma}{d\tau} &= (r - q)\phi + \kappa\theta B \\ \frac{d\Lambda}{d\tau} &= \lambda\left\{E^Q[\exp(\phi J^S + BJ^V) - 1] - m\phi\right\}\end{aligned} \tag{2.124}$$

with the initial conditions:

$$B(0) = b, \ \Gamma(0) = \gamma \ \text{ and } \ \Lambda(0) = 0.$$

Here, $\mathbf{q} = (\phi \ b \ \gamma)^T$ and we use the symbol Θ to indicate the dependence of these parameter functions on the model parameters in the SVSJ model. The parameter functions are obtained below by solving the Riccati system (2.124):

$$\begin{aligned}
B(\tau;\Theta,\mathbf{q}) &= \frac{b(\xi_- e^{-\zeta\tau}+\xi_+)-(\phi-\phi^2)(1-e^{-\zeta\tau})}{(\xi_+ +\varepsilon^2 b)e^{-\zeta\tau}+\xi_- -\varepsilon^2 b}, \\
\Gamma(\tau;\Theta,\mathbf{q}) &= (r-q)\phi\tau+\gamma-\frac{\kappa\theta}{\varepsilon^2}\left[\xi_+\tau+2\ln\frac{(\xi_+ +\varepsilon^2 b)e^{-\zeta\tau}+\xi_- -\varepsilon^2 b}{2\zeta}\right], \\
\Lambda(\tau;\Theta,\mathbf{q}) &= -\lambda(m\phi+1)\tau+\lambda e^{\phi\nu+\delta^2\phi^2/2} \\
&\quad \left[\frac{k_2}{k_4}\tau-\frac{1}{\zeta}\left(\frac{k_1}{k_3}-\frac{k_2}{k_4}\right)\ln\frac{k_3 e^{-\zeta\tau}+k_4}{k_3+k_4}\right], \qquad (2.125)
\end{aligned}$$

with

$$\begin{aligned}
&\zeta=\sqrt{(\kappa-\rho\varepsilon\phi)^2+\varepsilon^2(\phi-\phi^2)}, \quad \xi_\pm=\zeta\mp(\kappa-\rho\varepsilon\phi), \\
&k_1=\xi_+ +\varepsilon^2 b, \quad k_2=\xi_- -\varepsilon^2 b, \\
&k_3=(1-\phi\rho_J\eta)k_1-\eta(\phi-\phi^2+\xi_- b), \\
&k_4=(1-\phi\rho_J\eta)k_2+\eta(\phi-\phi^2-\xi_+ b). \qquad (2.126)
\end{aligned}$$

The proof of the formulas of these parameter functions is presented in the Appendix.

2.6.2 Numerical valuation of complex algorithms and Heston trap

Recall that the logarithm of a complex number z is given by

$$\log z=\ln|z|+i\arg z. \qquad (2.127a)$$

Note that the multivalued nature of $\log z$ stems from $\arg z$, where

$$\arg z=\operatorname{Arg} z+2n\pi i. \qquad (2.127b)$$

Here, $\operatorname{Arg} z\in[-\pi,\pi)$ is called the principal argument and n is an integer. The branch cut of $\log z$ is taken to be $(-\infty,0]$ along the negative real axis.

Since the mgf of the Heston model involves the complex logarithm, the mgf may become discontinuous if we restrict the logarithm to its principal branch. This would lead to incorrect evaluation of the mgf formula. An interesting account and numerical examples on such difficulties in the numerical valuation of the Heston model, so called the Heston trap, can be found in Albrecher *et al.* (2007). Zhu (2010) proposes an ad hoc procedure to seek for any discontinuities in an integration algorithm and make correction for them. Kahl and Jäckel (2005) propose an easier implementable method based on the rotation count algorithm. Guo and Hung (2007) and Fahrner (2007) propose various formulations of the mgf formula that avoid the problem of discontinuities in the logarithm function. Lord and Kahl (2010) show that there exists one special formulation of the mgf in which the principal branch is always the correct one.

We consider the Heston stochastic volatility model, whose dynamics of the asset price S_t and its instantaneous variance V_t under a risk neutral measure Q are governed by

$$\begin{aligned} dS_t &= \mu(t)S_t\, dt + \sqrt{V_t}S_t\, dW_t^S \\ dV_t &= \kappa(\theta - V_t)dt + \varepsilon\sqrt{V_t}\, dW_t^V, \end{aligned} \tag{2.128}$$

where the two Brownian motions are correlated with $dW_t^S\, dW_t^V = \rho\, dt$. We assume $|\rho| < 1$, $\kappa > 0$ and $\varepsilon > 0$. The drift $\mu(t)$ is determined by performing calibration of the forward curve of the underlying asset. The characteristic function of the logarithm of the asset price is exponentially affine in f, where f is the logarithm of the forward price, such that

$$E_t^Q\left[e^{iu\ln S_T}\right] = \exp\left(iuf + A(\tau;u) + VB(\tau;u)\right), \quad \tau = T - t.$$

The parameter functions $A(\tau;u)$ and $B(\tau;u)$ are seen to satisfy the following Riccati system of ordinary differential equations:

$$\begin{aligned} \frac{dA(\tau;u)}{d\tau} &= \kappa\theta B(\tau;u), \\ \frac{dB(\tau;u)}{d\tau} &= \alpha(u) - \beta(u)B(\tau;u) + \frac{\varepsilon^2}{2}B^2(\tau;u), \end{aligned} \tag{2.129}$$

with initial conditions: $A(0;u) = 0$ and $B(0;u) = 0$. Here, the parameters in (2.129) are given by

$$\alpha(u) = -\frac{u(i+u)}{2} \quad \text{and} \quad \beta(u) = \kappa - \rho\varepsilon ui. \tag{2.130}$$

Heston (1993) obtains the following solution to the Riccati system (2.129):

$$\begin{aligned} A(\tau;u) &= \frac{\kappa\theta}{\varepsilon^2}\{[\beta(u) - D(u)]\tau - 2\ln\psi_H(\tau;u)\}, \\ B(\tau;u) &= \frac{\beta(u) - D(u)}{\varepsilon^2}\,\frac{1 - e^{-D(u)\tau}}{1 - G(u)e^{-D(u)\tau}}, \end{aligned} \tag{2.131}$$

where

$$\begin{aligned} D(u) &= \sqrt{\beta(u)^2 - 2\alpha(u)\varepsilon^2}, \\ G(u) &= \frac{\beta(u) - D(u)}{\beta(u) + D(u)}, \\ \psi_H(\tau;u) &= \frac{G(u)e^{-D(u)\tau} - 1}{G(u) - 1}. \end{aligned} \tag{2.132}$$

Unfortunately, the formulation of $\ln\psi_H(\tau;u)$ in $A(\tau;u)$ causes discontinuities when the principal branch of the logarithm is used.

To remedy the difficulties of discontinuities in the complex logarithm function, Lord and Kahl (2010) propose an algebraic equivalent form $\psi_{LK}(\tau;u)$ to replace $\psi_H(\tau;u)$, where

$$\psi_{LK}(\tau;u) = \frac{c(u)e^{D(u)\tau} - 1}{c(u) - 1} \quad \text{with} \quad c(u) = \frac{1}{G(u)}, \tag{2.133a}$$

and $A(\tau;u)$ is modified accordingly to become

$$A(\tau;u) = \frac{\kappa\theta}{\varepsilon^2}\left\{[\beta(u) + D(u)\tau] - 2\ln\psi_{LK}(\tau;u)\right\}. \tag{2.133b}$$

Provided that $\text{Im}\{u\} \geq -\frac{\kappa}{\rho\varepsilon}$ and $\frac{\kappa}{\varepsilon} \leq \rho \leq \frac{2\kappa}{\varepsilon}$, they prove that no branch switching of the complex logarithm is required when $\psi_{LK}(\tau;u)$ is used in the Heston characteristic function. As a result, the rotation count algorithm is no longer needed.

2.6.3 Schöbel-Zhu model

The Schöbel-Zhu model uses the instantaneous volatility σ_t (instead of V_t as in the Heston model) to drive stochastic volatility. Under a risk neutral measure Q, the dynamics of the asset price S_t and its instantaneous volatility σ_t are governed by

$$\begin{aligned} dS_t &= \mu(t)S_t\, dt + \sigma_t S_t\, dW_t^S \\ d\sigma_t &= \kappa(\theta - \sigma_t)dt + \varepsilon\, dW_t^\sigma, \end{aligned} \tag{2.134}$$

where $dW_t^S\, dW_t^\sigma = \rho\, dt$. The other parameters have similar interpretation as in the affine model. By the transformation $V_t = \sigma_t^2$, the dynamic equation for V_t can be shown to be

$$dV_t = d\sigma_t^2 = 2\sigma_t\, d\sigma_t + \varepsilon^2\, dt = \left(\varepsilon^2 + 2\kappa\theta\sigma_t - 2\kappa V_t\right)dt + 2\varepsilon\sigma_t\, dW_t^\sigma. \tag{2.135}$$

Recall that f is the logarithm of the forward price. The Schöbel-Zhu model is seen to be affine in f, σ and V (also called the linear-quadratic model in f and σ). Similar to (2.6.2), the characteristic function of the logarithm of the asset price under the Schöbel-Zhu model is exponentially affine in f and takes an alternative form:

$$E_t^Q\left[e^{iu\ln S_T}\right] = \exp\left(iuf + A_\sigma(\tau;u) + \sigma B_\sigma(\tau;u) + VB_V(\tau;u)\right),\ \tau = T - t. \tag{2.136}$$

The parameter functions $A_\sigma(\tau;u)$, $B_\sigma(\tau;u)$ and $B_V(\tau;u)$ satisfy the following Riccati system of ordinary differential equations:

$$\begin{aligned} \frac{dA_\sigma(\tau;u)}{d\tau} &= \kappa\theta B_\sigma(\tau;u) + \frac{\varepsilon^2}{2}B_\sigma^2(\tau;u) + \varepsilon^2 B_V(\tau;u), \\ \frac{dB_\sigma(\tau;u)}{d\tau} &= 2\kappa\theta B_V(\tau;u) + B_\sigma(\tau;u)\left[2\varepsilon^2 B_V(\tau;u) - \frac{\beta(u)}{2}\right], \\ \frac{dB_V(\tau;u)}{d\tau} &= \alpha(u) - \beta(u)B_V(\tau;u) + 2\varepsilon^2 B_V^2(\tau;u), \end{aligned} \tag{2.137}$$

with initial conditions: $A_\sigma(0;u) = B_\sigma(0;u) = B_V(0;u) = 0$. The parameters $\alpha(u)$ and $\beta(u)$ are defined in (2.130).

Let $\phi_H(u)$ and $\phi_{SZ}(u)$ denote the respective characteristic function of the Heston model and Schöbel-Zhu model. The two characteristic functions are related by (Lord and Kahl, 2010)

$$\begin{aligned}&\phi_{SZ}(u;S,\sigma,\kappa,\varepsilon,\theta,\rho,\tau)\\&= \phi_H(u;S,\sigma^2,2\kappa,2\varepsilon,\frac{\varepsilon^2}{2\kappa},\rho,\tau)\exp(\widehat{A}_\sigma(\tau)+\sigma B_\sigma(\tau)),\end{aligned} \tag{2.138}$$

where $\widehat{A}_\sigma$ satisfies the following ordinary differential equation:

$$\frac{d\widehat{A}_\sigma(\tau;u)}{d\tau} = \kappa\theta B_\sigma(\tau;u) + 2\varepsilon^2 B_\sigma^2(\tau;u), \quad \widehat{A}_\sigma(0;u) = 0. \tag{2.139}$$

The parameter functions are found to be

$$\begin{aligned}B_V(\tau;u) &= \frac{\beta(u)-D_V(u)}{4\varepsilon^2}\,\frac{1-e^{-D_V(u)\tau}}{1-G(u)e^{-D_V(u)\tau}},\\ B_\sigma(\tau;u) &= \kappa\theta\frac{\beta(u)-D_V(u)}{D_V(u)\varepsilon^2}\,\frac{\left[1-e^{-\frac{D_V(u)\tau}{2}}\right]^2}{1-G(u)e^{-D_V(u)\tau}},\\ \widehat{A}_\sigma(\tau;u) &= \frac{\beta(u)-D_V(u)\kappa^2\theta^2}{2D_V(u)^3\varepsilon^2}\left\{\beta(u)[D_V(u)\tau-4]+D_V(u)[D_V(u)\tau-2]\right.\\ &\quad\left.+\frac{4e^{-\frac{D_V(u)\tau}{2}}\left[\frac{D_V(u)^2-2\beta(u)^2}{\beta(u)+D_V(u)}e^{-\frac{D_V(u)\tau}{2}}+2\beta(u)\right]}{1-G(u)e^{-D_V(u)\tau}}\right\}.\end{aligned} \tag{2.140}$$

where

$$D_V(u) = \sqrt{\beta(u)^2 - 8\alpha(u)\varepsilon^2}.$$

2.7 3/2 stochastic volatility model

Numerous empirical studies have reported inconsistency of the affine models in modeling asset return and its instantaneous variance with market observations. By fitting the S&P500 daily returns and implied volatilities over a 14-year period, and using an affine drift and constant-elasticity-of-variance (CEV) process for the instantaneous variance, Jones (2003) estimates the power γ (CEV parameter) of V_t in $\beta(V_t)$ [see (2.99)] to be 1.33. Chacko and Viceira (2003) employ the generalized method of moments on a 35-year period of weekly returns and a 71-year period of monthly returns, and estimate the CEV parameter to be 1.10 and 1.65, respectively. Using the same data as Jones (2003), Bakshi *et al.* (2006) test several stochastic volatility models on the time series of S&P100 implied volatilities and find that a linear drift model is rejected and the coefficient of the quadratic term is highly significant. Moreover, their estimate of the CEV parameter is 1.27. As deduced from these empirical findings, one should model the instantaneous variance process by using a diffusion process with nonaffine drift and CEV parameter greater than 1. With the choice of a quadratic drift and CEV parameter equal to 1.5, the 3/2 stochastic volatility model for modeling the instantaneous variance is more firmly supported by these empirical studies than the affine models.

The growing interest in the 3/2 model is attributed to the increasing concern in consistently modeling equity and volatility markets arising from the increasing popularity of the volatility derivatives market. Itkin and Carr (2010) introduce the 3/2 power time change process, Chan and Platen (2015) and Goard (2011) price variance swaps under the 3/2 stochastic volatility model. Yuen *et al.* (2015) derive closed-form pricing formulas for exotic variance swaps under the 3/2 model. By performing extensive numerical comparisons between the Heston model and 3/2 model, Drimus (2012) reports that the 3/2 model is superior to the Heston model since it is able to predict upward-sloping volatility of variance smiles, which exhibits better consistency with market observations. Most recently, Goard and Mazur (2013) report strong empirical evidence that VIX follows the 3/2 process rather than the square root process. In an effort to consistently modeling VIX and equity derivatives, Baldeaux and Badran (2014) consider the 3/2 plus jumps model for pricing VIX derivatives.

Analytic tractability of the underlying stochastic volatility model is essential for pricing variance and VIX derivatives. In order to facilitate pricing of variance derivatives, Carr and Sun (2007) obtain the joint characteristic function of the log asset price and its quadratic variation. Baldeaux and Badran (2014) extend their result to the 3/2 plus jumps model for pricing VIX derivatives. Lewis (2016) derives the joint transition density of the log asset price and the instantaneous variance for the 3/2 model with constant parameters.

In this section, we consider the Fourier transform of the joint density of the triple (X, I, V), where X is the log asset price, I is the continuous integrated variance and V is the instantaneous variance. We use the partial differential equation (PDE) approach that is similar to the approach used by Carr and Sun (2007) and Lewis (2016). Carr and Sun (2007) solve the governing PDE with a nice choice of transformation of variable so that the PDE is reduced to an ordinary differential equation (ODE). However, it is difficult to extend their technique with the inclusion of the instantaneous variance in the triplet. Alternatively, Lewis (2016) uses a sequence of partial Fourier transformations to reduce the original governing PDE to a one-dimensional linear PDE, which is then solvable by the standard method of characteristics. Our result extends that of Lewis (2016) to the 3/2 model with a time-dependent mean reversion. The closed-form Fourier transform formulas are useful for pricing a wide range of derivatives whose terminal payoffs depend on the terminal asset price and realized variance or both under the 3/2 model. We show how to use these Fourier transform formulas to price the forward target volatility options.

2.7.1 Model formulation

Under a pricing measure Q, the dynamic equations of the 3/2 stochastic volatility model are specified as follows:

$$\frac{\mathrm{d}S_t}{S_t} = (r-q)\,\mathrm{d}t + \sqrt{V_t}\,\mathrm{d}W_t^S,$$
$$\mathrm{d}V_t = V_t(\theta_t - \kappa V_t)\,\mathrm{d}t + \varepsilon V_t^{3/2}\,\mathrm{d}W_t^V, \tag{2.141}$$

where W_t^S and W_t^V are two correlated Brownian motions that observes $\mathrm{d}W_t^S\,\mathrm{d}W_t^V = \rho\,\mathrm{d}t$, where ρ is the correlation coefficient. Here, we assume constant riskless interest rate r and dividend yield q, though time-dependent deterministic riskless interest rate and dividend yield can be accommodated without much difficulty. In contrast to the square root process with an affine drift in the Heston model, the parameters in the 3/2 variance dynamics need to be interpreted differently. The speed of mean reversion of V_t to the long term mean level θ_t is now κV_t, which is linear in V_t and stochastic. Since $\kappa > 0$ under usual scenario, the speed of mean reversion is stronger when the instantaneous variance is higher. Also, ε should not be interpreted as the volatility of variance as in the Heston model. In fact, one needs to multiply it by a scaling factor V_t in order to make it comparable to its counterpart in the Heston model. The long-term mean reversion level is given by θ_t/κ. As pointed out by Itkin and Carr (2010), θ_t can be an independent stochastic process under the more general setting. By conditioning on the path of stochastic θ_t, analytic tractability of the 3/2 model with stochastic mean reversion level remains intact. In our discussion, we assume θ_t to be a deterministic function of time.

Similar to the Feller condition in the Heston model, Drimus (2012) shows that the parameters of the 3/2 model as specified by (2.141) are constrained by

$$\kappa - \rho\varepsilon \geq -\frac{\varepsilon^2}{2}. \tag{2.142}$$

Notice that under normal market condition, $\kappa > 0$ and $\rho \leq 0$, the above inequality is automatically satisfied.

The reciprocal of the 3/2 process of V_t is seen to be the CIR process. By letting $U_t = \frac{1}{V_t}$, U_t is given by the following inhomogeneous CIR process:

$$\mathrm{d}U_t = \left[(\kappa + \varepsilon^2) - \theta_t U_t\right]\mathrm{d}t - \varepsilon\sqrt{U_t}\,\mathrm{d}W_t^V. \tag{2.143}$$

The success of analytic tractability of the 3/2 model is derived from this analytic property. We explore analytic tractability of the 3/2 stochastic volatility model with time-dependent mean reversion parameter via the derivation of a closed-form formula for the partial Fourier transform of the triple joint transition density.

2.7.2 Partial Fourier transform of the triple joint density

Let $X_t = \ln S_t$ be the log asset price and we define the integrated variance of X_t by

$$I_t = \int_0^t V_s\,\mathrm{d}s, \tag{2.144}$$

Let $G(x,y,v,t;x',y',v',t')$ be the joint transition density of the triple (X,I,V) from state (x,y,v) at time t to state (x',y',v') at a later time t', $t' > t$. By the Feynman-Kac Theorem, G satisfies the following three-dimensional Kolmogorov backward equation:

$$-\frac{\partial G}{\partial t} = \left(r - q - \frac{v}{2}\right)\frac{\partial G}{\partial x} + \frac{v}{2}\frac{\partial^2 G}{\partial x^2} + v\frac{\partial G}{\partial y} + v(\theta_t - \kappa v)\frac{\partial G}{\partial v} + \frac{\varepsilon^2 v^3}{2}\frac{\partial^2 G}{\partial v^2} + \rho\varepsilon v^2\frac{\partial^2 G}{\partial x \partial v}, \tag{2.145}$$

subject to the terminal condition:

$$G(x,y,v,t';x',y',v',t') = \delta(x-x')\delta(y-y')\delta(v-v'),$$

where $\delta(\cdot)$ is the Dirac delta function. We allow the mean reversion level θ_t to be time dependent while all other parameters are assumed to be constant. Let $\check{G}$ denote the *partial Fourier transform* of G with respect to x and y, which is defined by

$$\check{G}(x,y,v,t;\omega,\eta,v',t') = \int_{-\infty}^{\infty}\int_{0}^{\infty} e^{\mathrm{i}\omega x' + \mathrm{i}\eta y'} G(x,y,v,t;x',y',v',t')\,\mathrm{d}y'\mathrm{d}x'. \tag{2.146}$$

Thanks to the affine structure of the pair (X,I), analytic evaluation of $\check{G}$ is more tractable than G itself. Also, $\check{G}$ is more relevant in pricing calculation of the target volatility option since the terminal payoff of the option depends on the terminal asset price S_T and realized integrated variance I_T. Obviously, $\check{G}$ also satisfies (2.145) with the terminal condition:

$$\check{G}(x,y,v,t';\omega,\eta,v',t') = e^{\mathrm{i}\omega x + \mathrm{i}\eta y}\delta(v-v').$$

Since (2.145) has no coefficient involving x or y, it is natural to consider the following solution form:

$$\check{G}(x,y,v,t;\omega,\eta,v',t') = e^{\mathrm{i}\omega x + \mathrm{i}\eta y} g(v,t;\omega,\eta,v',t'), \tag{2.147}$$

where g satisfies the following one-dimensional partial differential equation (PDE):

$$-\frac{\partial g}{\partial t} = \left[\mathrm{i}\omega\left(r - q - \frac{v}{2}\right) - \omega^2\frac{v}{2} + \mathrm{i}\eta v\right] g + \left[v(\theta_t - \kappa v) + \mathrm{i}\omega\rho\varepsilon v^2\right]\frac{\partial g}{\partial v} + \frac{\varepsilon^2 v^3}{2}\frac{\partial^2 g}{\partial v^2}, \tag{2.148}$$

subject to the terminal condition:

$$g(v,t';\omega,\eta,v',t') = \delta(v-v').$$

From (2.148), we see that g is essentially a function of v and t, with dependence on other parameters through the coefficients and auxiliary condition of the PDE. We manage to obtain closed-form analytic solution for $g(t,v;t',\omega,\eta,v')$, the details of which are summarized in Theorem 2.1.

Theorem 2.1 *Under the 3/2 stochastic volatility model as specified by (2.141), the partial Fourier transform of the triple joint transition density function defined by (2.146) is found to be*

$$\check{G}(x,y,v,t;\omega,\eta,v',t') \quad = \quad e^{\mathrm{i}\omega x+\mathrm{i}\eta y} g(v,t;\omega,\eta,v',t'),$$

where

$$g(v,t;,\omega,\eta,v',t')$$
$$= e^{a(t'-t)}\frac{A_t}{C_t}\exp\left(-\frac{A_t v+v'}{C_t v v'}\right)\frac{1}{(v')^2}\left(\frac{A_t v}{v'}\right)^{\frac{1}{2}+\frac{\tilde{\kappa}}{\varepsilon^2}} I_{2c}\left(\frac{2}{C_t}\sqrt{\frac{A_t}{vv'}}\right). \tag{2.149}$$

Here, the parameters are given by

$$a = \mathrm{i}\omega(r-q), \quad \tilde{\kappa} = \kappa - \mathrm{i}\omega\rho\varepsilon,$$
$$A_t = e^{\int_t^{t'}\theta_s\,\mathrm{d}s}, \quad C_t = \frac{\varepsilon^2}{2}\int_t^{t'} e^{\int_t^s \theta_{s'}\,\mathrm{d}s'}\,\mathrm{d}s,$$
$$c = \sqrt{\left(\frac{1}{2}+\frac{\tilde{\kappa}}{\varepsilon^2}\right)^2 + \frac{\mathrm{i}\omega+\omega^2-2\mathrm{i}\eta}{\varepsilon^2}}. \tag{2.150}$$

The proof of Theorem 2.1 is relegated to the Appendix.

Remarks

1. The approach of defining the partial transform of the triple joint density and proposing a solution form as in (2.147) works for other stochastic volatility models. For example, consider the following Heston model with time-dependent mean reversion parameter:

$$\frac{\mathrm{d}S_t}{S_t} = (r-q)\,\mathrm{d}t + \sqrt{V_t}\,\mathrm{d}W_t^S,$$
$$\mathrm{d}V_t = \kappa(\theta_t - V_t)\,\mathrm{d}t + \varepsilon\sqrt{V_t}\,\mathrm{d}W_t^V, \tag{2.151}$$

where $\mathrm{d}W_t^S\,\mathrm{d}W_t^V = \rho\,\mathrm{d}t$. Following a similar procedure, the governing equation for g can be found to be

$$-\frac{\partial g}{\partial t} = \left[\mathrm{i}\omega\left(r-q-\frac{v}{2}\right) - \omega^2\frac{v}{2} + \mathrm{i}\eta v\right] g$$
$$+\left[\kappa(\theta_t - v) + \mathrm{i}\omega\rho\varepsilon v\right]\frac{\partial g}{\partial v} + \frac{\varepsilon^2 v}{2}\frac{\partial^2 g}{\partial v^2}, \tag{2.152}$$

subject to the terminal condition:

$$g(v,t';\omega,\eta,v',t') = \delta(v-v').$$

Note that (2.152) has a similar form as (2.149), so it can be solved using similar procedure. However, for the Heston model under which the triple (X,I,V)

is joint affine, we would prefer to consider the full Fourier transform (joint characteristic function of the triple) since it admits an exponential affine form. Note that one may not be able to solve the corresponding Riccati system of ODEs explicitly in closed form when θ_t is a deterministic function of t (Itkin and Carr, 2010).

2. There have been sufficient amount of empirical studies that show pure diffusion processes, accompanied by either square root or 3/2 stochastic variance, cannot adequately capture the short-end implied volatility structure. With the inclusion of jump dynamics in the stochastic volatility models, the derivation methods can be extended to derive the partial Fourier transform under the compound Poisson jumps. In the square root model, we allow the inclusion of simultaneous jumps in both the asset price and instantaneous variance processes. However, due to limitation of analytic tractability, we only allow jump in the asset price process but not the instantaneous variance process in the $3/2$ model. Recall that we transform the $3/2$ process into the square root process by taking the reciprocal of the variance variable in order to take advantage of the affine structure of the square root model. This transformation of state variable does not work when the jump term is included in the $3/2$ stochastic variance process.

2.7.3 Partial Fourier transform of the joint density function of (X,V)

Once we have obtained the partial Fourier transform of the joint transition density of the triple (X,I,V), we can deduce the corresponding partial Fourier transform of the marginal transition density. Let the joint transition density function of (X,V) be denoted by $G_1(x,v,t;x',v',t')$ and define its partial Fourier transform by

$$\check{G}_1(x,v,t;\omega,v',t') = \int_{-\infty}^{\infty} e^{\mathrm{i}\omega x'} G_1(x,v,t;x',v',t')\,\mathrm{d}x'. \tag{2.153}$$

By virtue of (2.146), $\check{G}_1$ can be obtained by simply setting $\eta = 0$ in (2.149). The corresponding result is summarized in Theorem 2.2.

Theorem 2.2 *Under the 3/2 stochastic volatility model as specified by (2.141), the partial Fourier transform defined by (2.153) is given by (Zheng and Zeng, 2016)*

$$\check{G}_1(x,v,t;\omega,v',t') = e^{\mathrm{i}\omega x} g_1(v,t;\omega,v',t'), \tag{2.154}$$

where

$$\begin{aligned} & g_1(v,t;\omega,v',t') \\ &= g(v,t;\omega,0,v',t') \\ &= e^{a(t'-t)}\frac{A_t}{C_t}\exp\left(-\frac{A_t v+v'}{C_t v v'}\right)(v')^{-2}\left(\frac{A_t v}{v'}\right)^{\frac{1}{2}+\frac{\tilde{\kappa}}{\varepsilon^2}} I_{2c}\left(\frac{2}{C_t}\sqrt{\frac{A_t}{v v'}}\right), \end{aligned} \tag{2.155}$$

where $I_q(\cdot)$ is the modified Bessel function of order q. The parameters a, $\tilde{\kappa}$, A_t, and C_t are the same as those given in (2.150), *but with a modified parameter c as given by*

$$c = \sqrt{\left(\frac{1}{2} + \frac{\tilde{\kappa}}{\varepsilon^2}\right)^2 + \frac{\mathrm{i}\omega + \omega^2}{\varepsilon^2}}.$$

Remarks

1. One can also use Theorem 2.1 to derive the partial Fourier transform of the joint transition density of (I,V).

2. By further setting $\omega = 0$ in (2.154), we can deduce the transition density of V:

$$G_V(v,t;v',t') = g(v,t;0,0,v',t').$$

The conditional transition density of V_t can be inferred to be

$$p_V(V_{t'}|V_t) = \frac{A_t}{C_t}\exp\left(-\frac{A_tV_t + V_{t'}}{C_tV_tV_{t'}}\right)\frac{1}{V_{t'}^2}\left(\frac{A_tV_t}{V_{t'}}\right)^{\frac{1}{2}+\frac{\kappa}{\varepsilon^2}} I_{1+\frac{2\kappa}{\varepsilon^2}}\left(\frac{2}{C_t}\sqrt{\frac{A_t}{V_{t'}V_t}}\right). \tag{2.156}$$

One may obtain the same analytic formula using an alternative approach that is based on the conditional transition density of the CIR process [see (2.113)] and the transformation relation (2.143) between the CIR process and $3/2$ process.

2.7.4 Joint characteristic function of (X,I)

The valuation of the European target volatility option whose terminal payoff depends on both the terminal asset price and realized integrated variance can be performed using the Fourier inversion method with the use of the joint characteristic function of (X,I). Another application of this joint characteristic function is shown in pricing finite maturity timer options (see Sec. 6.3). It is seen that this joint characteristic function can be obtained from $\breve{G}$ by integrating with respect to v'.

Theorem 2.3 *Under the 3/2 stochastic volatility model as specified by (2.141), the joint characteristic function of the pair (X,I) is given by*

$$E_t[e^{\mathrm{i}\omega X_{t'} + \mathrm{i}\eta I_{t'}}] = e^{\mathrm{i}\omega X_t + \mathrm{i}\eta I_t} h(V_t,t;\omega,\eta,t'), \tag{2.157}$$

where

$$h(v,t;\omega,\eta,t') = e^{a(t'-t)}\frac{\Gamma(\tilde{\beta}-\tilde{\alpha})}{\Gamma(\tilde{\beta})}\left(\frac{1}{C_tv}\right)^{\tilde{\alpha}} {}_1F_1\left(\tilde{\alpha},\tilde{\beta},-\frac{1}{C_tv}\right). \tag{2.158}$$

Here,

$$\tilde{\alpha} = -\frac{1}{2} - \frac{\tilde{\kappa}}{\varepsilon^2} + c, \quad \tilde{\beta} = 1 + 2c,$$

Γ *is the gamma function [see* (2.63)*], and* ${}_1F_1$ *is the confluent hypergeometric function of the first kind defined by*

$$ {}_1F_1(z;\alpha,\gamma) = \sum_{n=0}^{\infty} \frac{(\alpha)_n}{(\gamma)_n} \frac{z^n}{n!} \tag{2.159}$$

with $(x)_n = \prod_{j=0}^{n-1}(x+j)$.

The proof of Theorem 2.3 is presented in the Appendix.

Remarks

1. The result presented in Theorem 2.3 agrees with that given in Theorem 3 in Carr and Sun (2007).
2. The corresponding marginal characteristic function of X and I can be deduced by simply setting $\eta = 0$ and $\omega = 0$ in (2.158), respectively.

Pricing of European target volatility option

Once we obtain the joint characteristic function of (X, I), we may use the Fourier inversion method to price a European target volatility call option whose terminal payoff is given by

$$V_T(S_T, I_T) = \overline{\sigma}\sqrt{\frac{T}{I_T}}(S_T - K)^+. \tag{2.160}$$

Here, K is the strike price, T is the maturity date, $\overline{\sigma}$ is the target annualized volatility, and $\sqrt{I_T/T}$ is the realized volatility over the life of the option $[0,T]$. The two-dimensional Fourier transform of $V_T(S_T, I_T)$ is found to be

$$\widehat{V}_T(\omega,\eta) = \overline{\sigma}(1+i)\sqrt{\frac{\pi T}{2\eta}} \frac{K^{1+i\omega}}{(i\omega - \omega^2)}, \quad \text{Im}\{\omega\} > 1 \text{ and } \text{Im}\{\eta\} > 0. \tag{2.161}$$

Here, $\text{Im}\{\omega\}$ denotes the imaginary part of the complex Fourier transform variable ω, and similar notation for $\text{Im}\{\eta\}$. Using the generalized Fourier transform method (Carr and Madan, 1999; Lewis, 2016), the time-t price function $V(S,I,V,t)$ of the target volatility call option is given by the following two-dimensional Fourier inversion integral (Torricelli, 2013)

$$V(S,I,V,t) = \frac{e^{-r(T-t)}}{4\pi^2} \int_{ik_1-\infty}^{ik_1+\infty} \int_{ik_2-\infty}^{ik_2+\infty} e^{-i\omega(r-q)(T-t)} S^{-i\omega} e^{-i\eta I} h(V,t;\omega,\eta,T)\widehat{V}_T(\omega,\eta)\, d\eta\, d\omega, \tag{2.162}$$

where the joint characteristic function $h(V,t;\omega,\eta,T)$ is given by (2.158). The horizontal Bromwich lines: $\omega = u + ik_1$ and $\eta = s + ik_2$, $u \in \mathbb{R}$ and $s \in \mathbb{R}$, are chosen so that the generalized Fourier inversion integral exists.

Appendix

Derivation of the parameter functions of $U(X,V,\tau)$ in (2.125)

To find the solution for the parameter function $B(\tau)$, we introduce the following transformation:

$$B(\tau) = -\frac{2}{\varepsilon^2}\frac{E'(\tau)}{E(\tau)}.$$

The differential equation for $B(\tau)$ is transformed into the following linear differential equation for $E(\tau)$:

$$E''(\tau) + (\kappa - \rho\varepsilon\phi)E'(\tau) - \frac{\varepsilon^2}{4}(\phi - \phi^2)E(\tau) = 0.$$

The initial condition, $B(0) = b$, can be transformed into the corresponding initial condition for $E(\tau)$ as follows:

$$E'(0) = -\frac{\varepsilon^2 b}{2}E(0).$$

Solving the equation for $E(\tau)$, we obtain

$$E(\tau) = E(0)\left(\frac{\xi_+ + \varepsilon^2 b}{2\zeta}e^{-\frac{\xi_- \tau}{2}} + \frac{\xi_- - \varepsilon^2 b}{2\zeta}e^{\frac{\xi_+ \tau}{2}}\right).$$

The solution to $B(\tau)$ is then found to be

$$B(\tau) = \frac{b(\xi_- e^{-\zeta\tau} + \xi_+) - (\phi - \phi^2)(1 - e^{-\zeta\tau})}{(\xi_+ + \varepsilon^2 b)e^{-\zeta\tau} + \xi_- - \varepsilon^2 b}.$$

Once the solution $B(\tau)$ has been determined, we obtain $\Gamma(\tau)$ by direct integration as shown below:

$$\begin{aligned}
\Gamma(\tau) &= \gamma + (r-q)\phi\tau + \kappa\theta\int_0^\tau B(u)\,du \\
&= \gamma + (r-q)\phi\tau - \frac{2\kappa\theta}{\varepsilon^2}\ln\frac{E(\tau)}{E(0)} \\
&= (r-q)\phi\tau + \gamma - \frac{\kappa\theta}{\varepsilon^2}\left[\xi_+\tau + 2\ln\frac{(\xi_+ + \varepsilon^2 b)e^{-\zeta\tau} + \xi_- - \varepsilon^2 b}{2\zeta}\right].
\end{aligned}$$

The solution of $\Lambda(\tau)$ requires an expectation calculation followed by integration, where

$$\Lambda(\tau) = -\lambda(m\phi + 1)\tau + \lambda\int_0^\tau E^Q[\exp(\phi J^S + B(u)J^V)]\,du.$$

We employ the iterated expectation as follows:

$$\begin{aligned}
E^Q[\exp(\phi J^S + B(u)J^V)] &= E^Q[E^Q[\exp(\phi J^S + B(u)J^V)]|J^V] \\
&= E^Q\left[\frac{e^{B(u)J^V}}{\sqrt{2\pi}\delta}\int_{-\infty}^{\infty}\exp\left(\phi x - \frac{(x-\nu-\rho_J J^V)^2}{2\delta^2}\right)\mathrm{d}x\right] \\
&= \exp\left(\phi\nu + \frac{\delta^2\phi^2}{2}\right)E^Q\left[\exp\left([\rho_J\phi + B(u)]J^V\right)\right] \\
&= \exp\left(\phi\nu + \frac{\delta^2\phi^2}{2}\right)\int_0^{\infty}\frac{1}{\eta}\exp\left([\rho_J\phi + B(u)]y - \frac{y}{\eta}\right)\mathrm{d}y \\
&= \exp\left(\phi\nu + \frac{\delta^2\phi^2}{2}\right)\frac{1}{1-[\rho_J\phi + B(u)]\eta}.
\end{aligned}$$

Lastly, we perform the integration of the last term as follows:

$$\begin{aligned}
\int_0^{\tau}\frac{1}{1-[\rho_J\phi + B(u)]\eta}\,\mathrm{d}u &= \int_0^{\tau}\frac{k_1e^{-\zeta u}+k_2}{k_3e^{-\zeta u}+k_4}\,\mathrm{d}u \\
&= \frac{k_2}{k_4}\tau - \frac{1}{\zeta}\left(\frac{k_1}{k_3} - \frac{k_2}{k_4}\right)\ln\frac{k_3e^{-\zeta\tau}+k_4}{k_3+k_4},
\end{aligned}$$

where

$$\begin{aligned}
k_1 &= \xi_+ + \varepsilon^2 b, \quad k_2 = \xi_- - \varepsilon^2 b, \\
k_3 &= (1-\phi\rho_J\eta)k_1 - \eta(\phi - \phi^2 + \xi_- b), \\
k_4 &= (1-\phi\rho_J\eta)k_2 + \eta(\phi - \phi^2 - \xi_+ b).
\end{aligned}$$

Note that the final integration step requires the technical condition:

$$\mathrm{Re}\{\rho_J\phi + B(u)\}\eta < 1.$$

Since η is generally small, this requirement is usually fulfilled. We consider the following two degenerate cases.

(i) Suppose $k_3 = 0$, the integral reduces to

$$\int_0^{\tau}\frac{1}{1-[\rho_J\phi + B(u)]\eta}\,\mathrm{d}u = \frac{k_2}{k_4}\tau - \frac{k_1}{k_4}\frac{e^{-\zeta\tau}-1}{\zeta}.$$

(ii) When $k_4 = 0$, the integral reduces to

$$\int_0^{\tau}\frac{1}{1-[\rho_J\phi + B(u)]\eta}\,\mathrm{d}u = \frac{k_1}{k_3}\tau + \frac{k_2}{k_3}\frac{e^{\zeta\tau}-1}{\zeta}.$$

Proof of Theorem 2.1

Let $\tilde{g} = ge^{a(t-t')}$, where $a = \mathrm{i}\omega(r-q)$. From (2.148), the governing equation for $\tilde{g}$ is given by

$$\frac{\partial\tilde{g}}{\partial t} + (\theta_t v - \tilde{\kappa}v^2)\frac{\partial\tilde{g}}{\partial v} + \frac{\varepsilon^2 v^3}{2}\frac{\partial^2\tilde{g}}{\partial v^2} - \left(\frac{\mathrm{i}\omega+\omega^2}{2} - \mathrm{i}\eta\right)v\tilde{g} = 0.$$

The terminal condition for $\tilde{g}$ remains to be

$$\tilde{g}(t',v;t',\omega,\eta,v') = \delta(v-v').$$

To solve the above partial differential equation, we perform the transformation of the variables from (v,v') to (u,u'), with $u = 1/v$ and $u' = 1/v'$. We define $f(t,u;t',\omega,\eta,u')$ via the following transformation relation:

$$\tilde{g}(t,v;t',\omega,\eta,v') = \frac{u^{\alpha}}{(u')^{\alpha-2}} f(t,u;t',\omega,\eta,u'),$$

where the parameter α is to be determined in the subsequent procedure. The governing equation for f can be found to be

$$\begin{aligned} -\frac{\partial f}{\partial t} &= \frac{\varepsilon^2 u}{2}\frac{\partial^2 f}{\partial u^2} + [\varepsilon^2(\alpha+1)+\tilde{\kappa}-\theta_t u]\frac{\partial f}{\partial u} - \alpha\theta_t f \\ &\quad + \left[\frac{\varepsilon^2}{2}(\alpha^2+\alpha)+\tilde{\kappa}\alpha - \frac{\mathrm{i}\omega+\omega^2}{2} + \mathrm{i}\eta\right]\frac{f}{u}, \end{aligned}$$

subject to the terminal condition:

$$f(t',u;t',\omega,\eta,u') = \delta(u-u').$$

To reduce the complexity of the above differential equation, we make the ingenious choice: $\alpha = \alpha(\omega,\eta)$ such that α satisfies the following quadratic equation:

$$\frac{\varepsilon^2}{2}(\alpha^2+\alpha)+\tilde{\kappa}\alpha - \frac{\mathrm{i}\omega+\omega^2}{2} + \mathrm{i}\eta = 0,$$

The governing PDE of f is then reduced to

$$-\frac{\partial f}{\partial t} = \frac{\varepsilon^2 u}{2}\frac{\partial^2 f}{\partial u^2} + [\varepsilon^2(\alpha+1)+\tilde{\kappa}-\theta_t u]\frac{\partial f}{\partial u} - \alpha\theta_t f,$$

where all the coefficients are affine in u. The two roots of the quadratic equation in α are given by

$$\alpha = -\left(\frac{1}{2}+\frac{\tilde{\kappa}}{\varepsilon^2}\right) \pm c, \tag{A.1}$$

where $c = c(\omega,\eta)$ is given by

$$c(\omega,\eta) = \sqrt{\left(\frac{1}{2}+\frac{\tilde{\kappa}}{\varepsilon^2}\right)^2 + \frac{\mathrm{i}\omega+\omega^2-2\mathrm{i}\eta}{\varepsilon^2}}.$$

The choice of which root of α is discussed later. We consider the Laplace transform of f with respect to u', which is defined as follows:

$$\hat{f}(t,u;t',\omega,\eta,\xi) = \int_0^{\infty} e^{-\xi u'} f(t,u;t',\omega,\eta,u')\, du'.$$

Indeed, $\hat{f}$ satisfies the same PDE as that of f, where

$$-\frac{\partial \hat{f}}{\partial t} = \frac{\varepsilon^2 u}{2}\frac{\partial^2 \hat{f}}{\partial u^2} + [\varepsilon^2(\alpha+1) + \tilde{\kappa} - \theta_t u]\frac{\partial \hat{f}}{\partial u} - \alpha\theta_t \hat{f}, \tag{A.2}$$

while the terminal condition is modified to become

$$\hat{f}(t', u; t', \omega, \eta, \xi) = e^{-u\xi}.$$

By virtue of the affine structure of the coefficients in the PDE, (A.2) admits the following exponential affine solution:

$$\hat{f}(t, u; t', \omega, \eta, \xi) = \exp(B(t;\xi)u + D(t;\xi)),$$

where $B(t;\xi)$ and $D(t;\xi)$ are parameter functions that can be determined by solving the following Riccati system of ODEs:

$$\begin{aligned}
\frac{\mathrm{d}B}{\mathrm{d}t} &= \theta_t B - \frac{\varepsilon^2}{2}B^2, \\
\frac{\mathrm{d}D}{\mathrm{d}t} &= \alpha\theta_t - [\varepsilon^2(\alpha+1) + \tilde{\kappa}]B,
\end{aligned}$$

with terminal conditions:

$$B(t';\xi) = -\xi \quad \text{and} \quad D(t';\xi) = 0.$$

The parameter functions are found to be

$$B(t;\xi) = -\frac{\xi}{A_t + C_t\xi},$$

where

$$A_t = e^{\int_t^{t'} \theta_s \,\mathrm{d}s}, \quad C_t = \frac{\varepsilon^2}{2}\int_t^{t'} e^{\int_t^s \theta_u \,\mathrm{d}u}\,\mathrm{d}s.$$

Thanks to the identity:

$$\frac{\mathrm{d}}{\mathrm{d}t}\ln(A_t + C_t\xi) = -\theta_t - \frac{\varepsilon^2}{2}\frac{\xi}{A_t + C_t\xi},$$

we obtain

$$D(t;\xi) = -2\left(\alpha + 1 + \frac{\tilde{\kappa}}{\varepsilon^2}\right)\ln(A_t + C_t\xi) + \left(\alpha + 2 + \frac{2\tilde{\kappa}}{\varepsilon^2}\right)\int_t^{t'} \theta_s \,\mathrm{d}s.$$

Putting these relations together yields

$$\hat{f}(t, u; t', \omega, \eta, \xi) = A_t^{\alpha+2+\frac{2\tilde{\kappa}}{\varepsilon^2}} \exp\left(-\frac{\xi u}{A_t + C_t\xi}\right)(A_t + C_t\xi)^{-2\alpha-2-\frac{2\tilde{\kappa}}{\varepsilon^2}}.$$

Finally, f can be obtained by taking the inverse Laplace transform of $\hat{f}$ with respect to ξ. Recall that for the function

$$h(p) = p^{-\nu-1}e^{\gamma/p} \text{ for } \nu > -1,$$

the inverse Laplace transform of $h(p)$ is given by

$$\mathcal{L}^{-1}\{h(p)\}(x) = \frac{1}{2\pi\mathrm{i}}\int_{\tau-\mathrm{i}\infty}^{\tau+\mathrm{i}\infty} e^{xp}h(p)\,\mathrm{d}p = \left(\frac{x}{\gamma}\right)^{\nu/2} I_\nu(2\sqrt{\gamma x}),$$

where $I_\nu(\cdot)$ is the modified Bessel function of the first kind. Closed-form expression of the inverse Laplace transform $\mathcal{L}^{-1}[\hat{f}]$ exists provided that

$$\mathrm{Re}\left\{\alpha + 1 + \frac{\tilde{\kappa}}{\varepsilon^2}\right\} > 0,$$

where $\mathrm{Re}\{\cdot\}$ denotes taking the real part. It can be shown [see Appendix A in Zheng and Zeng (2016) for details] that the satisfaction of the above technical condition requires the choice of the plus sign for α in (A.1) so that

$$\alpha = -\left(\frac{1}{2} + \frac{\tilde{\kappa}}{\varepsilon^2}\right) + c.$$

Writing $p = A_t + C_t\xi$ and applying the inverse Laplace transform to $\hat{f}$, we have

$$\begin{aligned} f(t,u;t',\omega,\eta,u') &= \frac{A_t^{\frac{3}{2}+c+\frac{\tilde{\kappa}}{\varepsilon^2}}}{2\pi\mathrm{i}C_t}\int_{\tau-\mathrm{i}\infty}^{\tau+\mathrm{i}\infty} e^{\frac{u'(p-A_t)}{C_t}} p^{-2c-1} e^{-\frac{u(p-A_t)}{C_t p}}\,\mathrm{d}p \\ &= \frac{A_t^{\frac{3}{2}+c+\frac{\tilde{\kappa}}{\varepsilon^2}} e^{-\frac{u+A_t u'}{C_t}}}{2\pi\mathrm{i}C_t}\int_{\tau-\mathrm{i}\infty}^{\tau+\mathrm{i}\infty} e^{\frac{u'p}{C_t}} p^{-2c-1} e^{\frac{uA_t}{C_t p}}\,\mathrm{d}p \\ &= \frac{A_t^{\frac{3}{2}-c+\frac{\tilde{\kappa}}{\varepsilon^2}}}{C_t} e^{-\frac{u+A_t u'}{C_t}} \left(\frac{A_t u'}{u}\right)^{c} I_{2c}\left(\frac{2}{C_t}\sqrt{A_t u u'}\right). \end{aligned}$$

When expressed in terms of v and v', g is found to be

$$g(t,v;t',\omega,\eta,v') = e^{a(t'-t)}\frac{A_t}{C_t}\exp\left(-\frac{A_t v + v'}{C_t v v'}\right)\frac{1}{(v')^2}\left(\frac{A_t v}{v'}\right)^{\frac{1}{2}+\frac{\tilde{\kappa}}{\varepsilon^2}} I_{2c}\left(\frac{2}{C_t}\sqrt{\frac{A_t}{vv'}}\right).$$

Proof of Theorem 2.3

Recall that

$$\begin{aligned} E_t[e^{\mathrm{i}\omega X_{t'}+\mathrm{i}\omega I_{t'}}] &= \int_0^\infty \check{G}(t,X_t,I_t,V_t;t',\omega,\eta,v')\,\mathrm{d}v' \\ &= e^{\mathrm{i}\omega X_t+\mathrm{i}\eta I_t}\int_0^\infty g(t,V_t;t',\omega,\eta,v')\,\mathrm{d}v'. \end{aligned}$$

It suffices to compute the following integral:

$$\begin{aligned}
h(t,v;t',\omega,\eta) &= \int_0^\infty g(t,v;t',\omega,\eta,v')\,\mathrm{d}v' \\
&= e^{a(t'-t)}\frac{A_t}{C_t}\int_0^\infty e^{-\frac{u+A_tu'}{C_t}}\left(\frac{A_tu'}{u}\right)^{\frac{1}{2}+\frac{\tilde{\kappa}}{\varepsilon^2}} I_{2c}\left(\frac{2}{C_t}\sqrt{A_tuu'}\right)\mathrm{d}u' \\
&= \frac{e^{a(t'-t)-\frac{u}{C_t}}}{C_tu^{\frac{1}{2}+\frac{\tilde{\kappa}}{\varepsilon^2}}}\int_0^\infty e^{-\frac{z}{C_t}}z^{\frac{1}{2}+\frac{\tilde{\kappa}}{\varepsilon^2}} I_{2c}\left(\frac{2\sqrt{uz}}{C_t}\right)\mathrm{d}z,
\end{aligned}$$

where $z = A_tu'$. To evaluate the integral, we observe the relation:

$$\int_0^\infty e^{-st}t^{\iota}I_{\varsigma}(\lambda\sqrt{t})\,\mathrm{d}t = \frac{\Gamma(\phi)}{\Gamma(\psi)}\frac{X^{\varsigma/2}}{s^{1+\iota}}\,{}_1F_1(\phi,\psi,X),$$

where $I_q(\cdot)$ is the modified Bessel function of order q and ${}_1F_1$ is the confluent hypergeometric function of the first kind. Here, $\phi = 1+\iota+\varsigma/2$, $\psi = 1+\varsigma$, $X = \frac{\lambda^2}{4s}$ and $\mathrm{Re}\{\phi,s\} > 0$. Combining these results together, we obtain

$$\begin{aligned}
h(t,v;t',\omega,\eta) &= e^{a(t'-t)-\frac{u}{C_t}}\frac{\Gamma(1-\alpha)}{\Gamma(2c+1)}\left(\frac{u}{C_t}\right)^{\tilde{\alpha}}{}_1F_1\left(1-\alpha,2c+1,\frac{u}{C_t}\right) \\
&= e^{a(t'-t)}\frac{\Gamma(\tilde{\beta}-\tilde{\alpha})}{\Gamma(\tilde{\beta})}\left(\frac{u}{C_t}\right)^{\tilde{\alpha}}{}_1F_1\left(\tilde{\alpha},\tilde{\beta},-\frac{u}{C_t}\right) \\
&= e^{a(t'-t)}\frac{\Gamma(\tilde{\beta}-\tilde{\alpha})}{\Gamma(\tilde{\beta})}\left(\frac{1}{C_tv}\right)^{\tilde{\alpha}}{}_1F_1\left(\tilde{\alpha},\tilde{\beta},-\frac{1}{C_tv}\right).
\end{aligned}$$

Here, the parameters are given by

$$\tilde{\alpha} = -\frac{1}{2}-\frac{\tilde{\kappa}}{\varepsilon^2}+c \quad\text{and}\quad \tilde{\beta} = 1+2c.$$

Note that the second equality follows from the identity:

$${}_1F_1(a,b,z) = e^z\,{}_1F_1(b-a,b,-z).$$

Chapter 3

VIX Derivatives under Consistent Models and Direct Models

In Sec. 1.5.2, we have presented a preliminary discussion on the construction of VIX, the volatility index launched by the Chicago Board Options Exchange (CBOE). Formally, VIX is the square root of the risk neutral expectation of the integrated variance of the S&P 500 index over the next 30 calendar days, reported on an annualized basis and expressed in percentage point. Before the introduction of VIX, Brenner and Galai (1989) propose a volatility index (Sigma index) and introduce derivatives on this index. Whaley (1993) discusses the role of derivatives on market volatility as effective hedging tools. Market practitioners use VIX derivatives to hedge the risks of investments in the S&P 500 index and achieve exposure to the S&P 500 volatility without having to delta hedge their S&P 500 option positions (Dash and Moran, 2005). Besides, market data on VIX and prices of VIX derivatives provide information content of the S&P 500 index return's variance dynamics. Duan and Yeh (2010) discuss the extraction of the jump and volatility risk premium implied by VIX. By analyzing traded VIX options data, Kaeck and Alexander (2012) conclude that non-affine diffusion and variance jumps are necessary in the dynamics of the S&P 500 index and its variance in order to provide good fit to the VIX term structures and implied volatility distributions of traded VIX option prices.

In this chapter, we consider various pricing models of VIX, which can be quite different from those of volatility derivatives. Though VIX and volatility derivatives are volatility products, we note that VIX on the maturing date is the square root of the expected integrated variance of S&P 500 index over the next 30 calendar days while the discrete realized volatility on the maturity date in volatility derivative is calculated from the initiation date to maturity date. Under the affine model, the square of VIX can be expressed as a linear function of instantaneous variance, so analytic tractability of pricing VIX futures and options can be enhanced. However, this nice linear relation does not exist under the $3/2$ stochastic volatility model.

In Sec. 3.1.1, we start with the discussion of the relation between variance swap rate and VIX^2 under jumps. We then discuss the product specifications of different VIX derivatives. Besides the standard VIX futures and options, we also consider VVIX (volatility index on VIX) and VXX (exchange-traded note on VIX).

There are two approaches for pricing VIX derivatives. In the consistent model approach, one considers the joint dynamics of the index value process and stochastic volatility and derives the dynamics for VIX. The other approach directly models the

DOI: 10.1201/9781003263524-3

dynamics for VIX. Pricing formulas of VIX futures and options under the consistent models and direct models are derived in Secs. 3.2 and 3.3, respectively.

3.1 VIX, variance swap rate and VIX derivatives

In Sec. 1.5.2, we show that VIX (expressed as percentage points) is calculated as 100 times the model-free measure of the square root of the expected 30-day variance of the rate of return of the forward price of the S&P 500 index (SPX). For notational simplicity, we drop the factor 100 in our later discussion. We let S_t denote the SPX value process and $F_t(t+\tau)$ be the time-t forward price of the S&P 500 index maturing τ-period later, where $F_t(t+\tau) = e^{r\tau}S_t$. We let the riskless interest rate r to be constant. Let $p_t(K;t+\tau)$ and $c_t(K;t+\tau)$ denote the time-t value of the put and call option with strike K and maturing date $t+\tau$, respectively. Under a risk neutral measure Q, we recall the following theoretical definition of $\text{VIX}^2(\tau)$ [see (1.31)]:

$$\text{VIX}_t^2(\tau) = \frac{2}{\tau}e^{r\tau}\Pi_t(F_t(t+\tau), t+\tau) = -\frac{2}{\tau}E_t^Q\left[\ln\frac{S_{t+\tau}}{e^{r\tau}S_t}\right], \tag{3.1}$$

where

$$\Pi_t(F_t(t+\tau), t+\tau) = \int_0^{F_t(t+\tau)} \frac{p_t(K;t+\tau)}{K^2}\,\mathrm{d}K + \int_{F_t(t+\tau)}^{\infty} \frac{c_t(K;t+\tau)}{K^2}\,\mathrm{d}K.$$

From financial perspective, we prefer the replicating portfolio to include the out-of-the-money call and put options since they have higher liquidity.

As discussed in Sec. 1.5.2, since we only have availability of options with finite number of strikes so the option with strike price that exactly equals $F_t(t+\tau)$ is not available in general. In fact, the VIX formula used in actual implementation is given by the practical approximation formula (1.32). In our discussion of pricing of VIX derivatives in this chapter, we use the theoretical formula of $\text{VIX}_t^2(\tau)$ as depicted in (3.1) for better analytical tractability. As a remark, Wu and Liu (2018) discuss the scope of this VIX truncation errors in the emerging market; in particular, the truncation error that exists in iVX (a product that mimics VIX in China's options market). The error is not only significant but also volatile under different market conditions. They propose a new approach to estimating VIX truncation errors using corridor variance swaps.

3.1.1 Relation between variance swap rate and VIX^2 under jumps

Under the assumption of pure diffusion process of the underlying index value process, we have shown in Sec. 1.5.2 that VIX^2 equals the risk neutral expected cumulative diffusion variance [see (1.31)]. This relation does not hold when the index value has jumps. Carr and Wu (2009) present an insightful discussion on how to quantify the difference in $\text{VIX}_t^2(\tau)$ and variance swap rate when jumps occur in the index value process. In the subsequent exposition, we use the notations used in their paper.

Suppose the futures price F_t exhibits jump so that F_t evolves as

$$dF_t = F_t\sigma_t\, dW_t + \int_{\mathbb{R}\backslash\{0\}} F_{t^-}(e^x - 1)\widetilde{J}_F(dx, dt)\; dx\; dt \tag{3.2}$$

under a risk neutral measure Q. Here, W_t is Q-Brownian motion and F_{t^-} denotes the futures price just prior to a jump at time t. Suppose the futures price jumps from F_{t^-} to $F_t = e^x F_{t^-}$ at time t, the compensated jump measure $\widetilde{J}_F(dx, dt)$ realizes a nonzero value at the given x. The compensated jump measure compensates the jump process so that the integral term in (3.2) represents the increment of a Q-pure jump martingale. For simplicity of presentation, we assume that the jump process exhibits finite variation. Under the specification of F_t in (3.2), by Itô's lemma (see 2.28), we have

$$\begin{aligned} f(F_{t+\tau}) = f(F_t) &+ \int_t^{t+\tau} f'(F_{s^-})\, dF_s + \int_t^{t+\tau} f''(F_s)\frac{F_s^2\sigma_s^2}{2}\, ds \\ &+ \int_t^{t+\tau}\int_{\mathbb{R}\backslash\{0\}} \left[f(F_{s^-}e^x) - f(F_{s^-}) - f'(F_{s^-})F_{s^-}(e^x - 1)\right]\widetilde{J}_F(dx, ds). \end{aligned}$$

Taking $f(F) = \ln F$, we obtain

$$\begin{aligned} \ln F_{t+\tau} = \ln F_t &+ \int_t^{t+\tau} \frac{1}{F_{s^-}}\, dF_s - \int_t^{t+\tau} \frac{\sigma_s^2}{2}\, ds \\ &+ \int_t^{t+\tau}\int_{\mathbb{R}\backslash\{0\}} (x + 1 - e^x)\, \widetilde{J}_F(dx, ds). \end{aligned} \tag{3.3}$$

The quadratic variation QV on the futures return over $[t, t+\tau]$ consists of the continuous component and jump component, which is given by

$$QV_t^{t+\tau} = \int_t^{t+\tau} \sigma_s^2\, ds + \int_t^{t+\tau}\int_{\mathbb{R}\backslash\{0\}} x^2\, \widetilde{J}_F(dx, ds). \tag{3.4}$$

The last double integral gives the sum of squares of the jumps of the futures return [see (2.21b)]. By eliminating $\int_t^{t+\tau} \sigma_s^2\, ds$ in (3.3) and (3.4) and writing

$$\frac{F_{t+\tau}}{F_t} - 1 = \int_t^{t+\tau} \frac{1}{F_t}\, dF_s,$$

we obtain

$$\begin{aligned} QV_t^{t+\tau} &= 2\left(\frac{F_{t+\tau}}{F_t} - 1 - \ln\frac{F_{t+\tau}}{F_t}\right) + 2\int_t^{t+\tau}\left(\frac{1}{F_{s^-}} - \frac{1}{F_t}\right) dF_s \\ &\quad - 2\int_t^{t+\tau}\int_{\mathbb{R}\backslash\{0\}}\left(\frac{x^2}{2} + x - 1 - e^x\right)\widetilde{J}_F(dx, ds). \end{aligned}$$

The Taylor expansion of $\ln F_{t+\tau}$ about the point F_t gives

$$\ln F_{t+\tau} = \ln F_t + \frac{1}{F_t}(F_{t+\tau} - F_t) - \int_0^{F_t} \frac{1}{K^2}(K - F_{t+\tau})^+ \mathrm{d}K - \int_{F_t}^{\infty} \frac{1}{K^2}(F_{t+\tau} - K)^+ \mathrm{d}K.$$

Recall that the variance swap rate SW over $[t, t+\tau]$ is given by

$$SW_t^{t+\tau} = E_t^Q \left[\frac{1}{\tau} QV_t^{t+\tau} \right].$$

We observe the following relations:

$$E_t^Q \left[\int_0^{F_t} \frac{(K - F_{t+\tau})^+}{K^2} \mathrm{d}K \right] = e^{r\tau} \int_0^{F_t} \frac{p_t(K; t+\tau)}{K^2} \mathrm{d}K,$$
$$E_t^Q \left[\int_0^{F_t} \frac{(F_{t+\tau} - K)^+}{K^2} \mathrm{d}K \right] = e^{r\tau} \int_0^{F_t} \frac{c_t(K; t+\tau)}{K^2} \mathrm{d}K,$$

and the martingale property:

$$E_t^Q \left[\int_t^{t+\tau} \left(\frac{1}{F_{s-}} - \frac{1}{F_t} \right) \mathrm{d}F_s \right] = 0.$$

Combining all results together, we finally obtain

$$SW_t^{t+\tau} - \mathrm{VIX}_t^2(\tau) = \frac{2}{\tau} E_t^Q \int_t^{t+\tau} \int_{\mathbb{R}\setminus\{0\}} \left(\frac{x^2}{2} + x + 1 - e^x \right) \widetilde{J}_F(\mathrm{d}x, \mathrm{d}s). \tag{3.5}$$

Recall that

$$\frac{x^2}{2} + x + 1 - e^x = -\frac{x^3}{6} + O(x^4).$$

The difference in $SW_t^{t+\tau}$ and $\mathrm{VIX}_t^2(\tau)$ arises from the jumps in F_t, which is $O(x^3)$. Since the difference is proportional to $E_t^Q \left[\int_t^{t+\tau} \int_{\mathbb{R}\setminus\{0\}} -x^3 \widetilde{J}_F(\mathrm{d}x, \mathrm{d}s) \right]$, $SW_t^{t+\tau}$ is larger than $\mathrm{VIX}_t^2(\tau)$ if the price jumps of the futures have negative skewness under Q.

Ait-Sahalia *et al.* (2020) performed empirical studies on the difference of $SW_t^{t+\tau}$ and $\mathrm{VIX}_t^2(\tau)$ for varying maturities. They observe that the differences are mostly positive, statistically significant, larger during market turmoils but sizable even in quiet times. These observations are consistent with the presence of a significant price jump component. There are other attributes that may contribute to the difference of $SW_t^{t+\tau}$ and $\mathrm{VIX}_t^2(\tau)$. First, a larger liquidity risk premium is embedded in $SW_t^{t+\tau}$ than in SPX options since SPX options are more liquid. The higher the liquidity risk premium, the lower the $SW_t^{t+\tau}$ so liquidity should bias downward. That is, without liquidity issue, the difference $SW_t^{t+\tau} - \mathrm{VIX}_t^2(\tau)$ is even larger in positive value. Second, the SPX and variance swaps are traded somewhat in disconnected markets. It is not uncommon that VIX derivatives and SPX options give opposing information on volatility dynamics.

3.1.2 VIX derivatives

VIX derivatives can be used to hedge the volatility risk in the S&P 500 index and/or achieve exposure to S&P 500 volatility. Indeed, investors can invest directly in volatility as an asset class by means of VIX futures and options. Since 2015, the VIX becomes the second most traded underlying in the CBOE options market, only after the S&P 500 index itself.

VIX futures and options

VIX futures and options are among the most popular contracts traded in the CBOE. Trading in VIX futures was started in 2004 while that of the VIX options in 2006. The contract multiplier for each VIX futures contract is $1,000 while that of the VIX option is $100. The popularity of trading on VIX futures and options has been growing over the years. On June 24, 2016, in reaction to the Brexit referendum, over 721,000 VIX futures contracts were traded. On November 9, 2016, the volume was 644,892 in reaction to the surprise outcome of the US election. For VIX options at CBOE, reported record of 2,562,477 contracts was traded on August 10, 2017. The bid-ask spreads of traded VIX derivatives remain to be wide, possibly due to lack of reliable pricing models.

For VIX futures maturing at T, the time-t futures price is given by

$$F_t = E_t^Q[\text{VIX}_T(\tau)], \quad t < T. \tag{3.6}$$

Similarly, the time-t price of the European and American VIX call options maturing at T with strike price K is given by

$$c_t = e^{-r(T-t)} E_t^Q[(\text{VIX}_T(\tau) - K)^+], \tag{3.7a}$$

$$C_t = \sup_{u \in [t,T]} E_t^Q[e^{-r(u-t)}(\text{VIX}_u(\tau) - K)^+], \tag{3.7b}$$

respectively.

VVIX

In 2012, the CBOE introduced a new volatility index VVIX into the VIX market. In a similar spirit to VIX, VVIX is a risk neutral forward looking measure of the 30-day implied volatility of the VIX, where VVIX is implied by a portfolio of VIX options while VIX is implied by a portfolio of S&P index options. Zang *et al.* (2017) compare the historical time series of VIX and VVIX. They detect co-jump of VIX and VVIX and show that VVIX provides additional information content on the volatility of VIX beyond VIX itself. Like VIX, VVIX also exhibits the mean reversion feature. Also, both VIX and VVIX share some of their peak values, especially during the 2008 financial tsunami. Park (2016) presents a scatter plot of VVIX log returns against VIX log returns and reveals positively correlated changes in the VVIX and VIX. The volatility of VIX, as measured by VVIX, tends to increase as VIX increases.

VXX: Exchange-traded note on VIX

Besides VIX futures and options, exchange-traded notes (ETN) on the VIX become more popularly traded. The first such product VXX was launched by Barclays in 2009 and there have been more than thirty of these types of ETNs since then, with several billion dollars in market caps and daily volumes. A comprehensive empirical study on VIX ETN can be found in Alexander and Korovilas (2013). The VXX is a non-securitized debt obligation, similar to a zero coupon bond. Its redemption value depends on the nearest and second-nearest maturing VIX futures contract, which is rebalanced daily to create a nearly constant one-month maturity.

Let I_t be the time-t value of VXX and $F(t,T)$ be the time-t VIX futures price with expiry T. Let T_1 be the nearest month expiry and T_2 be the second next-month expiry. The weight factor $w(t)$ is defined by

$$w(t)(T_1-t)+[1-w(t)](T_2-t)=30 \text{ days}$$

since the VXX tracks a 30-day VIX futures. The return of the VXX between dates t and $t+1$ is given by

$$\begin{aligned}&\frac{I_{t+1}-I_t}{I_t}\\ &=r+\frac{w(t)[F(t+1,T_1)-F(t,T_1)]+[1-w(t)][F(t+1,T_2)-F(t,T_2)]}{w(t)F(t,T_1)+[1-w(t)]F(t,T_2)}. \qquad (3.8)\end{aligned}$$

Taking the differential limit, the dynamic equation of the VXX is given by

$$\frac{\mathrm{d}I_t}{I_t}=r\,\mathrm{d}t+\frac{w(t)\mathrm{d}F(t,T_1)+[1-w(t)]\mathrm{d}F(t,T_2)}{w(t)F(t,T_1)+[1-w(t)]F(t,T_2)}. \qquad (3.9)$$

The theoretical benchmark of the 30-day maturity VIX futures should be given by

$$w(t)F(t,T_1)+[1-w(t)]F(t,T_2).$$

The VIX ETN is created to help investors on VIX futures to reduce the roll-over risk, which is the risk that the futures price for the next maturity may be unfavorable compared to the futures price with the nearest maturity. The rolling cost of the ETN can be quantified as the difference between its return and the return of the theoretical benchmark. Grasselli and Wagalath (2020) show that the roll yield for the VXX is given by

$$\begin{aligned}&\frac{\mathrm{d}I_t}{I_t}-\left\{r\,\mathrm{d}t+\frac{\mathrm{d}[w(t)F(t,T_1)+[1-w(t)]F(t,T_2)]}{w(t)F(t,T_1)+[1-w(t)]F(t,T_2)}\right\}\\ &=\frac{F(t,T_1)-F(t,T_2)}{(T_2-T_1)\{w(t)F(t,T_1)+[1-w(t)]F(t,T_2)\}}. \qquad (3.10)\end{aligned}$$

To interpret the above result, the second term on the left side is the rate of return of the theoretical benchmark (with inclusion of the risk-free rate of return r) while the first term gives the actual rate of return. Using (3.9), the difference of the two terms can be simplified to give the right-hand term.

Since the term structure of the VIX is generally in contango (futures price rolls down to the spot price as maturity approaches), the roll yield is negative. This imposes rolling cost for the ETN, so investors on VXX lose quite significantly over the years. More detailed discussion on the roll yield and pricing models of VXX and VXX derivatives can be found in Gehricke and Zhang (2018) and Grasselli and Wagalath (2020).

3.2 Pricing VIX derivatives under consistent models

The consistent model specifies the joint dynamics of the index value process S_t and its instantaneous variance process V_t. This pricing approach has the advantage that it allows combined pricing of the index derivatives and VIX products in the equity and VIX markets, thus enabling analysis of variance risks of both markets consistently. Some of the earliest works that adopt the Heston model (without jump) include Zhang and Zhu (2006), Zhu and Zhang (2007), and Luo and Zhang (2012). Baldeaux and Badran (2014) argue that stochastic volatility model for S_t without jumps may not produce implied volatilities of VIX options that exhibit volatility skew structures that are consistent with market observed implied skew observed in traded VIX options. Wang and Wang (2020) consider VIX valuation and its futures pricing through a generalized affine realized volatility model with hidden components and jump. Jing *et al.* (2021) present valuation of VIX options under the Hawkes jump-diffusion model that captures the clustering pattern of jumps observed in the asset price process. In this section, we consider pricing of VIX futures and options under the affine stochastic volatility models (with simultaneous jumps on S_t and V_t), 3/2-model (with jumps on S_t), affine GARCH type model and Lévy model with stochastic volatility (Barndorff-Nielsen and Shephard model).

3.2.1 Affine stochastic volatility models

To price VIX futures and options, it is desirable to express $\text{VIX}_t^2(\tau)$ in terms of V_t and other relevant model parameters that characterize the formulation of joint process of S_t and V_t. By virtue of the linear property of the affine stochastic volatility model, we can establish a linear relation between $\text{VIX}_T^2(\tau)$ and V_t.

Under a risk neutral measure Q, the joint process of S_t and V_t of the affine jump-diffusion model is specified by the dynamic equations:

$$\frac{\mathrm{d}S_t}{S_t} = (r - m\lambda)\,\mathrm{d}t + \sqrt{V_t}\,\mathrm{d}W_t^1 + (e^{J^S} - 1)\,\mathrm{d}N_t \tag{3.11a}$$

$$\mathrm{d}V_t = \kappa(\theta - V_t)\,\mathrm{d}t + \sigma_V\sqrt{V_t}\,\mathrm{d}W_t^2 + J^V\,\mathrm{d}N_t \tag{3.11b}$$

where the Brownian motions W_t^1 and W_t^2 observe $dW_t^1 dW_t^2 = \rho\, dt$, θ is the constant mean reversion level of V_t, r is the constant risk-free interest rate, σ_V is the constant volatility of V_t and κ is the constant multiplier on the mean reversion drift rate.

In the above joint process of S_t and V_t, the arrivals of simultaneous jumps on S_t and V_t are modeled by the Poisson process N_t with constant intensity λ. Let J^S and J^V denote the respective random jump size on S_t and V_t, where J^S and J^V are independent of N_t. Furthermore, we assume J^V to be exponentially distributed with mean μ_V, where

$$J^V \sim \text{Exp}(1/\mu_V); \tag{3.12a}$$

and $J^S|J^V$ is normally distributed with mean $\mu_S + \rho_J \mu_V$ and variance σ_S^2 where

$$J^S|J^V \sim N(\mu_S + \rho_J \mu_V, \sigma_S^2). \tag{3.12b}$$

Recall from (2.101) that

$$m = E^Q[e^{J^S} - 1] = E^Q[E^Q[e^{J^S}|J_V] - 1] = \frac{e^{\frac{\sigma_S^2}{2} + \mu_S}}{1 - \rho_J \mu_V} - 1, \tag{3.13}$$

where ρ_J and μ_V satisfy $\rho_J \mu_V < 1$.

Using Itô's lemma, (3.11a) can be rewritten in terms of $\ln S_t$ as follows:

$$d\ln S_t = \left(r - m\lambda - \frac{V_t}{2}\right) dt + \sqrt{V_t}\, dW_t^1 + J^S\, dN_t.$$

Integrating both sides from t to $t+\tau$ gives

$$\ln \frac{S_{t+\tau}}{S_t} = (r - m\lambda)\tau - \int_t^{t+\tau} \frac{V_u}{2}\, du + \int_t^{t+\tau} \sqrt{V_u}\, dW_u^1\, du + \int_t^{t+\tau} J^S\, dN_u.$$

By virtue of independence of the Poisson jump process and jump distribution, we obtain

$$E_t^Q\left[\int_t^{t+\tau} J^S\, dN_u\right] = (\mu_S + \rho_J \mu_V)\lambda\tau.$$

Also, we observe

$$E_t^Q\left[\int_t^{t+\tau} \sqrt{V_u}\, dW_u^1\right] = 0.$$

We recall the result in (3.1) and combine with the above relations to obtain

$$\begin{aligned} \text{VIX}_t^2(\tau) &= -\frac{2}{\tau} E_t^Q\left[\ln \frac{S_{t+\tau}}{S_t e^{r\tau}}\right] \\ &= -\frac{2}{\tau}\left\{[(\mu_S + \rho_J \mu_V) - m]\lambda\tau - \frac{1}{2} E_t^Q\left[\int_t^{t+\tau} V_u\, du\right]\right\} \\ &= 2\lambda[m - (\mu_S + \rho_J \mu_V)] + \frac{1}{\tau} E_t^Q\left[\int_t^{t+\tau} V_u\, du\right]. \end{aligned} \tag{3.14}$$

By solving the dynamic equation of V_t under the affine jump-diffusion model, we obtain

$$\begin{aligned} V_u &= e^{-\kappa(u-t)}V_t + \left[1-e^{-\kappa(u-t)}\right]\theta + \int_t^u \sigma_V e^{-\kappa(u-s)}\sqrt{V_s}\,\mathrm{d}W_s^2 \\ &\quad + \int_t^u e^{-\kappa(u-s)}J^V\,\mathrm{d}N_s, \quad u>t. \end{aligned}$$

Taking expectation E_t^Q on both sides of the above equation gives

$$E_t^Q[V_u] = e^{-\kappa(u-t)}V_t + [1-e^{-\kappa(u-t)}]\theta + \frac{\lambda\mu_V}{\kappa}\left[1-e^{-\kappa(u-t)}\right], \quad u>t.$$

Substituting the above result into (3.14), we then obtain

$$\begin{aligned} \mathrm{VIX}_t^2(\tau) &= 2\lambda[m-(\mu_S+\rho_S\mu_V)] + \frac{1}{\tau}\int_t^{t+\tau} E_t^Q[V_u]\,\mathrm{d}u \\ &= 2\lambda[m-(\mu_S+\rho_S\mu_V)] + \frac{1-e^{-\kappa\tau}}{\kappa\tau}V_t + \left(1-\frac{1-e^{-\kappa\tau}}{\kappa\tau}\right)\left(\theta+\frac{\lambda\mu_V}{\kappa}\right). \end{aligned}$$

This is an elegant result under the affine type model since $\mathrm{VIX}_t^2(\tau)$ can be expressed as a linear function in V_t in the form (Ma *et al.*, 2020)

$$\mathrm{VIX}_t^2(\tau) = a_0(\tau) + a_1(\tau)V_t, \tag{3.15}$$

where the τ-dependent parameter functions are given by

$$a_0(\tau) = \left(1-\frac{1-e^{-\kappa\tau}}{\kappa\tau}\right)\left(\theta+\frac{\lambda\mu_V}{\kappa}\right) + 2\lambda[m-(\mu_S+\rho_J\mu_V)], \tag{3.16a}$$

$$a_1(\tau) = \frac{1-e^{-\kappa\tau}}{\kappa\tau}. \tag{3.16b}$$

These parameter functions depend on the parameter in the dynamics of V_t and the parameters in the jump component of S_t.

Under the case of zero jump, where $\lambda = 0$, $\mathrm{VIX}_t^2(\tau)$ is the weighted average of the instantaneous variance V_t and its long term mean θ (Luo and Zhang, 2012). The weight function $a_0(\tau)$ depends on κ and τ only. With the absence of the jump component, $\mathrm{VIX}_t^2(\tau)$ is independent of the parameters that characterize the dynamics of the index process S_t.

Extended models

Leippold *et al.* (2012) present a more direct derivation of the linear relation between $\mathrm{VIX}_t^2(\tau)$ and V_t, with an extension under the generalization of the intensity parameter λ. Suppose we assume λ_t to be state dependent, where $\lambda_t = \lambda_0 + \lambda_1 V_t$, such that the affine structure remains. The same dynamic equations for S_t and V_t are assumed and the jump dynamics remain the same. We consider

$$\begin{aligned}
\mathrm{VIX}_t^2(\tau) &= \frac{2}{\tau}E_t^Q\left[\int_t^{t+\tau}\frac{\mathrm{d}S_u}{S_u} - \mathrm{d}\ln S_u\right] \\
&= \frac{2}{\tau}E_t^Q\left[\int_t^{t+\tau}[\frac{V_u}{2} + (e^{J_S} - 1 - J_S)(\lambda_0 + \lambda_1 V_u)]\,\mathrm{d}u\right] \\
&= \frac{\xi_1}{\tau}E_t^Q\left[\int_t^{t+\tau} V_u\,\mathrm{d}u\right] + \xi_2 \\
&= \frac{\xi_1}{\tau}[\alpha(\tau)V_t + \beta(\tau)] + \xi_2,
\end{aligned} \tag{3.17}$$

where

$$\xi_1 = 1 + 2\lambda_1[\kappa - (\mu_S + \rho_J\mu_V)], \quad \xi_2 = 2\lambda_0[\kappa - (\mu_S + \rho_J\mu_V)],$$
$$\alpha(\tau) = \frac{1 - e^{-(\kappa-\lambda_1\mu_V)\tau}}{\kappa - \lambda_1\mu_V}, \qquad \beta(\tau) = \frac{\kappa\theta + \lambda_0\mu_V}{\kappa - \lambda_1\mu_V}[\tau - \alpha(\tau)].$$

Later research works assume stochastic mean reversion level θ_t with affine dynamics that is similar to the dynamics of V_t (Zhang *et al.*, 2010; Luo and Zhang, 2012; Kaeck and Alexander, 2013; Branger *et al.*, 2016; Bardgett *et al.*, 2019; Luo *et al.*, 2019). Bardgett *et al.* (2019) conduct a comprehensive analysis of VIX dynamics by combining the time series of SPX and VIX together with their options data. They show that a stochastic central tendency of volatility and jumps in volatility are crucial in capturing volatility smiles in both the SPX and VIX markets, and the tail of the risk neutral distribution of variance. Other extensions of the affine type models include regime switching (Papanicolaou and Sircar, 2014; Li, 2016), stochastic volatility-of-volatility (Fouque and Saporito, 2018) and two-variance models with jumps (Lo *et al.*, 2019).

As an illustration, Kaeck and Alexander (2012) include stochastic mean reversion level θ_t in the consistent model. Under a risk neutral measure Q, the joint dynamics of S_t, V_t and θ_t are governed by

$$\frac{\mathrm{d}S_t}{S_t} = (r - m\lambda)\mathrm{d}t + \sqrt{V_t}\,\mathrm{d}W_t^1 + (e^{J^S} - 1)\mathrm{d}N_t \tag{3.18a}$$

$$\mathrm{d}V_t = \kappa_V(\theta_t - V_t)\mathrm{d}t + \sigma_V\sqrt{V_t}\,\mathrm{d}W_t^2 + J^V\,\mathrm{d}N_t \tag{3.18b}$$

$$\mathrm{d}\theta_t = \kappa_\theta(\overline{\theta} - \theta_t)\mathrm{d}t + \sigma_\theta\sqrt{\theta_t}\,\mathrm{d}W_t^3, \tag{3.18c}$$

with $\mathrm{d}W_t^1\mathrm{d}W_t^2 = \rho\,\mathrm{d}t$ while W_t^3 is uncorrelated with W_t^1 and W_t^2. The definitions of the parameters and jump dynamics of J^S and J^V are the same as those in (3.12a,b).

By following a similar derivation procedure, we manage to obtain

$$\begin{aligned}
E_t^Q\left[\int_t^T V_u \, \mathrm{d}u\right] = & \frac{1-e^{-\kappa_V(T-t)}}{\kappa_V} V_t \\
& + \frac{[1-e^{-\kappa_V(T-t)}]\kappa_\theta - [1-e^{-\kappa_\theta(T-t)}]\kappa_V}{\kappa_\theta(\kappa_\theta-\kappa_V)}(\theta_t - \overline{\theta}) \\
& + \left(\overline{\theta} + \frac{\lambda\mu_V}{\kappa_V}\right)\left[T-t-\frac{1-e^{-\kappa_V(T-t)}}{\kappa_V}\right]. \qquad (3.19)
\end{aligned}$$

Pricing of VIX futures and options

By virtue of the linear relation between $\text{VIX}_t^2(\tau)$ and V_t, the evaluation of VIX futures and options can be performed effectively once the conditional density function $p(V_{t+\tau}|V_t)$ is available. For the dynamics of V_t specified in (3.11b) and (3.12a), the moment generating function $U(\phi;t,\tau,V_t)$ of $V_{t+\tau}$ can be found readily. We consider the affine form solution of the form $U(\phi;t,\tau,V_t)$, where

$$U(\phi;t,\tau,V_t) = e^{A(\phi;\tau,\lambda)+B(\phi;\tau)+D(\phi;\tau)V_t}. \qquad (3.20)$$

Recall that $U(\phi;t,\tau,V_t)$ in (3.20) is simply the marginal (dropping dependence on X) of the moment generating function defined in (2.123). By following a similar step of substituting $U(\phi;t,\tau,V_t)$ into the governing partial differential equation (2.122), we obtain a system of Riccati ordinary differential equations [simplified from (2.124) by dropping dependence on X]. By solving the corresponding Riccati system, the parameter functions in $U(\phi;t,\tau,V_t)$ are found to be (Zhu and Lian, 2012b)

$$\begin{aligned}
A(\phi;\tau,\lambda) &= \frac{2\mu_V\lambda}{2\mu_V\kappa-\sigma_V^2}\ln\left(1+(e^{-\kappa\tau}-1)\frac{\phi(\sigma_V^2-2\mu_V\kappa)}{2\kappa(1-\mu_V\phi)}\right), \\
B(\phi;\tau) &= -\frac{2\kappa\theta}{\sigma_V^2}\ln\left(1+(e^{-\kappa\tau}-1)\frac{\phi\sigma_V^2}{2\kappa}\right), \\
D(\phi;\tau) &= \frac{2\kappa\phi}{\sigma_V^2\phi+(2\kappa-\sigma_V^2\phi)e^{\kappa\tau}}.
\end{aligned}$$

Here, ϕ is complex valued and we write $\phi = \phi_R + i\phi_I$. To ensure $B(\phi;\tau)$ to be a well-defined continuous function of ϕ, it suffices to set

$$\mathrm{Re}\left\{\ln\left(1+(e^{-\kappa\tau}-1)\frac{\phi\sigma_V^2}{2\kappa}\right)\right\} > 0.$$

This is equivalent to set

$$\phi_R < \frac{2\kappa}{\sigma_V^2(1-e^{-\kappa\tau})}.$$

A similar technical requirement can be deduced for $A(\phi;\tau,\lambda)$. Combining these technical requirements, one should set (Lian and Zhu, 2013)

$$\phi_R < \min\left(\frac{1}{\mu_V}, \frac{2\kappa}{(1-e^{-\kappa\tau})\sigma_V^2+2\mu_V\kappa e^{-\kappa\tau}}\right). \qquad (3.21)$$

Since $\text{VIX}^2_{t+\tau}(\tau)$ has an explicit linear dependence on $V_{t+\tau}$, it is necessary to find the transition density function $p(V_{t+\tau}|V_t)$ of the instantaneous variance process V_t in order to price VIX futures and option. We can obtain $p(V_{t+\tau}|V_t)$ by taking the Laplace inversion of $U(\phi;t,\tau,V_t)$, where

$$p(V_{t+\tau}|V_t) = \frac{1}{\pi}\int_0^\infty \text{Re}\{e^{-\phi V_{t+\tau}}U(\phi;t,\tau,V_t)\}\,\text{d}\phi. \tag{3.22}$$

The time-t VIX futures price is given by

$$F_t(\tau) = E_t^Q[\text{VIX}_{t+\tau}(\tau)] = \int_0^\infty p(V_{t+\tau}|V_t)\sqrt{a_0(\tau)+a_1(\tau)V_{t+\tau}}\,\text{d}V_{t+\tau}, \tag{3.23a}$$

where $a_0(\tau)$ and $a_1(\tau)$ are given by (3.16a,b). Note that $p(V_{t+\tau}|V_t)$ is independent of the parameters in the dynamics of S_t while $a_0(\tau)$ depends on μ_S and ρ_J (parameters in the jump component of S_t).

Similarly, the time-t VIX call option price with strike price K is given by

$$c_t(\tau;K) = e^{-r\tau}\int_0^\infty p(V_{t+\tau}|V_t)(\sqrt{a_0(\tau)+a_1(\tau)V_{t+\tau}}-K)^+\,\text{d}V_{t+\tau}. \tag{3.23b}$$

We recall the following Laplace transform formulas:

$$\int_0^\infty e^{-\phi y}\sqrt{y}\,\text{d}y = \frac{\sqrt{\pi}}{2}\frac{1}{\phi^{3/2}}$$

and

$$\int_0^\infty e^{-\phi y}(\sqrt{y}-K)^+\,\text{d}y = \frac{\sqrt{\pi}}{2}\frac{1-\text{erf}(K\sqrt{\phi})}{\phi^{3/2}},$$

where the complex error function $\text{erf}(z)$ is defined by

$$\text{erf}(z) = \frac{2}{\sqrt{\pi}}\int_0^z e^{-u^2}\,\text{d}u.$$

Combining all the above relations and writing $\text{VIX}_t = x$, one can easily deduce the following integral price formulas for the VIX futures and European call option with strike price K (Lian and Zhu, 2013):

$$F_t(x,\tau) = \frac{1}{2a_1(\tau)\sqrt{\pi}}\int_0^\infty \text{Re}\left\{e^{\frac{a_0(\tau)}{a_1(\tau)}\phi}U(\phi;t,\tau,\frac{x^2-a_0(\tau)}{a_1(\tau)})\left(\frac{\phi}{a_1(\tau)}\right)^{-3/2}\right\}\text{d}\phi_I, \tag{3.24a}$$

$$c_t(x,\tau) = \frac{e^{-r\tau}}{2a_1(\tau)\sqrt{\pi}}$$
$$\int_0^\infty \text{Re}\left\{e^{\frac{a_0(\tau)}{a_1(\tau)}\phi}U(\phi;t,\tau,\frac{x^2-a_0(\tau)}{a_1(\tau)})\frac{1-\text{erf}(K\sqrt{\phi/a_1(\tau)})}{[\phi/a_1(\tau)]^{3/2}}\right\}\text{d}\phi_I, \tag{3.24b}$$

where $\phi = \phi_R + i\phi_I$ with ϕ_R satisfying (3.21).

The price formulas of the VIX futures and European call option can be simplified if we assume no jump on the instantaneous variance V_t. With zero jump on V_t, we manage to obtain the density function of V_t [see (2.113)] without resort to finding the Laplace inversion of the moment generating function using (3.22). The τ-dependent parameter function $a_0(\tau)$ in (3.16a) is modified by setting $\rho_J \mu_V = 0$ while $a_1(\tau)$ in (3.16b) remains the same. Writing $y = V_{t+\tau}$, we recall from (2.113) that the density function of $V_{t+\tau}$ in (3.22) admits the following closed-form representation

$$p_{V_{t+\tau}}(y|V_t) = ce^{-w-cy}\left(\frac{cy}{w}\right)^{q/2} I_q(2\sqrt{wcy}), \tag{3.25}$$

where I_q is the modified Bessel function of the first kind of order q,

$$c = \frac{2\kappa}{\sigma_V^2(1-e^{-\kappa\tau})}, \quad w = ce^{-\kappa\tau}V_t \quad \text{and} \quad q = \frac{2\kappa\theta}{\sigma_V^2} - 1.$$

In terms of $u = \mathrm{VIX}_{t+\tau}$, $x = \mathrm{VIX}_t$, the density function of $\mathrm{VIX}_{t+\tau}$ can be expressed as

$$p_{\mathrm{VIX}_{t+\tau}}(u|x) = \frac{2u}{a_1(\tau)} p_{V_{t+\tau}}\left(\frac{u^2 - a_0(\tau)}{a_1(\tau)} \middle| \frac{x^2 - a_0(\tau)}{a_1(\tau)}\right), \quad u \geq 0.$$

The corresponding integral formulas for the VIX futures and European call option are given by (Kokholm and Stisen, 2015)

$$F_t(x,\tau) = \int_0^\infty u p_{\mathrm{VIX}_{t+\tau}}(u|x)\, \mathrm{d}u, \tag{3.26a}$$

$$c_t(x,\tau) = e^{-r\tau}\int_K^\infty (u-K) p_{\mathrm{VIX}_{t+\tau}}(u|x)\, \mathrm{d}u. \tag{3.26b}$$

A brief account on hedging VIX futures and options can be found in Sepp (2008). He argues that the standard delta-gamma hedging is not quite appropriate for the VIX options due to infrequent and large jumps observed in VIX dynamics. To hedge the jump risks, he proposes to follow the delta-jump-hedging strategy that uses two futures contracts, aiming at eliminating the expected jump impact. The two hedge ratios of the hedging futures are determined by setting the hedging position and its variance delta to be both zero.

Numerical evaluation of the price functions of VIX derivatives

The integral price formulas of VIX futures and options can be evaluated by direct numerical integration, though special precautions are required in searching for the appropriate damping factors in the numerical evaluation of the Laplace integrals in (3.24a,b). Alternatively, one may use the Markov chain type algorithm to construct the lattice tree for the dynamics of V_t [see the willow tree algorithm in Ma *et al.* (2020)] and perform the usual time stepping numerical calculations as in the standard lattice tree algorithms.

In Fig. 3.1, we show the plots of the European VIX call option prices with strike price $K = 13$ and varying values of volatility of instantaneous variance σ_V and jump intensity parameter λ under different maturities (2-month, 3-month and 4-month). The other model parameters of the joint dynamics of S_t and V_t are listed in Table 3.1. The plots reveal that the European VIX call option price function is an increasing function of σ_V, λ, and maturity. All these observations agree with financial intuition.

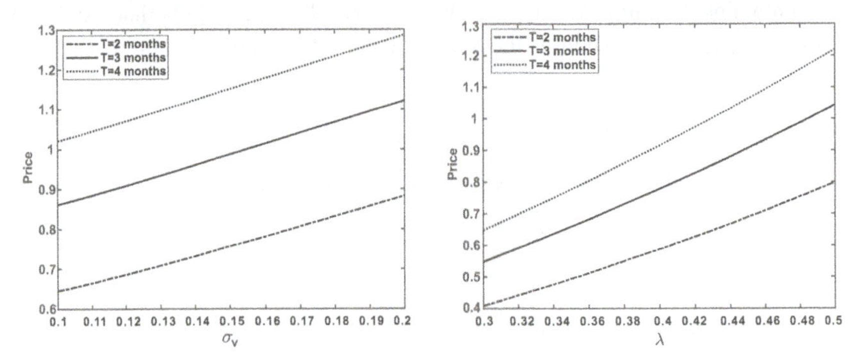

FIGURE 3.1: Plots of the European VIX call option prices with $K = 13$ and varying values of σ_V and λ under different maturities.

r	V_0	κ	θ	σ_V	μ_S	σ_S	λ	ρ_J	μ_V
0.0319	0.0076	3.46	0.008	0.14	−0.0865	0.0001	0.47	−0.38	0.05

TABLE 3.1: Model parameter values of the affine stochastic volatility model with simultaneous jumps in S_t and V_t.

Analytic approximation methods

As an alternative approach to compute prices of VIX derivatives without resort to the direct numerical integration of the integral price formulas, one may use various analytic approximation methods for deriving approximate pricing formulas. Kwok and Zheng (2018) present a full account of the use of the saddlepoint approximation for pricing VIX derivatives (see Sec. 4.2 in their book). Their numerical tests demonstrate high level of accuracy and computational efficiency in computing VIX derivative prices. Barletta and Nicolato (2018) derive orthogonal expansions for pricing VIX options under the affine jump-diffusion models by approximating the density function with an orthogonal expansion of polynomials weighted by a kernel. Instead of using the common Edgeworth expansion or Gram-Charlier expansion, they use a flexible family of distributions, which generalize the Gamma kernel associated with the classical Laguerre expansion. Cui *et al.* (2021) consider approximate valuation of VIX derivatives under various stochastic volatility models using combined Itô-Taylor expansion and Markov chain approximation. Tong and Huang (2021) develop an approximation approach for pricing VIX options under the log-linear Realized GARCH model.

3.2.2 $3/2$-model with jumps in index value

The affine stochastic volatility model enjoys nice analytic tractability due to its affine structure such that closed-form solutions for the parameter functions can be found by solving the system of Riccati ordinary differential equations. However, based on various empirical studies on modeling asset price dynamics, we witness inconsistency of the affine stochastic volatility models with the market data (see the detailed discussion in Sec. 2.7). The $3/2$-model chooses the variance exponent in the instantaneous variance V_t that comes closer to the empirical observed values while retains some level of analytic tractability. In this subsection, we consider pricing VIX futures and options under the $3/2$-model and some of its extensions.

Under a risk neutral measure Q, the joint process of S_t and V_t of the $3/2$-model is specified by the dynamic equations:

$$\frac{\mathrm{d}S_t}{S_t} = (r - m\lambda)\mathrm{d}t + \sqrt{V_t}\,\mathrm{d}W_t^1 + (e^{J^S} - 1)\mathrm{d}N_t, \tag{3.27a}$$

$$\mathrm{d}V_t = V_t(\theta - \kappa V_t)\mathrm{d}t + \sigma_V V_t^{3/2}\,\mathrm{d}W_t^2, \tag{3.27b}$$

where the parameters κ, θ, and σ_V have similar interpretation as those of the $1/2$-model, $\mathrm{d}W_t^1\mathrm{d}W_t^2 = \rho\,\mathrm{d}t$, N_t is the Poisson process with constant intensity λ and the jump component J^S is normally distributed with mean μ_S and variance σ_S^2. Note that S_t takes the same dynamics as in (3.11a) with compound Poisson jumps while V_t does not have jump [unlike (3.11b)]. Correspondingly, by setting $\rho_J\mu_V = 0$ in (3.13), we have

$$m = E[e^{J^S} - 1] = e^{\frac{\sigma_S^2}{2} + \mu_S} - 1.$$

Following the same derivation procedure, we can obtain a similar relation between $\mathrm{VIX}_t^2(\tau)$ and V_t as shown in (3.14), where

$$\mathrm{VIX}_t^2(\tau) = 2\lambda(m - \mu_S) + \frac{1}{\tau}E_t^Q\left[\int_t^{t+\tau} V_u\,\mathrm{d}u\right]. \tag{3.28}$$

Unlike the $1/2$-model, there is no simple analytic solution for the expectation of the instantaneous variance.

We consider an alternative pricing approach by relating $\int_t^{t+\tau} E_t^Q[V_u]\,du$ to the Laplace transform of the integrated realized variance $\int_t^{t+\tau} V_u\,du$, which is defined by

$$L(V_t;\ell) = E\left[e^{-\ell\int_t^{t+\tau} V_u\,\mathrm{d}u}\,\middle|\,V_t\right]. \tag{3.29}$$

Here, ℓ is the Laplace transform variable. For the dynamics of V_t under the 3/2-model as governed by (3.27b), $L(V_t;\ell)$ is found to be (Zhao *et al.*, 2018)

$$L(V_t;\ell) = \frac{\Gamma(\gamma-\alpha)}{\Gamma(\gamma)}\left(\frac{2}{\sigma_V^2 U_t}\right)^{\alpha} {}_1F_1\left(-\frac{2}{\sigma_V^2 U_t};\alpha,\gamma\right), \tag{3.30}$$

where

$$U_t = \frac{e^{\theta\tau}-1}{\theta}V_t,\ \alpha = -\left(\frac{\kappa}{\sigma_V^2}+\frac{1}{2}\right)+\sqrt{\left(\frac{\kappa\theta}{\sigma_V^2}+\frac{1}{2}\right)^2+\frac{2\ell}{\sigma_V^2}},\ \gamma = 2\left(\alpha+1+\frac{\kappa}{\sigma_V^2}\right),$$

Γ is the Gamma function [see (2.63)] and ${}_1F_1$ is the confluent hypergeometric function of the first kind defined in (2.159). In fact, one can obtain $L(V_t;\ell)$ by setting $\omega = 0$ and $\eta = i\ell$ in $h(V_t;\omega,\eta)$ in (2.158). By taking the negative of the derivative of the above Laplace transform $L(V_t;\ell)$ with respect to ℓ and setting $\ell = 0$, we obtain

$$g(V_t;\tau) = -\frac{\partial}{\partial \ell}L(V_t;\ell)\bigg|_{\ell=0} = E\left[\int_t^{t+\tau} V_u\,\mathrm{d}u \,\bigg|\, V_t\right] = \int_t^{t+\tau} E_t^Q[V_u]\,\mathrm{d}u. \tag{3.31}$$

Together with (3.28), we obtain the following formal representation of $\mathrm{VIX}_t^2(\tau)$ in terms of the Laplace transform $L(V_t;\ell)$ and jump parameters in the dynamics of S_t:

$$\mathrm{VIX}_t^2(\tau) = 2\lambda(m-\mu_S)+\frac{g(V_t;\tau)}{\tau}. \tag{3.32}$$

The valuation of $g(V_t;\tau)$ can be affected by numerical differentiation of $L(V_t;\ell)$, with closed-form formula that can be expressed in terms of the confluent hypergeometric function [see (3.30)]. Let $f_{V_{t+\tau}|V_t}(y;x)$ denote the transition density function of $V_{t+\tau}$ with $V_{t+\tau} = y$ and $V_t = x$, whose analytic formula is stated in (2.156). The integral formulas of the time-t price for the VIX futures and European call option are given by (Baldeaux and Badran, 2014)

$$F_t = E[\mathrm{VIX}_{t+\tau}(\tau)] = \int_0^\infty \sqrt{\frac{g(y;\tau)}{\tau}+2\lambda(m-\mu_S)}\, f_{V_{t+\tau}|V_t}(y;x)\,\mathrm{d}y \tag{3.33a}$$

and

$$\begin{aligned} c_t &= e^{-r\tau}E[(\mathrm{VIX}_{t+\tau}(\tau)-K)^+] \\ &= e^{-r\tau}\int_0^\infty \left(\sqrt{\frac{g(y;\tau)}{\tau}+2\lambda(m-\mu_S)}-K\right)^+ f_{V_{t+\tau}|V_t}(y;x)\,\mathrm{d}y, \end{aligned} \tag{3.33b}$$

respectively.

Note that the inclusion of jumps in the index value does not have any effect on the analytic tractability of the 3/2-model since only one extra term $2\lambda(m-\mu_S)$ is added in the integral price formulas. Baldeaux and Badran (2014) show that the 3/2-model with jumps in index value produces a better short-term fit to the implied volatility of S&P index options than its pure diffusion counterpart.

Extended models

Another stochastic volatility model that exhibits analytic tractability is the 4/2-model plus jumps (Grasselli, 2017). Under a risk neutral measure Q, the joint process of asset price S_t and instantaneous variance V_t of the 4/2-model plus jumps is governed by

$$\frac{dS_t}{S_t} = (r - m\lambda)dt + \left(a\sqrt{V_t} + \frac{b}{\sqrt{V_t}}\right)dW_t^1 + (e^{J^S} - 1)dN_t, \tag{3.34a}$$

$$dV_t = \kappa(\theta - V_t)dt + \sigma_V\sqrt{V_t}\,dW_t^2, \tag{3.34b}$$

where a and b are constant parameters. Similar to the 3/2-model, we have to assume zero jump in V_t in order to achieve analytic tractability. Other parameter values and jump distribution follow similar interpretation as those of the 3/2-model. The 4/2-model reduces to 1/2-model and 3/2-model when $b = 0$ and $a = 0$, respectively. The distribution of J^S is assumed to be normal with mean μ_S and variance σ_S^2.

Lin *et al.* (2017) manage to derive the closed-form formula for the Laplace transform of the integrated realized variance under the 4/2-model. Together with the known transition density function of V_t, one can compute VIX futures price and call option price using the same set of integral price formulas (3.33a,b) by substituting the corresponding $g(V_t;\tau)$ and $f_{V_{t+\tau}|V_t}(y;x)$ for the 4/2-model. The integral formula for $\text{VIX}_t^2(\tau)$ under the 4/2-model is found to be

$$\begin{aligned}
&\text{VIX}_t^2(\tau)\\
&= 2\lambda(\widetilde{\mu} - \mu_S) + 2ab + \frac{a^2}{\tau}\left[\theta\tau + \frac{x-\theta}{\kappa}(1 - e^{-\kappa\tau})\right]\\
&+ \int_0^\tau \frac{b^2\kappa}{\sigma_V^2\tau}\frac{\Gamma\left(\frac{2\kappa\theta}{\sigma_V^2} - 1\right)}{\Gamma\left(\frac{2\kappa\theta}{\sigma_V^2}\right)}\left(\sinh\frac{\kappa u}{2}\right)^{-\frac{2\kappa\theta}{\sigma_V^2}} \exp\left(\frac{\kappa}{\sigma_V^2}\left(\kappa\theta u + x - x\coth\frac{\kappa u}{2}\right)\right)\\
&\quad \left(1 + \coth\frac{\kappa u}{2}\right)^{1-\frac{2\kappa\theta}{\sigma_V^2}} {}_1F_1\left(\frac{2\kappa x}{\sigma_V^2(e^{\kappa u} - 1)}; \frac{2\kappa\theta}{\sigma_V^2} - 1, \frac{2\kappa\theta}{\sigma_V^2}\right) du,
\end{aligned} \tag{3.35}$$

where $x = V_t$ and

$$\widetilde{\mu} = \exp\left(\mu_S + \frac{\sigma_S^2}{2}\right) - 1.$$

Here, Γ and ${}_1F_1$ denote the Gamma function and hypergeometric confluent function of the first kind, respectively. As evidenced by their numerical studies, Lin *et al.* (2017) conclude that the 4/2-model provides better overall performance in pricing VIX futures and options.

Lin *et al.* (2019) also propose the free stochastic volatility model in pricing VIX derivatives. The free stochastic volatility model allows the more generalized power index α in the volatility term of the index value process, which reduces to the affine 1/2-model when $\alpha = 1/2$. The joint dynamics of asset price S_t and instantaneous variance V_t under a risk neutral measure Q are assumed to be governed by

$$\begin{aligned}
\frac{dS_t}{S_t} &= (r - \lambda_1\widetilde{\mu}_1 - \lambda_2\widetilde{\mu}_2)dt + V_t^\alpha\,dW_t^1\\
&\quad + (e^{J_1} - 1)dN_{1t} + (e^{J_2} - 1)dN_{2t}, \qquad (3.36)\\
dV_t &= \kappa(\theta - V_t)dt + \sigma_V\sqrt{V_t}\,dW_t^2,
\end{aligned}$$

where W_t^1 and W_t^2 are correlated Q-Brownian motions with correlation coefficient ρ, κ and θ are the mean reversion speed and reversion level of V_t, respectively, σ_V is the volatility of V_t, and r is the riskless interest rate. The volatility of S_t takes the form V_t^α, $\alpha \in \left[-\frac{1}{2}, \frac{3}{2}\right]$, which is called the *free* stochastic volatility. The freedom on the choice of α aims to capture better fluctuations of volatility implied in the S&P 500 index. The free volatility model assumes more general form of asymmetric jumps, where the upward and downward jumps are driven by independent compound Poisson processes. Here, N_{1t} and N_{2t} denote the risk neutral Poisson processes driving upward and downward jumps with jump intensities λ_1 and λ_2, respectively. The upward jump sizes J_1 are assumed to follow an independent exponential distribution with a positive mean, $\mu_1 > 0$ and probability density function $\frac{1}{\mu_1}e^{-\frac{x}{\mu_1}}$, $x > 0$. The two parameters μ_1 and $\widetilde{\mu}_1$ are related by

$$\widetilde{\mu}_1 = E_Q[e^{J_1} - 1] = \frac{1}{1-\mu_1} - 1. \tag{3.37a}$$

Similarly, the downward jump size J_2 is assumed to follow an independent exponential distribution with a negative mean, $\mu_2 < 0$, and probability density function $\frac{1}{|\mu_2|}e^{\frac{x}{\mu_2}}$, $x < 0$. The two parameters μ_2 and $\widetilde{\mu}_2$ are related by

$$\widetilde{\mu}_2 = E_Q[e^{J_2} - 1] = \frac{1}{1-\mu_2} - 1. \tag{3.37b}$$

Similar to $\mathrm{VIX}_t^2(\tau)$ under the 4/2-model, Lin *et al.* (2019) manage to obtain the following integral formula for VIX_t^2 at $V_t = x$ under the free stochastic volatility model:

$$\begin{aligned}\mathrm{VIX}_t^2(\tau) &= 2[\lambda_1(\widetilde{\mu}_1 - \mu_1) + \lambda_2(\widetilde{\mu}_2 - \mu_2)] \\ &+ \int_0^\tau \left[\frac{\sigma_V^{4\alpha}}{\kappa^{2\alpha}\tau} \frac{\Gamma\left(\frac{2\kappa\theta}{\sigma_V^2} + 2\alpha\right)}{\Gamma\left(\frac{2\kappa\theta}{\sigma_V^2}\right)} \left(\sinh\frac{\kappa u}{2}\right)^{-\frac{2\kappa\theta}{\sigma_V^2}} \right. \\ &\exp\left(\frac{\kappa}{\sigma_V^2}\left(\kappa\theta u + x - x\coth\frac{\kappa u}{2}\right)\right)\left(1 + \coth\frac{\kappa u}{2}\right)^{-2\alpha - \frac{2\kappa\theta}{\sigma_V^2}} \\ &\left. {}_1F_1\left(\frac{2\kappa x}{\sigma_V^2(e^{\kappa u} - 1)}; \frac{2\kappa\theta}{\sigma_V^2} + 2\alpha, \frac{2\kappa\theta}{\sigma_V^2}\right)\right] du. \end{aligned} \tag{3.38}$$

Here, ${}_1F_1$ is the confluent hypergeometric function of the first kind.

3.2.3 Barndorff-Nielsen and Shephard model

The Barndorff-Nielsen and Shephard model is a non-Gaussian stochastic volatility model with instantaneous variance governed by an Ornstein-Uhlenbeck process driven by a subordinator without drift. Arai (2019) manages to derive a linear relation between $\mathrm{VIX}_t^2(\tau)$ and V_t under the Barndorff-Nielsen and Shephard model, similar to that of the $1/2$-model.

Under the Barndorff-Nielsen and Shephard model, the dynamics of S_t under a risk neutral measure Q is given by

$$S_t = S_0 \exp\left((r+\mu)t - \int_0^t \sigma_s^2 \, \mathrm{d}s + \int_0^t \sigma_s \, \mathrm{d}W_s + \rho H_{\lambda t} \right), \tag{3.39}$$

where the leverage coefficient $\rho \le 0$ and intensity parameter $\lambda > 0$. Here, W_t is a standard Brownian motion and H is a subordinator with drift. Also, σ_t^2 is governed by an Ornstein-Uhlenbeck process

$$\mathrm{d}\sigma_t^2 = -\lambda \sigma_t^2 \, \mathrm{d}t + \mathrm{d}H_{\lambda t}. \tag{3.40}$$

The drift parameter μ is determined so that the discounted asset price process is a martingale. It can be shown that

$$\mu = \int_0^\infty (1 - e^{\rho x}) \, \Pi(\mathrm{d}x), \tag{3.41}$$

where $\Pi(\mathrm{d}x)$ is the Lévy measure of $H_{\lambda t}$. By following a similar proof as in (3.17), Arai (2019) shows that

$$\mathrm{VIX}_t^2(\tau) = \frac{1}{\tau} E_t^Q \left[\int_t^{t+\tau} \sigma_s^2 \, \mathrm{d}s \right] - 2 \int_0^\infty (1 + \rho x - e^{\rho x}) \, \Pi(\mathrm{d}x). \tag{3.42}$$

By solving the dynamic equation for σ_t^2 and taking expectation, we have

$$\frac{1}{\tau} E_t^Q \left[\int_t^{t+\tau} \sigma_s^2 \, \mathrm{d}s \right] = \frac{1 - e^{-\lambda\tau}}{\lambda\tau} \sigma_t^2 + \frac{1}{\lambda} \left(1 - \frac{1 - e^{-\lambda\tau}}{\lambda\tau} \right) \int_0^\infty x \, \Pi(\mathrm{d}x). \tag{3.43}$$

Combining the above results together, we obtain a similar linear relation between $\mathrm{VIX}_t^2(\tau)$ and σ_t^2, where

$$\begin{aligned} \mathrm{VIX}_t^2(\tau) = {} & \frac{1 - e^{-\lambda\tau}}{\lambda\tau} \sigma_t^2 + \frac{1}{\lambda} \left(1 - \frac{1 - e^{-\lambda\tau}}{\lambda\tau} \right) \int_0^\infty x \, \Pi(\mathrm{d}x) \\ & - 2 \int_0^\infty (1 + \rho x - e^{\rho x}) \, \Pi(\mathrm{d}x). \end{aligned} \tag{3.44}$$

For the choice of the subordinator $H_{\lambda t}$, Arai (2019) consider the two most common choices in the Barndorff-Nielsen and Shephard model, namely, the inverse Gaussian process and Gamma process. The corresponding Lévy measures are given by

$$\Pi_{\mathrm{IG}}(\mathrm{d}x) = \frac{\lambda a}{2\sqrt{2\pi}} x^{-3/2} (1 + b^2 x) e^{-\frac{b^2}{2} x} \mathbf{1}_{\{(0,\infty)\}}(x) \, \mathrm{d}x, \quad a > 0 \text{ and } b > 0, \tag{3.45a}$$

$$\Pi_{\mathrm{G}}(\mathrm{d}x) = \lambda a b e^{-bx} \mathbf{1}_{\{(0,\infty)\}}(x) \, \mathrm{d}x, \quad a > 0 \text{ and } b > 0, \tag{3.45b}$$

respectively.

3.2.4 GARCH type models

Compared to the stochastic volatility models, there is a relatively thinner literature on pricing VIX derivatives using the GARCH type models. This may be due to the concern that the conventional locally risk neutral valuation relationship in the GARCH type models (Duan, 1995) may compensate for the equity risk premium only. Later research works (Hao and Zhang, 2013; Kanniainen *et al.*, 2014) show that the joint estimation of the underlying asset combined with options and VIX data can account for the variance risk premium. This would improve the pricing performance of the GARCH type models in pricing volatility and VIX derivatives.

Heston-Nandi GARCH model

We consider VIX futures pricing under the affine-type discrete time Heston-Nandi GARCH model (Heston and Nandi, 2000). Similar to the earlier affine-type stochastic volatility models, we derive the moment generating function of the conditional variance. Thanks to the affine structure of the Heston-Nandi model, we manage to obtain a linear relationship between $\mathrm{VIX}_t^2(\tau)$ and the instantaneous variance V_t. For the Heston-Nandi GARCH model, the dynamics of the returns R_{t+1} of the S&P 500 index S_t under the statistical measure is governed by

$$R_{t+1} = \ln\frac{S_{t+1}}{S_t} = r_{t+1} + \lambda h_{t+1} - \frac{h_{t+1}}{2} + \sqrt{h_{t+1}}\varepsilon_{t+1}h_{t+1}, \tag{3.46}$$

where ε_t follows a standard normal distribution, r_t is the time-t risk-free interest rate, h_t is the time-t instantaneous daily variance of the return of the S&P 500 index, and λ is the equity risk premium parameter associates with the conditional variance. We perform risk neutralization using the locally risk neutral valuation relationship (Duan, 1995) to obtain the following dynamics of R_{t+1} under the risk neutral measure Q:

$$R_{t+1} = \ln\frac{S_{t+1}}{S_t} = r_{t+1} - \frac{h_{t+1}}{2} + \sqrt{h_{t+1}}\varepsilon_{t+1}^{*}h_{t+1}, \tag{3.47a}$$

where

$$h_{t+1} = \omega + \beta h_t + \alpha(\varepsilon_t^* - \delta^*\sqrt{h_t})^2, \tag{3.47b}$$

$$\varepsilon_t^* = \varepsilon_t + \lambda\sqrt{h_t} \quad \text{and} \quad \delta^* = \delta + \lambda. \tag{3.47c}$$

The long-term variance as quantified by the unconditional expectation of h_t under the Q measure is given by

$$\overline{h} = E^Q[h_t] = \frac{\widetilde{\omega}}{1-\widetilde{\beta}}, \tag{3.48}$$

where

$$\widetilde{\omega} = \omega + \alpha \quad \text{and} \quad \widetilde{\beta} = \beta + \alpha(\delta+\lambda)^2.$$

The time-t implied variance term structure $V_t(n)$ is defined to be

$$V_t(n) = \frac{1}{n}\sum_{k=1}^{n} E_t^Q[h_{t+k}], \tag{3.49}$$

which can be shown to be a weighted average of the conditional variance at the next time point $t+1$ and the long-run variance under the risk neutral measure Q. More precisely, we have

$$V_t(n) = \Gamma(n)h_{t+1} + [1-\Gamma(n)]\overline{h}, \tag{3.50}$$

where $\overline{h}$ is given by (3.48) and

$$\Gamma(n) = \frac{1-\widetilde{\beta}^n}{n(1-\widetilde{\beta})}.$$

To establish (3.50), we consider

$$E_t^Q[h_{t+2}] - \overline{h} = \widetilde{\omega} + \widetilde{\beta}h_{t+1} - \frac{\widetilde{\omega}}{1-\widetilde{\beta}} = \widetilde{\beta}(h_{t+1} - \overline{h}),$$

and recursively, we obtain

$$E_t^Q[h_{t+k}] = \overline{h} + \widetilde{\beta}^{k-1}(h_{t+1} - \overline{h}), \quad k = 1, 2, \ldots, n.$$

Recall the definition of $V_t(n)$, where

$$V_t(n) = \frac{1}{n}\sum_{k=1}^{n} E_t^Q[h_{t+k}] = \overline{h} + \frac{1}{n}\sum_{k=1}^{n}\widetilde{\beta}^{k-1}(h_{t+1} - \overline{h}) = \overline{h} + \frac{1}{n}\frac{1-\widetilde{\beta}^n}{1-\widetilde{\beta}}(h_{t+1} - \overline{h}),$$

and hence the result in (3.50). As a result, based on the convention of 252 trading days in one calendar year and 22 trading days in one calendar month, we have

$$\text{VIX}_t = \sqrt{252V_t(22)} = \sqrt{252\{\Gamma(22)h_{t+1} + [1-\Gamma(22)]\overline{h}\}}. \tag{3.51}$$

By virtue of the linear relation between VIX_T^2 and h_{T+1}, where

$$\text{VIX}_T^2 = a_0 + a_1 h_{T+1}. \tag{3.52}$$

By observing the formula

$$E_t^Q[\sqrt{x}] = \frac{1}{2\sqrt{\pi}}\int_0^{\infty} \frac{1 - E_t^Q[e^{-\phi x}]}{\phi^{3/2}}\, \mathrm{d}\phi,$$

we obtain the following integral price formula for the VIX futures as follows:

$$\begin{aligned} F_t = E_t^Q[\text{VIX}_T] &= E_t^Q[a_0 + a_1 h_{T+1}] \\ &= \frac{1}{2\sqrt{\pi}}\int_0^{\infty} \frac{1 - e^{-\phi a_0} E_t^Q[e^{-\phi a_1 h_{T+1}}]}{\phi^{3/2}}\, \mathrm{d}\phi, \end{aligned} \tag{3.53}$$

where $E_t^Q[e^{-\phi a_1 h_{T+1}}]$ is the moment generating function of the conditional variance at time $T+1$. Thanks to the affine structure of the Heston-Nandi GARCH model,

Wang *et al.* (2017) show that the moment generating function admits the following exponential affine representation:

$$E_t^Q[e^{\phi h_{t+n+1}}] = e^{C(\phi,n)+H(\phi,n)h_{t+1}}, \tag{3.54}$$

where the functions $C(\phi,n)$ and $H(\phi,n)$ can be computed based on the following recursive relations:

$$C(\phi,n+1) = C(\phi,n) - \frac{1}{2}\ln(1-2\alpha H(\phi,n)) + \omega H(\phi,n), \tag{3.55a}$$

$$H(\phi,n+1) = \frac{\alpha\delta^* H(\phi,n)}{1-2\alpha H(\phi,n)} + \beta H(\phi,n), \tag{3.55b}$$

with initial conditions:

$$C(\phi,0) = 0 \quad \text{and} \quad H(\phi,0) = \phi.$$

The proof of (3.54) and (3.55a,b) is presented in the Appendix.

Wang *et al.* (2017) performed empirical studies on the pricing performance of the affine Heston-Nandi GARCH model. They consider various estimation methods of the model parameters based on different sets of data used and conclude that the model estimated by the joint use of the VIX and VIX futures prices yields the best pricing performance and provides a good balance in fitting both the VIX and VIX futures.

Inverse Gaussian GARCH model

Analytic integral price formulas for the VIX futures can also be derived for various extensions of the GARCH-type models. Yang and Wang (2018) consider the inverse Gaussian (IG) GARCH model, whose dynamics of daily returns R_{t+1} and conditional variance h_{t+1} under the statistical measure P are governed by

$$R_{t+1} = \ln\frac{S_{t+1}}{S_t} = r_t + \xi h_{t+1} + \eta y_{t+1}, \tag{3.56a}$$

$$h_{t+1} = \omega + bh_t + cy_t + a\frac{h_t^2}{y_t}, \tag{3.56b}$$

where ξ, ω, a, b, and c are constant parameters. Shocks to returns are y_{t+1}, which is assumed to be an independent inverse Gaussian distribution with degree of freedom $\frac{h_{t+1}}{\eta^2}$; that is,

$$y_{t+1} \sim \text{IG}\left(\frac{h_{t+1}}{\eta^2}\right), \tag{3.57}$$

where IG is the inverse Gaussian distribution and η is a scale parameter. As η approaches zero, the inverse Gaussian distribution converges to the normal distribution. Hence, the affine Heston-Nandi model is nested within the IG GARCH model. More importantly, Yang and Wang (2018) show that a similar linear relation between VIX_t^2 and h_t can be obtained and the conditional moment generating function assumes similar exponential affine form. As a result, analytic integral price formula for the VIX futures can also be obtained under the IG GARCH model.

As later works, Cao, H. *et al.* (2020) manage to derive price formulas for VIX options under the above two affine GARCH model. Also, Yang *et al.* (2019c) extend the Heston-Nandi GARCH model and IG GARCH model by adding jumps as parts of the shocks to asset returns. They manage to derive integral price formulas for the VIX futures similar to those for the no-jump counterparts.

Discrete time stochastic volatility model

Hitaj *et al.* (2017) consider a more systematic framework to price VIX call options under a class of discrete time stochastic volatility models that combine the virtues of analytic tractability of continuous time affine stochastic volatility models and ease of estimation of parameters in discrete time models. This is done by substituting the mixing random variable with an affine GARCH process. Their model assumes that the dynamic of the discrete time log return under the statistical measure P is governed by

$$R_t = \ln \frac{S_{t+1}}{S_t} = r_t + \lambda_0 h_t + \lambda_1 V_t + \sigma \sqrt{V_t} Z_t, \tag{3.58a}$$

where $Z_t \sim N(0,1)$ and V_t is independent of Z_t. The process of h_t is assumed to be

$$h_t = \alpha_0 + \alpha_1 V_{t-1} + \beta h_{t-1}. \tag{3.58b}$$

The parameters λ_0, λ_1, σ, α_0, α_1, and β are assumed to be constant, and $\sigma \geq 0$. The conditional distribution of log return belongs to the family of Normal Variance Mean Mixture models since Z_t is normally distributed. Hitaj *et al.* (2017) consider three special choices of the distribution to the affine GARCH process V_t, namely, the Gamma distribution, Inverse Gaussian distribution, and Normal Tempered Stable distribution (the first two distributions are nested within the last distribution). Similar to the GARCH models considered in the above, these discrete time stochastic volatility models exhibit similar analytic tractability of admitting an exponential affine form for the conditional moment generating function of V_t. Similarly, we can express VIX_t^2 as a linear function of h_{t+1} with time-dependent coefficient functions. However, since VIX_t^2 is an autoregressive process, the conditional expected value of VIX is not available in closed-form formula.

As ongoing research, there are a variety of affine type GARCH models that exhibit similar analytic tractability. Two such examples are the extended LHARG model analyzed by Huang *et al.* (2019) and GJR-GARCH model studied by Xie *et al.* (2020).

3.3 Direct modeling of VIX

There are earlier pricing models on continuous realized volatility that model the volatility process directly based on (i) mean reverting square root process for the underlying volatility (Grünbichler and Longstaff, 1996), (ii) mean reverting process in the logarithm of volatility (Detemple and Osakwe, 2000). In this section, we follow similar approach on pricing VIX derivatives based on direct modeling of the VIX process.

We start with the class of one-factor models, where the state variable X_t can be either the VIX index or its logarithm. Empirical data on VIX clearly reveal that VIX index exhibits the mean reverting feature and mostly positive spikes. The one-factor model for either $X_t = \mathrm{VIX}_t$ or $X_t = \ln \mathrm{VIX}_t$ takes the form (Kaeck and Alexander, 2013):

$$dX_t = \kappa(\theta - X_t)dt + \sigma X_t^\alpha \, dW_t + J_t \, dN_t, \tag{3.59}$$

where κ is the speed of mean reversion, θ is the long term value of the process, σ is a constant for the diffusion term, W_t is the standard Brownian motion, J_t is the jump size and N_t is a Poisson process with varying intensity $\lambda_0 + \lambda_1 X_t$. Psychoyios *et al.* (2010) discuss various choices of the jump dynamics. The exponent α may be set to be $1/2$ or $3/2$ for the VIX process, and α equals 0 or 1 for the log VIX process. Besides the above assumed form, Yan and Zhao (2019) explore some other alternative stochastic volatility models that also admit nice analytic tractability.

Later direct models assume the above form for VIX_t or $\ln \mathrm{VIX}_t$, plus additional stochastic state variables for the volatility of volatility and/or mean reversion level. For example, Wang *et al.* (2016) propose the following two-factor affine jump-diffusion model whose dynamics of VIX_t and volatility of volatility V_t under a risk neutral measure Q are governed by

$$d\ln \mathrm{VIX}_t = \kappa(\theta - \ln \mathrm{VIX}_t)dt + \sqrt{V_t}\, dW_t^1 + J_t \, dN_t, \tag{3.60a}$$

$$dV_t = \kappa_V(\theta_V - V_t)dt + \sigma_V \sqrt{V_t}\, dW_t^2, \tag{3.60b}$$

where $dW_t^1 dW_t^2 = \rho \, dt$. Alternatively, Mencía and Sentana (2013) propose another two-factor affine diffusion model whose dynamics of $\ln \mathrm{VIX}_t$ and mean reversion level θ_t under a risk neutral measure Q are governed by

$$d\ln \mathrm{VIX}_t = \kappa(\theta - \ln \mathrm{VIX}_t)dt + \sigma \, dW_t^1 \tag{3.61a}$$

$$d\theta_t = \kappa_\theta(\overline{\theta} - \theta_t)dt + \sigma_\theta \, dW_t^2, \tag{3.61b}$$

where W_t^1 and W_t^2 are uncorrelated Brownian motions. Park (2016) further proposes the three-factor direct model with stochastic state variables on VIX_t, volatility of volatility V_t, and mean reversion level θ_t, as well as allowing asymmetric positive and negative jumps on VIX_t. His model claims to reflect reality better since investors react differently to good and bad news, so these positive and negative shocks arrive independently with different rates and sizes. Kaeck and Seeger (2020) present studies on the empirical hedging performance of alternative VIX option pricing models. They show that while sophisticated models with stochastic mean-reversion and jumps have superior pricing performance, however their hedging performance is inferior to a simple Black model hedge.

While most direct models of VIX use the stochastic volatility models, Li *et al.*'s (2017) model the VIX dynamics as a pure jump semimartingale with infinite jump activity and infinite variation. They apply time change on the 3/2 diffusion by a class of additive subordinators with infinite activity. Their model takes the initial term structure of VIX futures as input. Pricing formulas for VIX futures and European VIX options can be derived via eigenfunction expansions. Their direct model for VIX manages to achieve good fit for the VIX implied volatility surface, which typically shows very steep skews.

In the next two subsections, we consider pricing VIX derivatives under various multifactor affine jump-diffusion models and one-factor 3/2-model with jumps.

3.3.1 Multifactor affine jump-diffusion models

Wang *et al.* (2016) propose the following two-factor affine jump-diffusion direct model for $\ln \mathrm{VIX}_t$, whose joint dynamics of $\ln \mathrm{VIX}_t$ and its instantaneous variance V_t under a risk neutral measure Q is governed by

$$\mathrm{d}\ln \mathrm{VIX}_t = \kappa(\theta - \ln \mathrm{VIX}_t)\mathrm{d}t + \sqrt{V_t}\,\mathrm{d}W_t^1 + J\,\mathrm{d}N_t, \tag{3.62a}$$

$$\mathrm{d}V_t = \kappa_V(\theta_V - V_t)\mathrm{d}t + \sigma_V\sqrt{V_t}\,\mathrm{d}W_t^2, \tag{3.62b}$$

where the Brownian motions W_t^1 and W_t^2 observe $\mathrm{d}W_t^1\mathrm{d}W_t^2 = \rho\ \mathrm{d}t$, κ and κ_V are the respective mean reversion speed of $\ln \mathrm{VIX}_t$ and V_t, θ and θ_V are the respective long term mean of $\ln \mathrm{VIX}_t$ and V_t, N_t is the Poisson portion with jump intensity λ, and the jump size variables $\{J_k\}$, $k = 1,2,\ldots$, σ_V is the volatility parameter of V_t. We assume the jumps to follow the double exponential jump distribution $\mathrm{Exp}(\eta_1, \eta_2, p)$ with $\eta_1 > 1$, $\eta_2 > 0$ and $0 < p < 1$ [see (2.46)]. The jump in stochastic volatility V_t is not included since earlier empirical studies show that jumps in the instantaneous variance of the underlying does not provide significant improvement in fitting the implied volatility surface.

For notational convenience, we write $X_t = \ln \mathrm{VIX}_t$. Thanks to the affine structure of the two-factor direct model, closed-form solution of the conditional characteristic function

$$\Phi(\tau;\phi) = E_t^Q[e^{i\phi X_T}], \quad \tau = T - t, \tag{3.63}$$

can be found. Once $\Phi(\tau;\phi)$ is known, the VIX futures price is given by

$$F_t = E_t^Q[\mathrm{VIX}_T] = \Phi(\tau; -i). \tag{3.64}$$

Also, the price of the European call option on VIX_t with strike price K is given by

$$c(\mathrm{VIX}_t, \tau; K) = e^{-r\tau}(F_t\Pi_1 - K\Pi_2), \tag{3.65}$$

where

$$\Pi_j = \frac{1}{2} + \frac{1}{\pi}\int_0^\infty \mathrm{Re}\left\{\frac{\Phi_j(\tau;\phi)e^{-i\phi\ln K}}{i\phi}\right\}\mathrm{d}\phi, \quad j = 1,2,$$

$$\Phi_1(\tau;\phi) = \frac{\Phi(\tau;\phi - i)}{\Phi(\tau;-i)}, \quad \Phi_2(\tau;\phi) = \Phi(\tau;\phi).$$

Next, we seek for closed-form solution of $\Phi(\tau;\phi)$. Solving the dynamic equation of (3.62a), we obtain

$$X_T = e^{-\kappa\tau}X_t + \int_t^T e^{-\kappa(T-s)}\kappa\theta \, ds + \int_t^T e^{-\kappa(T-s)}\sqrt{V_s}\, dW_s^1 + \int_t^T e^{-\kappa(T-s)}J \, dN_s.$$

Recall that

$$\Phi(\tau;\phi) = E_t^Q[e^{i\phi X_T}] = \exp(e^{-\kappa\tau}X_t + \theta(1-e^{-\kappa\tau})i\phi) \, E_t^Q\left[\exp\left(i\phi\int_t^T e^{-\kappa(T-s)}\sqrt{V_s}\, dW_s^1\right)\right] E_t^Q\left[\exp\left(i\phi\int_t^T e^{-\kappa(T-s)}J \, dN_s\right)\right]. \tag{3.66}$$

Since $dW_t^1 dW_t^2 = \rho \, dt$, we use the Cholesky decomposition to write

$$W_t^1 = \rho W_t^2 + \sqrt{1-\rho^2}W_t,$$

where W_t is independent of W_t^2. We write the second term in (3.66) as follows:

$$\begin{aligned} &E_t^Q\left[\exp\left(i\phi\int_t^T e^{-\kappa(T-s)}\sqrt{V_s}\, dW_s^1\right)\right] \\ &= E_t^Q\left[\exp\left(i\phi\rho\int_t^T e^{-\kappa(T-s)}\sqrt{V_s}\, dW_s^2\right)\right] \\ &\quad E_t^Q\left[\exp i\phi\sqrt{1-\rho^2}\int_t^T e^{-\kappa(T-s)}\sqrt{V_s}\, dW_s \middle| W_s^2,\, t\le s\le T\right]. \end{aligned}$$

From the dynamics of V_t as governed by (3.62b), we obtain

$$d(e^{-\kappa(T-s)}V_s) = e^{-\kappa(T-s)}[\kappa_V\theta_V + (\kappa-\kappa_V)V_s]ds + e^{-\kappa(T-s)}\sigma_V\sqrt{V_s}\, dW_s^2$$

so that

$$\begin{aligned} &\int_t^T e^{-\kappa(T-s)}\sqrt{V_s}\, dW_s^2 \\ &= \frac{1}{\sigma_V}\left[V_T - e^{-\kappa\tau}V_t - \frac{\kappa_V\theta_V}{\kappa}(1-e^{-\kappa\tau}) - (\kappa-\kappa_V)\int_t^T e^{-\kappa(T-s)}V_s \, ds\right]. \end{aligned}$$

Also, note that W_t and W_t^2 are independent so that the following stochastic integral

$$\int_t^T e^{-\kappa(T-s)}\sqrt{V_s}\, dW_s$$

is conditionally normal with zero mean and variance

$$(1-\rho^2)\int_t^T e^{-2\kappa(T-s)}V_s \, ds.$$

Combining these results together, we obtain

$$\begin{aligned}
&E_t^Q\left[\exp\left(i\phi\int_t^T e^{-\kappa(T-s)}\sqrt{V_s}\,\mathrm{d}W_s^1\right)\right]\\
&=\exp\left(i\phi\left[-\frac{\rho e^{-\kappa\tau}V_t}{\sigma_V}-\frac{\rho\kappa_V\theta_V}{\sigma_V\kappa}(1-e^{-\kappa\tau})\right]\right)\\
&\quad E_t^Q\left[\exp\left(i\phi\left[\frac{\rho}{\sigma_V}V_T-\frac{\rho(\kappa-\kappa_V)}{\sigma_V}\int_t^T e^{-\kappa(T-s)}V_s\,\mathrm{d}s\right]\right.\right.\\
&\qquad\left.\left.-\frac{\phi^2}{2}(1-\rho^2)\int_t^T e^{-2\kappa(T-s)}V_s\,\mathrm{d}s\right)\right].
\end{aligned}$$

Putting all the above results together, the conditional characteristic function $\Phi(\tau;\phi)$ can be expressed as product of two expectations:

$$\Phi(\tau;\phi)=e^{u_0(\tau;\phi)}E_V(\tau;\phi)E_J(\tau;\phi), \tag{3.67}$$

where

$$\begin{aligned}
u_0(\tau;\phi)&=i\phi\left[e^{-\kappa\tau}X_t+\left(\theta-\frac{\rho\kappa_V\theta_V}{\sigma_V-\kappa}\right)(1-e^{-\kappa\tau})-\frac{\rho e^{-\kappa\tau}V_t}{\sigma_V}\right],\\
E_V(\tau;\phi)&=E_t^Q\left[\exp\left(u_1V_T-\int_t^T u_2(\tau;\phi)V_s\,\mathrm{d}s\right)\right],\\
E_J(\tau;\phi)&=E_t^Q\left[\exp\left(i\phi\int_t^T e^{-\kappa(T-s)}J\,\mathrm{d}N_s\right)\right],
\end{aligned}$$

with

$$u_1=i\phi\frac{\rho}{\sigma_V},\ \text{and}\ u_2(\tau;\phi)=i\phi\frac{\rho(\kappa-\kappa_V)}{\sigma_V}e^{-\kappa\tau}+\frac{\phi^2}{2}(1-\rho^2)e^{-2\kappa\tau}.$$

Based on the dynamics of V_t as specified by (3.62b) under a risk neutral measure Q, by virtue of the Feynman-Kac Theorem, $E_V(\tau;\phi)$ is seen to be governed by the following partial differential equation:

$$\frac{\partial E_V}{\partial\tau}=\frac{\sigma_V^2}{2}V\frac{\partial^2E_V}{\partial V^2}+\kappa_V(\theta_V-V)\frac{\partial E_V}{\partial V}-u_2(\tau)VE_V \tag{3.68}$$

with initial condition:

$$E_V(0;\phi)=e^{i\phi\frac{\rho}{\sigma_V}V}.$$

Thanks to the affine structure of V_t, $E_V(\tau;\phi)$ admits solution of the exponential affine form:

$$E_V(\tau;\phi)=\exp(C(\tau;\phi)+D(\tau;\phi)V), \tag{3.69}$$

where the two parameter functions $C(\tau;\phi)$ and $D(\tau;\phi)$ satisfy the following system of Riccati ordinary differential equations:

$$\frac{\mathrm{d}C}{\mathrm{d}\tau}=\kappa_V\theta_VD, \tag{3.70a}$$

$$\frac{\mathrm{d}D}{\mathrm{d}\tau}=\frac{\sigma_V^2}{2}D^2-\kappa_VD-i\phi\frac{\rho(\kappa-\kappa_V)}{\sigma_V}e^{-\kappa\tau}-\frac{\phi^2}{2}(1-\rho^2)e^{-2\kappa\tau}. \tag{3.70b}$$

Wang *et al.* (2016) manage to solve for $C(\tau;\phi)$ and $D(\tau;\phi)$ in terms of the Kummer functions of the first and second kind. Also, they obtain the following closed-form solution for $E_J(\tau;\phi)$ when the jump size J follows the double exponential distribution function $\text{Exp}(\eta_1,\eta_2,p)$, where

$$E_J(\tau;\phi)=\frac{\lambda p}{\kappa}\ln\frac{\eta_1-i\phi e^{-\kappa\tau}}{\eta_1-i\phi}+\frac{\lambda(1-p)}{\kappa}\ln\frac{\eta_2+i\phi e^{-\kappa\tau}}{\eta_2+i\phi}. \tag{3.71}$$

Combining all the above results together, we finally obtain

$$\Phi(\tau;\phi)=\exp(A(\tau;\phi)+B(\tau;\phi)V_t+i\phi e^{-\kappa\tau}\ln\text{VIX}_t), \tag{3.72}$$

where

$$A(\tau;\phi)=C(\tau;\phi)+i\phi\left(\theta-\frac{\rho\kappa_V\theta_V}{\sigma_V\kappa}\right)(1-e^{-\kappa\tau})+E_J(\tau;\phi),$$
$$B(\tau;\phi)=D(\tau;\phi)-i\phi\frac{\rho e^{-\kappa\tau}}{\sigma_V}.$$

Though closed-form expressions of $C(\tau;\phi)$ and $D(\tau;\phi)$ can be found, Wang *et al.* (2016) comment that the numerical evaluation of the Kummer functions can be very time-consuming. They comment that the numerical solution of the Riccati system of ordinary differential equations (3.70a,b) using the standard Runge-Kutta algorithm can be more computationally efficient.

Extended model

The jump term in $\ln\text{VIX}_t$ can be generalized to the Lévy type jump instead of the double exponential jump or Gaussian jump. Cao, J. *et al.* (2020) consider the more general affine type jump-diffusion model with stochastic mean reversion and Lévy type jump L_t^J. The dynamic equations for $\ln\text{VIX}_t$, its instantaneous variance V_t and mean reversion θ_t under a risk neutral measure Q are given by

$$\begin{aligned}
\mathrm{d}\ln\text{VIX}_t &= \kappa(\theta_t-\ln\text{VIX}_t)+\rho\sqrt{V_t}\,\mathrm{d}W_t^1+\sqrt{1-\rho^2}\sqrt{V_t}\,\mathrm{d}W_t^2+\mathrm{d}L_t^J,\\
\mathrm{d}\theta_t &= \kappa_\theta(\overline{\theta}-\theta_t)\,\mathrm{d}t+\sigma_\theta\,\mathrm{d}W_t^3,\\
\mathrm{d}V_t &= \kappa_V(\theta_V-V_t)\,\mathrm{d}t+\sigma_V\,\mathrm{d}W_t^1,
\end{aligned} \tag{3.73}$$

where ρ is the correlation coefficient, κ, κ_θ, and κ_V are the respective constant mean reversion speeds, σ_θ and σ_V are the respective constant volatility parameters. Given the affine structure of the model, one can derive the corresponding Riccati system of ordinary differential equations to solve for the parameter functions in the exponential affine form of the moment generating function of the joint dynamics of $\ln\text{VIX}_t$, θ_t and V_t. For general Lévy type jumps, there will be no closed-form solution for the parameter functions. However, the solution of the Riccati system of ordinary differential equation can be found numerically.

3.3.2 3/2 plus models

Goard and Mazur (2013) performed empirical studies on the pricing performance of various models in capturing the behavior of the VIX. The tested models are nested within the class of models that take the general form:

$$\mathrm{dVIX}_t = (c_1 + \frac{c_2}{\mathrm{VIX}_t} + c_3\mathrm{VIX}_t \ln \mathrm{VIX}_t + c_4\mathrm{VIX}_t + c_5\mathrm{VIX}_t^2)\mathrm{d}t + k\mathrm{VIX}_t^{\gamma}\,\mathrm{d}W_t. \quad (3.74)$$

They find that the model with γ equals 1.5 and nonzero c_4 and c_5 provides the best performance compared to other models with γ chosen to be $1/2$ or 1. Also, an analytic price formula can be derived for the European call option on the VIX when VIX_t under the statistical measure follows

$$\mathrm{dVIX}_t = (c_4\mathrm{VIX}_t + c_5\mathrm{VIX}_t^2)\mathrm{d}t + \varepsilon\mathrm{VIX}_t^{3/2}\mathrm{d}W_t. \quad (3.75)$$

Let $\lambda(\mathrm{VIX}_t, t)$ denote the market price of risk associated with the VIX, and assume that it takes the form

$$\lambda(\mathrm{VIX}_t, t) = a\mathrm{VIX}_t^{-1/2} + b\mathrm{VIX}_t^{1/2}. \quad (3.76)$$

Under the risk neutral measure Q, VIX_t follows

$$\begin{aligned}\mathrm{dVIX}_t &= [c_4\mathrm{VIX}_t + c_5\mathrm{VIX}_t^2 - \lambda(\mathrm{VIX}_t, t)\varepsilon\mathrm{VIX}_t^{3/2}]\mathrm{d}t + \varepsilon\mathrm{VIX}^{3/2}\,\mathrm{d}\widetilde{W}_t \\ &= (\alpha\mathrm{VIX}_t + \beta\mathrm{VIX}_t^2)\mathrm{d}t + \varepsilon\mathrm{VIX}_t^{3/2}\,\mathrm{d}\widetilde{W}_t, \quad (3.77)\end{aligned}$$

where $\widetilde{W}_t$ is the Brownian motion under Q, $\alpha = c_4 - a\varepsilon$ and $\beta = c_5 - b\varepsilon$, $c_5 \le 0$ and $\beta \le 0$. Thanks to the observation that the reciprocal $1/\mathrm{VIX}_t$ follows the CIR process, Goard and Mazur (2013) manage to find the integral price formula of European call option on the VIX with strike price K as follows:

$$\begin{aligned}c(\mathrm{VIX}_t, \tau; K) = &\frac{2\alpha e^{-r\tau}}{\varepsilon^2 q}\exp\left(-\frac{2\alpha e^{-\alpha\tau}}{\varepsilon^2\mathrm{VIX}_t q}\right)\mathrm{VIX}_t^{-\frac{\beta}{\varepsilon^2}+\frac{1}{2}}\exp\left(\alpha\tau\left(-\frac{\beta}{\varepsilon^2}+\frac{1}{2}\right)\right) \\ &\int_0^{1/K}\phi^{\frac{1}{2}-\frac{\beta}{\varepsilon^2}}\left(\frac{1}{\phi}-K\right)e^{-\frac{2\alpha\phi}{\varepsilon^2 q}}I_\nu\left(\frac{4\alpha\sqrt{\phi}e^{-\frac{\alpha\tau}{2}}}{\varepsilon^2\sqrt{\mathrm{VIX}_t q}}\right)\mathrm{d}\phi, \quad (3.78)\end{aligned}$$

where $\tau = T - t$, $\nu = 1 - \frac{2\beta}{\varepsilon^2}$, $q = 1 - e^{-\alpha\tau}$ and $I_\nu(\cdot)$ is the modified Bessel function of order ν. The proof of (3.78) is presented in the Appendix.

Tan *et al.* (2018) consider the extension of the 3/2-model with jumps, where the VIX_t process under Q is governed by

$$\begin{aligned}\mathrm{dVIX}_t = &\left[\kappa\mathrm{VIX}_t(\theta - \mathrm{VIX}_t) - \lambda(\mathrm{VIX}_t, t)\varepsilon\mathrm{VIX}_t^{3/2}\right]\mathrm{d}t \\ &+ \varepsilon\mathrm{VIX}_t^{3/2}\,\mathrm{d}\widetilde{W}_t + (e^J - 1)\mathrm{VIX}_t\,\mathrm{d}N_t. \quad (3.79)\end{aligned}$$

Compared with (3.77), there is an additional jump term characterized by the Poisson process N_t and jump size J. Suppose the jump sizes $\{J_k\}$, $k = 1, 2, \ldots$ are assumed to follow iid exponential distribution $\mathrm{Exp}(\eta_1, \eta_2, p)$, one can derive the conditional characteristic function in closed form based on the knowledge of the characteristic function of $1/\mathrm{VIX}_t$ and $E_J(\tau;\phi)$ [see (3.71)]. Once the conditional characteristic function of VIX_t is available, we can find the price of the European call option on the VIX in a similar form as shown in (3.65).

Detemple and Kitapbayev (2018) consider pricing of European and American call options on the VIX under the generalized 3/2-model, where the process of VIX_t is governed by some power function of the CIR process. Under the physical measure, VIX_t is governed by

$$\mathrm{VIX}_t = f(Y_t), \tag{3.80a}$$

where

$$\mathrm{d}Y_t = (\beta - \alpha Y_t)\mathrm{d}t + \varepsilon\sqrt{Y_t}\,\mathrm{d}W_t. \tag{3.80b}$$

Here, α, β, and ε are parameters that assume positive constant values, and

$$f(y) = 1/y^{\nu}, \tag{3.81}$$

where $\nu > 0$ and $\beta > \frac{\varepsilon^2}{2}(\nu+1)$. The 3/2-model is obtained when $\nu = 1$. They show that the generalized 3/2-model produces a positive skew of implied volatility that is consistent with the VIX market dynamics.

Appendix

Proof of (3.54) and (3.55a,b)

The key step in the proof is to establish the recursive relations stated in (3.55a,b). When $n = 0$, by virtue of (3.54), we have

$$E_t^Q[e^{\phi h_{t+1}}] = e^{C(\phi,0)+H(\phi,0)h_{t+1}}$$

so we obtain the pair of initial conditions:

$$C(\phi,0) = 0 \quad \text{and} \quad H(\phi,0) = \phi.$$

We consider the following judicious choice of the form of the recursive relations:

$$\begin{aligned} C(\phi,n+1) &= C(\phi,n) + F_C(H(\phi,n)) \\ H(\phi,n+1) &= F_H(H(\phi,n)), \end{aligned}$$

where the functions F_C and F_H are to be determined based on the dynamic equation of h_{t+1} in (3.47b). We deduce the following pair of relations:

(i) $E_t^Q[e^{\phi h_{t+n+2}}] = e^{C(\phi,n+1)+H(\phi,n+1)h_{t+1}}$,

(ii) $E_t^Q[e^{\phi h_{t+n+2}}] = E_t^Q\left[E_{t+1}^Q\left[e^{\phi h_{t+n+2}}\right]\right] = E_t^Q\left[e^{C(\phi,n)+H(\phi,n)h_{t+2}}\right]$.

For notational simplicity, we write $u = H(\phi, n)$. By virtue of (3.47b), we observe

$$E_t^Q\left[e^{uh_{t+2}}\right] = e^{u\omega + u\beta h_{t+1} + u\alpha\delta^{*2}h_{t+1}} E_t^Q\left[e^{u\alpha\varepsilon_{t+1}^{*2} - 2u\alpha\delta^*\sqrt{h_{t+1}}\varepsilon_{t+1}^*}\right].$$

After performing the expectation calculation, we obtain

$$E_t^Q[e^{uh_{t+2}}] = e^{u\omega - \frac{\ln(1-2u\alpha)}{2}} e^{\left(u\beta + \frac{u\alpha\delta^{*2}}{1-2u\alpha}\right)h_{t+1}}.$$

Lastly, we compare the above pair of relations in (i) and (ii) to obtain

$$F_C(u) = u\omega - \frac{\ln(1-2u\alpha)}{2},$$
$$F_H(u) = u\beta + \frac{u\alpha\delta^{*2}}{1-2u\alpha}.$$

Proof of (3.78)

Suppose VIX_t is governed by (3.77), where

$$\text{dVIX}_t = (\alpha\text{VIX}_t + \beta\text{VIX}_t^2)\text{d}t + \varepsilon\text{VIX}_t^{3/2}\,\text{d}\widetilde{W}_t.$$

By Itô's lemma, the reciprocal $Y_t = 1/\text{VIX}_t$ follows the process

$$\text{d}Y_t = (\varepsilon^2 - \beta - \alpha Y_t)\text{d}t - \varepsilon\sqrt{Y_t}\,\text{d}\widetilde{Z}_t.$$

Provided that the Feller condition: $\beta < \frac{\varepsilon^2}{2}$ is satisfied, Y_t and VIX_t would remain positive. The transition density function of Y_t is known to be

$$f(Y_T|Y_t) = Ce^{-u-z}\left(\frac{z}{u}\right)^{q/2} I_\nu(\delta\sqrt{uz}),$$

where

$$C = \frac{2\alpha}{\varepsilon^2(1-e^{-\alpha\tau})},\; u = CY_te^{-\alpha\tau},\; z = CY_t,\; \nu = 1 - \frac{2\beta}{\varepsilon^2},\; \tau = T - t.$$

The time-t price of the call option on VIX with strike price K is given by

$$c(\text{VIX}_t, t; K) = e^{-r\tau}E_t^Q\left[\max\left(\frac{1}{Y_T} - K, 0\right)\right].$$

By using the transition density function $f(Y_T|Y_t)$, we obtain the price formula (3.78).

Chapter 4

Swap Products on Discrete Variance and Volatility

In this chapter, we focus on pricing of swap products on discrete realized variance and volatility under stochastic volatility models (affine type models and 3/2-model) and Lévy models in finding their fair strikes. The usual market convention uses the annualized sum of daily squared log returns (typically daily closing prices) over the period. Specifically, the tenor of the discrete realized variance of an asset price process over $[0,T]$ is characterized by the set of monitoring dates $0 = t_0 < t_1 < \cdots < t_N = T$, where T is the maturity date and N is the total number of monitoring dates. The discrete realized variance of S_t over $[t_0, t_N]$ is defined by

$$I(0,T;N) = \frac{A}{N} \sum_{k=1}^{N} \left(\ln \frac{S_{t_k}}{S_{t_{k-1}}} \right)^2, \tag{4.1}$$

where A is the annualization factor. The definition of $I(0,T;N)$ exhibits high level of path dependency of the asset price process and poses challenges to pricing of derivative products on discrete variance and volatility.

As discussed in Chapter 1, the more generalized form of discrete realized variance is defined by

$$I_w(0,T;N) = \frac{A}{N} \sum_{k=1}^{N} w_k r_k^2, \tag{4.2}$$

where r_k can be the logarithm of return $\ln \frac{S_{t_k}}{S_{t_{k-1}}}$ or rate of return $\frac{S_{t_k}}{S_{t_{k-1}}} - 1$, and w_k is the weight that is dependent on the asset prices. Some examples that have been considered earlier include

(i) gamma swap: $w_k = \frac{S_{t_k}}{S_{t_{k-1}}}$,

(ii) self-quantoed swap: $w_k = \frac{S_{t_N}}{S_{t_0}}$,

(iii) corridor variance swap: $w_k = \mathbf{1}_{\{L \le S_{k-1} \le U\}}$, where $[L,U]$ is the corridor.

The discrete realized volatility of S_t over $[t_0, t_N]$ can be defined naturally by taking the square root of some chosen discrete realized volatility. This square root formulation poses technical difficulties in expectation calculations. There exists no closed

DOI: 10.1201/9781003263524-4

form formula for the fair strike of swap on discrete realized volatility. We may employ analytic approximation method in finding the fair strike, like the use of the saddlepoint approximation method (Zheng and Kwok, 2014c). On the other hand, Barndorff-Nielsen and Shephard (2003) propose another measurement of discrete realized volatility as defined by

$$V(0,T;N) = \sqrt{\frac{\pi}{2NT}} \sum_{k=1}^{N} \left| \frac{S_{t_i} - S_{t_{i-1}}}{S_{t_{i-1}}} \right|, \tag{4.3}$$

where $T = t_N - t_0$ is the length of the sampling period. Nice analytic tractability of pricing volatility swaps is achieved under this choice of discrete realized volatility.

The challenges on pricing swap products on discrete variance and volatility lie on nonlinearity of the discrete realized variance and volatility in terms of the asset prices monitored at discrete time points. A wide variety of versatile analytic methods have been developed in the literature. The success of analytic tractability relies on availability of closed-form formulas for the moment generating functions and transition density functions of the asset price process under stochastic volatility models, Lévy models and GARCH type models. In this chapter, we focus on pricing swaps on discrete realized variance and volatility under stochastic volatility models and Lévy models. There have been new successes on pricing swap products under the GARCH-type models. Readers may be interested to consult Badescu *et al.* (2020) for pricing variance and volatility swaps under the general affine GARCH model and Badescu *et al.* (2019) for the non-affine GARCH models. Their methodologies rely on solving differential recursions for the coefficients of the joint cumulant generating function of log price and conditional variance processes.

This chapter is organized as follows: The earlier three sections are relegated to different analytic methods for pricing various variance swap products and volatility swap under stochastic volatility models (affine models and 3/2-model). These include the direct expectation approach in Sec. 4.1, nested expectation via partial integro-differential equation in Sec. 4.2 and moment generating function method in Sec. 4.3. Pricing methods for various variance swap products under the Lévy models are presented in Sec. 4.4. We also consider the continuous limit of deducing the fair strike formulas of variance swap products under continuous sampling of realized variance from those under discrete realized variance.

4.1 Direct expectation of square of log return

In this section, we price vanilla swap on discrete realized variance under the affine stochastic volatility models with jumps by taking direct expectation of the individual terms, each involving the square of log return. For convenience, we assume the notional of the variance swap to be unity. By virtue of linearity of the terminal payoff, the fair strike K_v of the vanilla swap on discrete realized variance is given by termwise expectation, where

$$K_v = E_0^Q[I(0,T;N)] = \frac{A}{N}\sum_{k=1}^{N} E_0^Q\left[\left(\ln\frac{S_{t_k}}{S_{t_{k-1}}}\right)^2\right] \tag{4.4}$$

under a risk neutral measure Q. It suffices to consider valuation of individual expectation of the square of log return of S_t over $[t_{k-1}, t_k]$, $k = 1, 2, \ldots, N$. Based on certain assumption of the dynamics of the instantaneous variance V_t, it is possible to expand $\left(\ln\frac{S_{t_k}}{S_{t_{k-1}}}\right)^2$ in terms of quadratic products of different variance quantities. Broadie and Jain (2008) perform the direct expectation calculations of these quadratic variance terms and manage to derive closed-form formulas for the fair strike of discrete vanilla variance swap under the Merton jump-diffusion model and Heston stochastic volatility model. Later, Bernard and Cui (2014) show that the tedious procedure can be simplified by exploring an enhanced representation of $\left(\ln\frac{S_{t_k}}{S_{t_{k-1}}}\right)^2$ in terms of dynamics of V_t. Their analytic procedure can be applied to wider class of stochastic volatility models, like the Hull-White model and Schöbel-Zhu model. Here, we consider the extension of their approach to stochastic volatility model with jumps on S_t.

Under a risk neutral measure Q, suppose the dynamics of asset price S_t and instantaneous variance V_t are governed by

$$\begin{aligned}
\frac{dS_t}{S_t} &= (r - q - \lambda m)\,dt + \sqrt{V_t}\left(\rho\, dW_t^1 + \sqrt{1-\rho^2}\, dW_t^2\right) + (e^{J^S} - 1)\,dN_t, \\
dV_t &= \alpha(V_t)\,dt + \beta(V_t)\,dW_t^1,
\end{aligned} \tag{4.5}$$

where r is the riskless interest rate, q is the dividend yield, J^S is the random jump component on S_t, and $m = E[e^{J^S} - 1]$. The Brownian motions W_t^1 and W_t^2 are taken to be uncorrelated. The Poisson jump process N_t with constant intensity λ is assumed to be independent of the two Brownian motions. The drift $\alpha(V_t)$ and diffusion coefficient $\beta(V_t)$ of the instantaneous variance process V_t are adapted to the natural filtration generated by W_t^1. The two stochastic processes S_t and V_t are correlated with constant correlation coefficient ρ.

We would like to express $\ln\frac{S_{t_k}}{S_{t_{k-1}}}$ in terms of the coefficient functions $\alpha(V_t)$ and $\beta(V_t)$. Accordingly, we define the two auxiliary functions

$$f(x) = \int_0^x \frac{\sqrt{z}}{\beta(z)}\,dz \quad \text{and} \quad g(x) = \alpha(x)f'(x) + \frac{\beta^2(x)}{2}f''(x). \tag{4.6}$$

Applying Itô's lemma to $f(V_t)$, we have

$$\begin{aligned}
df(V_t) &= f'(V_t)[\alpha(V_t)\,dt + \beta(V_t)\,dW_t^1] + \frac{\beta^2(V_t)}{2}f''(V_t)\,dt \\
&= g(V_t)\,dt + f'(V_t)\beta(V_t)\,dW_t^1 \\
&= g(V_t)\,dt + \sqrt{V_t}\,dW_t^1.
\end{aligned}$$

Integrating the above equation from t_{k-1} to t_k and rearranging the terms, we have

$$\int_{t_{k-1}}^{t_k} \sqrt{V_t}\, dW_t^1 = f(V_{t_k}) - f(V_{t_{k-1}}) - \int_{t_{k-1}}^{t_k} g(V_t)\, dt. \tag{4.7}$$

By Itô's lemma, the dynamic equation of $\ln S_t$ is given by

$$d\ln S_t = \left(r - q - \lambda m - \frac{V_t}{2}\right) dt + \sqrt{V_t}(\rho\, dW_t^1 + \sqrt{1-\rho^2} dW_t^2) + J^S\, dN_t.$$

Combining the above equation with (4.7), we obtain the following integral representation of log return:

$$\begin{aligned}\ln \frac{S_{t_k}}{S_{t_{k-1}}} &= \int_{t_{k-1}}^{t_k} \left(r - q - m\lambda - \frac{V_t}{2}\right) dt + \sqrt{1-\rho^2} \int_{t_{k-1}}^{t_k} \sqrt{V_t}\, dW_t^2 + \sum_{n=1}^{\Delta N_{t_k}} J_n^s \\ &\quad + \rho \left[f(V_{t_k}) - f(V_{t_{k-1}}) - \int_{t_{k-1}}^{t_k} g(V_t)\, dt \right],\end{aligned} \tag{4.8}$$

where $\Delta N_{t_k} = N_{t_k} - N_{t_{k-1}}$ is the number of Poisson jumps within $[t_{k-1}, t_k]$.

It suffices to consider the expectation of a typical term $\left(\ln \frac{S_{t_k}}{S_{t_{k-1}}}\right)^2$ under Q. By observing the martingale property of $\int_{t_{k-1}}^{t_k} \sqrt{V_t}\, dW_t^2$ and independence between W_t^2 and N_t, we obtain

$$\begin{aligned}E_0^Q\left[\left(\ln \frac{S_{t_k}}{S_{t_{k-1}}}\right)^2\right] &= \mu^2 \Delta t_k^2 + \frac{1}{4} E_0^Q\left[\left(\int_{t_{k-1}}^{t_k} V_t\, dt\right)^2\right] + (1-\rho^2) E_0^Q\left[\int_{t_{k-1}}^{t_k} V_t\, dt\right] \\ &\quad + \rho^2 E_0^Q[f(V_{t_k}) - f(V_{t_{k-1}})]^2 + \rho^2 E_0^Q\left[\left(\int_{t_{k-1}}^{t_k} g(V_t)\, dt\right)^2\right] \\ &\quad + E_0^Q\left[\left(\sum_{n=1}^{\Delta N_{t_k}} J_n^s\right)^2\right] + \mu \Delta t_k E_0^Q\left[2\sum_{n=1}^{\Delta N_{t_k}} J_n^s - \int_{t_{k-1}}^{t_k} V_t\, dt\right] \\ &\quad + \rho E_0^Q\left[\int_{t_{k-1}}^{t_k} g(V_t)\, dt \int_{t_{k-1}}^{t_k} V_t\, dt\right] - E_0^Q\left[\int_{t_{k-1}}^{t_k} V_t\, dt \sum_{n=1}^{\Delta N_{t_k}} J_n^s\right] \\ &\quad - \rho E_0^Q\left[[f(V_{t_k}) - f(V_{t_{k-1}})] \int_{t_{k-1}}^{t_k} [2\rho g(V_t) + V_t]\, dt\right],\end{aligned} \tag{4.9}$$

where $\mu = r - q - \lambda m$ and $\Delta t_k = t_k - t_{k-1}$.

Affine stochastic volatility model with jumps

We consider pricing of the discrete vanilla variance swap under the affine model with jumps in asset return, where the coefficient functions in the instantaneous variance process are given by

$$\alpha(V_t) = \kappa(\theta - V_t) \text{ and } \beta(V_t) = \varepsilon\sqrt{V_t}.$$

The auxiliary functions f and g admit the linear functional forms in V_t, where

$$f(V_t) = V_t/\varepsilon \text{ and } g(V_t) = \frac{\kappa}{\varepsilon}(\theta - V_t).$$

When the jump size distribution J^S follows the Gaussian distribution with mean ν and variance δ^2 so that

$$m = E[J^S - 1] = e^{\nu+\delta^2/2} - 1,$$

each term in (4.9) can be evaluated in closed form.

We assume uniform time interval so that $\Delta t_k = t_k - t_{k-1} = T/N$, $k = 1, 2, \ldots, N$. The jump terms can be found to be

$$E_0^Q\left[\left(\sum_{n=1}^{\Delta N_{t_k}} J_n^S\right)^2\right] = \left[\lambda\frac{T}{N} + \left(\frac{\lambda T}{N}\right)^2\right](\delta^2 + \nu^2) \quad \text{and} \quad E_0^Q\left[\sum_{n=1}^{\Delta N_{t_k}} J_n^S\right] = \frac{\lambda T \nu}{N}.$$

Substituting these results into (4.9), we obtain

$$\begin{aligned}
&E_0^Q\left[\left(\ln\frac{S_{t_k}}{S_{t_{k-1}}}\right)^2\right] \\
&= \left[(\mu+\nu\lambda)^2 + \frac{\rho^2\kappa^2\theta^2}{\varepsilon^2}\right]\left(\frac{T}{N}\right)^2 + \lambda(\delta^2+\nu^2)\frac{T}{N} + \lambda^2\delta^2\frac{T^2}{N^2} \\
&\quad + \left\{1 - \rho^2 - \left[\mu + \nu\lambda + \frac{2\rho\kappa\theta}{\varepsilon}\left(\frac{\rho\kappa}{\varepsilon} - \frac{1}{2}\right)\right]\frac{T}{N}\right\} E_0^Q\left[\int_{t_{k-1}}^{t_k} V_t \,\mathrm{d}t\right] \\
&\quad - \frac{2\rho^2\kappa\theta}{\varepsilon^2}\frac{T}{N} E_0^Q[V_{t_k} - V_{t_{k-1}}] \\
&\quad + \frac{\rho^2}{\varepsilon^2} E_0^Q[V_{t_k} - V_{t_{k-1}}]^2 + \left(\frac{\rho\kappa}{\varepsilon} - \frac{1}{2}\right)^2 E_0^Q\left[\left(\int_{t_{k-1}}^{t_k} V_t \,\mathrm{d}t\right)^2\right] \\
&\quad + \frac{2\rho}{\varepsilon}\left(\frac{\rho\kappa}{\varepsilon} - \frac{1}{2}\right) E_0^Q\left[(V_{t_k} - V_{t_{k-1}})\int_{t_{k-1}}^{t_k} V_t \,\mathrm{d}t\right].
\end{aligned} \tag{4.10}$$

From the dynamic equation of V_t, we deduce the identity

$$\frac{1}{\kappa} E_0^Q[V_{t_k} - V_{t_{k-1}}] + E_0^Q\left[\int_{t_{k-1}}^{t_k} V_t \,\mathrm{d}t\right] = \frac{\theta T}{N}.$$

Further simplification of (4.10) gives

$$\begin{aligned}
E_0^Q\left[\left(\ln\frac{S_{t_k}}{S_{t_{k-1}}}\right)^2\right] &= \frac{a^2T^2}{N^2} + \lambda(\delta^2+\nu^2)\frac{T}{N} + \left(1 - \rho^2 + \frac{2abT}{N}\right) E_0^Q\left[\int_{t_{k-1}}^{t_k} V_t \,\mathrm{d}t\right] \\
&\quad + \lambda^2\delta^2\frac{T^2}{N^2} - \frac{2a\rho T}{\varepsilon N} E_0^Q[V_{t_k} - V_{t_{k-1}}] + \frac{\rho^2}{\varepsilon^2} E_0^Q[(V_{t_k} - V_{t_{k-1}})^2] \\
&\quad + b^2 E\left[\left(\int_{t_{k-1}}^{t_k} V_t \,\mathrm{d}t\right)^2\right] + \frac{2b\rho}{\varepsilon} E_0^Q\left[(V_{t_k} - V_{t_{k-1}})\int_{t_{k-1}}^{t_k} V_t \,\mathrm{d}t\right],
\end{aligned} \tag{4.11}$$

where

$$a = \mu + \nu\lambda - \frac{\rho\kappa\theta}{\varepsilon} \quad \text{and} \quad b = \frac{\rho\kappa}{\varepsilon} - \frac{1}{2}.$$

To evaluate the remaining expectation terms, we employ several technical results. Recall that (see Sec. 2.5.2)

$$E_0^Q[V_t] = (1 - e^{-\kappa t})\theta + e^{-\kappa t} V_0$$

so that

$$E_0^Q\left[\int_{t_{k-1}}^{t_k} V_t \, \mathrm{d}t\right] = \frac{\theta T}{N} + (V_0 - \theta)\frac{e^{-\kappa t_{k-1}} - e^{-\kappa t_k}}{\kappa}.$$

By using the conditional expectation and tower rule, with $s < t$, we have

$$\begin{aligned}
E_0^Q[V_t V_s] &= E_0^Q[E_s^Q[V_t | V_s] V_s] = E_0^Q[(1 - e^{-\kappa(t-s)})\theta V_s + e^{-\kappa(t-s)} V_s^2] \\
&= [1 - e^{-\kappa(t-s)}]\theta[\theta + e^{-\kappa s}(V_0 - \theta)] \\
&\quad + e^{-\kappa(t+s)}\left[(V_0 - \theta)^2 - \frac{\varepsilon^2}{\kappa}\left(V_0 - \frac{\theta}{2}\right)\right] \\
&\quad + e^{-\kappa(t-s)}\left(\theta + \frac{\varepsilon^2}{2\kappa}\right)[\theta + 2e^{-\kappa s}(V_0 - \theta)].
\end{aligned}$$

We then have

$$\begin{aligned}
E_0^Q\left[\left(\int_{t_{k-1}}^{t_k} V_t \, \mathrm{d}t\right)^2\right] &= \int_{t_{k-1}}^{t_k} \int_{t_{k-1}}^{t_k} E_0^Q[V_t V_s] \, \mathrm{d}s\mathrm{d}t \\
&= \int_{t_{k-1}}^{t_k} \int_{t_{k-1}}^{t} E_0^Q[V_t V_s] \, \mathrm{d}s\mathrm{d}t + \int_{t_{k-1}}^{t_k} \int_{t}^{t_k} E_0^Q[V_t V_s] \, \mathrm{d}s\mathrm{d}t.
\end{aligned}$$

The first term can be computed as follows:

$$\begin{aligned}
\int_{t_{k-1}}^{t_k} \int_{t_{k-1}}^{t} E_0^Q[V_t V_s] \, \mathrm{d}s\mathrm{d}t = &\frac{\theta^2 T^2}{2N^2} + \left[\frac{e^{-\kappa t_{k-1}} - e^{-\kappa t_k}}{\kappa^2} - \frac{e^{-\kappa t_k} T}{\kappa N}\right](V_0 - \theta)\left(\theta + \frac{\varepsilon^2}{\kappa}\right) \\
&+ \lambda^2 \delta^2 \frac{T}{N} + \left[\frac{e^{-\kappa t_{k-1}} T}{\kappa N} - \frac{e^{-\kappa t_{k-1}} - e^{-\kappa t_k}}{\kappa^2}\right]\theta(V_0 - \theta) \\
&+ \left[\frac{e^{-\kappa t_{k-1}}(e^{-\kappa t_{k-1}} - e^{-\kappa t_k})}{\kappa^2} - \frac{e^{-2\kappa t_{k-1}} - e^{-2\kappa t_k}}{2\kappa^2}\right] \\
&\left[(V_0 - \theta)^2 - \frac{\varepsilon^2}{\kappa}\left(V_0 - \frac{\theta}{2}\right)\right] \\
&+ \left(\frac{T}{\kappa N} - \frac{1 - e^{-\kappa T/N}}{\kappa^2}\right)\frac{\varepsilon^2 \theta}{2\kappa}.
\end{aligned}$$

The expectation of the other cross terms in (4.11) can be computed in a similar manner. Finally, by summing (4.11) from $k = 1$ to N, dividing both sides by T and simplifying the expression, one can obtain

$$K_v(0, T; N) = K_v(0, T; \infty) + R_v(0, T; N), \tag{4.12}$$

where

$$K_v(0,T;\infty) = E_0^Q[I(0,T;\infty)] = \theta + (V_0 - \theta)\frac{1-e^{-\kappa T}}{\kappa T} + \lambda(\delta^2 + \nu^2), \tag{4.13a}$$

and

$$\begin{aligned} R_v(0,T;N) = \frac{T}{N}&\left[\left(\frac{\theta}{2} - \mu - \nu\lambda\right)^2 + (V_0 - \theta)\left(\frac{\theta}{2} - \mu - \nu\lambda\right)\frac{1-e^{-\kappa T}}{\kappa T}\right] \\ &+ \theta\left(\frac{\varepsilon^2}{4\kappa^2} - \frac{\rho\varepsilon}{\kappa}\right)\left[1 - \frac{N(1-e^{-\kappa T/N})}{\kappa T}\right] \\ &+ (V_0 - \theta)\frac{1-e^{-\kappa T}}{\kappa T}\left(\frac{\varepsilon^2}{2\kappa^2} - \frac{\rho\varepsilon}{\kappa}\right)\left[1 + \frac{\kappa T}{N(1-e^{\kappa T/N})}\right] \\ &- \left[\varepsilon^2(\theta - 2V_0) + 2\kappa(V_0 - \theta)^2\right]\frac{1-e^{-2\kappa T}}{8\kappa^3 T}\frac{1-e^{-\kappa T/N}}{1+e^{-\kappa T/N}}. \end{aligned} \tag{4.13b}$$

The fair strike formula (4.12) derived from the direct expectation approach possesses the nice representation as the sum of two terms: strike price formula under continuous monitoring plus a correction term arising from discrete monitoring. It is quite straightforward to show that the correction term becomes zero when the number of monitoring instants becomes infinite. In fact, $R_v(0,T;N)$ admits the following asymptotic expansion in powers of $\frac{1}{N}$. We have

$$R_v(0,T;N) = \frac{C}{N} + O\left(\frac{1}{N^2}\right), \tag{4.14}$$

where

$$\begin{aligned} C \quad = \quad & T\left(\frac{\theta}{2} - \mu - \nu\lambda\right)^2 + (V_0 - \theta)\left(\frac{\theta}{2} - \mu - \nu\lambda\right)\frac{1-e^{-\kappa T}}{\kappa} \\ &+ \frac{\theta\varepsilon T}{2}\left(\frac{\varepsilon}{4\kappa} - \rho\right) + (V_0 - \theta)(1 - e^{-\kappa T})\left(\frac{\varepsilon^2}{4\kappa^2} - \frac{\rho\varepsilon}{2\kappa}\right) \\ &+ \left[\varepsilon^2(\theta - 2V_0) + 2\kappa(V_0 - \theta)^2\right]\frac{1-e^{-2\kappa T}}{16\kappa^2}. \end{aligned}$$

As a verification, $K_v(0,T;\infty)$ in (4.12) agrees with the fair strike formula (2.120) derived in Sec. 2.5 under continuous sampling of realized variance by setting $\eta = \rho_J = 0$ accordingly.

Furthermore, we may consider the asymptotic limit of the fair strike formula (4.12) under the small time limit in powers of maturity T. First, we consider the expansion of $K_v(0,T;N)$ in powers of T as follows:

$$K_v(0,T;N) = V_0 + \lambda(\delta^2 + \nu^2) + B_1 T + B_2 T^2 + O(T^3), \tag{4.15}$$

where

$$\begin{aligned} B_1 &= -\frac{\kappa(V_0-\theta)}{2}+\frac{1}{4N}[(V_0-2\mu-2\nu\lambda)^2-2\varepsilon\rho V_0]+\frac{\lambda^2\delta^2}{N}, \\ B_2 &= \frac{\kappa^2(V_0-\theta)}{6}+\frac{\kappa(V_0-\theta)(\varepsilon\rho+2\mu+2\nu\lambda-V_0)+\varepsilon^2V_0/2}{4N} \\ &\quad +\frac{\varepsilon\kappa\rho(V_0+\theta)-\varepsilon^2V_0/2}{12N^2}. \end{aligned}$$

The expansion of $K_\nu(0,T;\infty)$ in (4.13a) in powers of T is given by

$$K_\nu(0,T;\infty)=V_0+\lambda(\delta^2+\nu^2)-\frac{\kappa}{2}(V_0-\theta)T+\frac{\kappa^2}{6}(V_0-\theta)T^2+O(T^3). \quad (4.16a)$$

As a result, one obtains the following expansion in powers of T for the correction term:

$$\begin{aligned} & K_\nu(0,T;N)-K_\nu(0,T;\infty) \\ = & \frac{T}{4N}[(V_0-2\mu-2\nu\lambda)^2-2\varepsilon\rho V_0+4\lambda^2\delta^2]+O(T^2). \end{aligned} \quad (4.16b)$$

The correction term tends to zero as $T\to 0$, so this implies that the small time limits of the discrete and continuous fair strikes are equal (Keller-Ressel and Muhle-Karle, 2013).

The success of analytic tractability in the above direct expectation calculations relies on the simple functional forms of $f(V_t)$ and $g(V_t)$, which are linear functions in V_t under the affine model. Under the Hull-White model, $f(V_t)$ and $g(V_t)$ involve square root of V_t only; while under the Schöbel-Zhu model, $f(V_t)$ and $g(V_t)$ are quadratic functions in V_t. By following similar procedures, we can derive similar fair strike formulas of discrete vanilla variance swap under the Hull White model and Schöbel-Zhu model (Bernard and Cui, 2014). This is possible since the direct expectation calculations of $E_0^Q[V_t^pV_t^q]$, where p and q are either $1/2$ or positive integers, are tractable. This direct expectation approach can be used to price discrete variance swaps when the stochastic volatility model is extended from the affine model to the Hawkes jump-diffusion model (Liu and Zhu, 2019). Rujivan and Zhu (2014) propose an alternative direct expectation approach by considering the expansion of the fair strike formula for variance swap under the Heston model (zero jump) in quadratic power of V_0.

The direct expectation approach is restrictive in two aspects. First, it does not provide a universal framework for pricing all different types of variance swap products. It is not quite straightforward to extend the approach to price variance swaps on generalized discrete realized variance. Though it is still manageable to derive the fair strike formula for the gamma swap under the Heston model (Rujivan, 2016), the derivation procedures are much tedious. However, it is not plausible to derive the fair strike for the corridor variance swap using this direct expectation approach. Second, this approach is not robust to accommodate generalization of the underlying asset price process, say, under the $3/2$-model.

4.2 Nested expectation via partial integro-differential equation

Zhu and Lian (2011) adopt the dimension reduction technique due to Little and Pant (2001) to show that evaluation of a typical term $E_0^Q\left[\left(\ln\frac{S_{t_i}}{S_{t_{i-1}}}\right)^2\right]$ can be achieved by solving two one-dimensional initial value problems. They manage to find the analytic solution, thanks to the closed-form analytic expressions of the marginal moment generating functions of the log asset price and instantaneous variance under the Heston stochastic volatility model. In this section, we illustrate the versatility of this nested expectation via solution of a partial integro-differential equation to find the fair strikes of various swap products on discrete realized variance under the Heston model and $3/2$ model with jumps. The return-type specification of the asset price can be either the log return or simple rate of return.

First, we consider the expectation calculation of square of log return:

$$E_0^Q\left[\left(\ln\frac{S_{t_i}}{S_{t_{i-1}}}\right)^2\right], \quad i=1,2,\ldots,N,$$

which may be visualized as a contingent claim pricing problem over $[0,t_i]$ with terminal payoff at t_i equals $\left(\ln\frac{S_{t_i}}{S_{t_{i-1}}}\right)^2$. Since both $S_{t_{i-1}}$ and S_{t_i} are not known at $t=0$, the pricing problem exhibits mild path dependence. By virtue of the property of the Dirac function, we define (Little and Pant, 2001)

$$I_t^i = \int_0^t \delta(u-t_{i-1})S_u\,\mathrm{d}u = \begin{cases} S_{t_{i-1}} & t_{i-1}<t \\ 0 & 0\le t<t_{i-1},\ i>1. \end{cases} \tag{4.17}$$

The variable I_t^i captures the realization of the asset price at t_{i-1}, where I_t^i starts at zero value, then jumps in value only at t_{i-1} by the amount $S_{t_{i-1}}$. It remains constant over $[0,t_{i-1}]$ and $[t_{i-1},t_i]$.

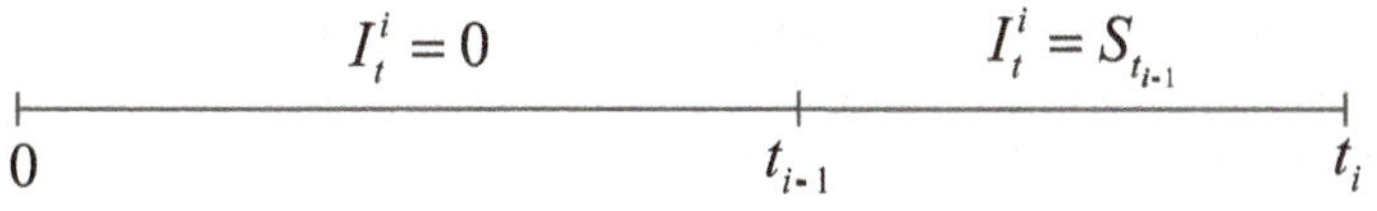

$I_t^i=0$ $\qquad$ $I_t^i=S_{t_{i-1}}$

0 $\qquad$ t_{i-1} $\qquad$ t_i

FIGURE 4.1: I_t^i records the realized asset price $S_{t_{i-1}}$ and exhibits jump in value of amount $S_{t_{i-1}}$ at t_{i-1}.

The pricing problem can be solved backward in time in a two-step procedure. We solve backward from t_i to t_{i-1}, apply an appropriate jump condition at t_{i-1}, then solve backward from t_{i-1} to t_0. First, we illustrate the procedure proposed by Zhu and Lian (2012a) to solve the governing two-dimensional equation in two steps using the

Fourier transform method under the Heston stochastic volatility model. Later, we show how to extend the approach to price generalized variance swaps under the 3/2 model with jumps (Yuen *et al.*, 2015).

4.2.1 Vanilla variance swaps under the Heston stochastic volatility model

Under a risk neutral pricing measure Q, the dynamics of the asset price S_t and instantaneous variance V_t under the Heston stochastic volatility model are governed by

$$\frac{dS_t}{S_t} = r\, dt + \sqrt{V_t}\, dW_t \tag{4.18a}$$

$$dV_t = \kappa(\theta - V_t)\, dt + \varepsilon\sqrt{V_t}\, dW_t^V, \tag{4.18b}$$

where $dW_t\, dW_t^V = \rho\, dt$. The other parameters r, κ, θ and ε have the usual meaning as stated in (2.121). We consider a pricing problem over $(0,t_i)$ of a contingent claim whose payoff at expiry time t_i is $\left(\ln \frac{S_{t_i}}{S_{t_{i-1}}}\right)^2$, $i > 1$. The special case $i = 0$ is trivial since S_0 is known at $t = 0$.

We write the price function of the above contingent claim as $U_i(S,V,I,t)$. The governing equation for $U_i(S,V,I,t)$ is the following partial differential equation:

$$\frac{\partial U_i}{\partial t} + \frac{V}{2}S^2\frac{\partial^2 U_i}{\partial S^2} + \rho\varepsilon VS\frac{\partial^2 U_i}{\partial S \partial V} + \frac{\varepsilon^2}{2}V\frac{\partial^2 U_i}{\partial V^2} + rS\frac{\partial U_i}{\partial S} + \kappa(\theta - V)\frac{\partial U_i}{\partial V} + \delta(t - t_{i-1})\frac{\partial U_i}{\partial I} - rU_i = 0,\ 0 < t < t_i, \tag{4.19a}$$

with terminal condition:

$$U(S,V,I,t_i) = \left(\ln \frac{S}{I}\right)^2. \tag{4.19b}$$

Here, $\delta(\cdot)$ is the Dirac function. At any time that does not fall on the time point t_{i-1}, the term $\delta(t - t_{i-1})\frac{\partial U_i}{\partial I}$ is zero since I remains constant over $(0,t_{i-1})$ and (t_{i-1},t_i). In the two-step procedure discussed below, we reduce the pricing formulation to be two-dimensional involving S and V while I serves as a dummy variable that captures the realization of the asset price $S_{t_{i-1}}$ at t_{i-1}.

(i)
$$\frac{\partial U_i}{\partial t} + \frac{V}{2}S^2\frac{\partial^2 U_i}{\partial S^2} + \rho\varepsilon VS\frac{\partial^2 U_i}{\partial S \partial V} + \frac{\varepsilon^2}{2}V\frac{\partial^2 U_i}{\partial V^2} + rS\frac{\partial U_i}{\partial S} + \kappa(\theta - V)\frac{\partial U_i}{\partial V} - rU_i = 0, \quad t_{i-1} < t < t_i, \tag{4.20a}$$

$$U_i(S,V,I,t_i) = \left(\ln \frac{S}{I}\right)^2. \tag{4.20b}$$

$$\text{(ii)} \qquad \frac{\partial U_i}{\partial t} + \frac{V}{2}S^2\frac{\partial^2 U_i}{\partial S^2} + \rho\varepsilon VS\frac{\partial^2 U_i}{\partial S\partial V} + \frac{\varepsilon^2}{2}V\frac{\partial^2 U_i}{\partial V^2} + rS\frac{\partial U_i}{\partial S}$$
$$+ \kappa(\theta - V)\frac{\partial U_i}{\partial V} - rU_i = 0, \quad 0 < t < t_{i-1}, \tag{4.21a}$$

$$U_i(S,V,0,t_{i-1}^-) = U_i(S,V,S,t_{i-1}^+). \tag{4.21b}$$

The terminal condition $U_i(S,V,I,t_i)$ has dependence on S but not on V, while I behaves like a parameter. We may use the one-dimensional Fourier transform with respect to $x = \ln S$ to solve $U_i(S,V,I,t)$ over (t_{i-1},t_i). Once $U_i(S,V,I,t_{i-1}^+)$ is obtained, we subsequently set $I = S$. Though I has jump of amount $S_{t_{i-1}}$ across t_{i-1}, the state variables S and V and the price function U_i remains to be continuous across t_{i-1}. Under the Heston model, the evaluation procedure over $(0,t_{i-1})$ is simplified to an one-dimensional pricing problem since the terminal condition $U_i(S,V,0,t_{i-1}^-)$ has dependence on V only.

Let $\mathcal{F}\{f(x)\}$ denote the Fourier transform of a function $f(x)$ with respect to $x = \ln S$ and ω is the Fourier transform variable. We solve the pricing problem backward in time, starting with the interval (t_{i-1},t_i). Accordingly, we define $\tau = t_i - t$. By taking the Fourier transform of the governing equation (4.20a) and terminal payoff (4.20b), the solution to $U_i(S,V,I,t)$, $t_{i-1} < t < t_i$, can be expressed as

$$U_i(S,V,I,t) = \mathcal{F}^{-1}\left\{e^{C(\omega,\tau)+D(\omega,\tau)V}\mathcal{F}\left\{\left(\ln\frac{S}{I}\right)^2\right\}\right\}, \tag{4.22}$$

where

$$C(\omega,\tau) = r(\omega i - 1)\tau + \frac{\kappa\theta}{\varepsilon^2}[(a+b)\tau - 2\ln\frac{1-ge^{b\tau}}{1-g}],$$
$$D(\omega,\tau) = \frac{a+b}{\varepsilon^2}\frac{1-e^{b\tau}}{1-ge^{b\tau}},$$
$$a = \kappa - \rho\varepsilon\omega i, \quad b = \sqrt{a^2 + \varepsilon^2(\omega^2 + \omega i)}, \quad g = \frac{a+b}{a-b}.$$

Here, the symbol i denotes $\sqrt{-1}$. The proof of (4.22) is presented in the Appendix.

Recall the Fourier transform formula:

$$\mathcal{F}\{x^n\} = 2\pi i^n\delta^{(n)}(\omega), \quad n = 0,1,2,\ldots, \tag{4.23a}$$

where $\delta^{(n)}(\omega)$ is the nth order derivative of $\delta(\omega)$ satisfying

$$\int_{-\infty}^{\infty}\delta^{(n)}(\omega)\Phi(\omega)\,\mathrm{d}\omega = (-1)^n\Phi^{(n)}(0). \tag{4.23b}$$

The Fourier transform of terminal payoff (4.20b) is given by

$$\mathcal{F}\{(x-\ln I)^2\} = 2\pi[-\delta^{(2)}(\omega) - 2i\delta^{(1)}(\omega)\ln I + \delta(\omega)(\ln I)^2].$$

Using formula (4.22), we obtain

$$\begin{aligned} & U_i(S,V,I,t) \\ = & \ \mathcal{F}^{-1}\{2\pi e^{C(\omega,\tau)+D(\omega,\tau)V}[-\delta^{(2)}(\omega)-2i\delta^{(1)}(\omega)\ln I+\delta(\omega)(\ln I)^2]\} \\ = & \int_{-\infty}^{\infty} e^{C(\omega,\tau)+D(\omega,\tau)V}[-\delta^{(2)}(\omega)-2i\delta^{(1)}(\omega)\ln I+\delta(\omega)(\ln I)^2]e^{\omega xi}\,d\omega \\ = & \ -f''(0)+2if'(0)\ln I+f(0)(\ln I)^2, \end{aligned} \tag{4.24}$$

where

$$f(\omega)=e^{C(\omega,\tau)+D(\omega,\tau)V+x\omega i}.$$

By computing $f''(0)$, $f'(0)$ and $f(0)$ and setting $I=S$ (equivalently, $x=\ln S$), we obtain

$$\begin{aligned} & U_i(S,V,S,t_{i-1}^+) \\ = & -e^{-r\Delta t_i}\{[D^{(1)}(\Delta t_i)]^2V^2+2C^{(1)}(\Delta t_i)D^1(\Delta t_i)V+[C^{(1)}(\Delta t_i)]^2+C^{(2)}(\Delta t_i)\}, \end{aligned} \tag{4.25}$$

where $\Delta t_i = t_i - t_{i-1}$, and

$$\begin{aligned} D^{(1)}(\Delta t_i) &= \left.\frac{\partial D(\omega,\Delta t_i)}{\partial \omega}\right|_{\omega=0}, \quad D^{(2)}(\Delta t_i) = \left.\frac{\partial^2 D(\omega,\Delta t_i)}{\partial \omega^2}\right|_{\omega=0}, \\ C^{(1)}(\Delta t_i) &= \left.\frac{\partial C(\omega,\Delta t_i)}{\partial \omega}\right|_{\omega=0}, \quad C^{(2)}(\Delta t_i) = \left.\frac{\partial^2 C(\omega,\Delta t_i)}{\partial \omega^2}\right|_{\omega=0}. \end{aligned}$$

We write

$$G_i(V)=U_i(S,V,0,t_{i-1}^+)=U_i(S,V,S,t_{i+1}^+). \tag{4.26}$$

Since the terminal payoff in the second step procedure has no dependence on I, we can express the price function $U_i(S,V,0,t)$ as

$$U_i(S,V,0,t)=\int_0^{\infty} e^{-r(t_{i-1}-t)}e^{-r\Delta t_i}G_i(V_{t_{i-1}})p(V_{t_{i-1}}|V_t)\,dV_{t_{i-1}}, \quad 0<t<t_{i-1}, \tag{4.27}$$

where the density function of the CIR process of V_t is given by [see (2.113)]

$$\begin{aligned} & p(V_{t_{i-1}}|V_t)=ce^{-w-x}\left(\frac{x}{w}\right)^{q/2} I_q(2\sqrt{w\kappa}), \\ & c=\frac{2\kappa}{\varepsilon^2[1-e^{-\kappa(t_{i-1}-t)}]}, \ w=cV_te^{-\kappa(t_{i-1}-t)}, \ x=cV_{t_{i-1}}, \ q=\frac{2\kappa\theta}{\varepsilon^2}-1, \end{aligned} \tag{4.28}$$

and $I_q(\cdot)$ is the modified Bessel function of the first kind of order q. Given that $G_i(V)$ is a quadratic function in V, Zhu and Lian (2012a) manage to show that the analytic integration of (4.27) admits the following nice closed-form solution:

$$\begin{aligned}
E_0^Q\left[\left(\ln\frac{S_{t_i}}{S_{t_{i-1}}}\right)^2\right] &= e^{rt_i}U_i(S,V,0,0)\\
&= \int_0^\infty G(V_{t_{i-1}})p(V_{t_{i-1}}|V_0)\,\mathrm{d}V_{t_{i-1}}\\
&= -[D^{(1)}(\Delta t_i)]^2\frac{\tilde{q}+2W_i+(\tilde{q}+W_i)^2}{c_i^2}\\
&\quad -[2C^{(1)}(\Delta t_i)D^1(\Delta t_i)+D^{(2)}(\Delta t_i)]\frac{\tilde{q}+W_i}{c_i}\\
&\quad -[C^{(1)}(\Delta t_{i-1})]^2-C^{(2)}(\Delta t_i), \qquad (4.29)
\end{aligned}$$

where

$$c_i=\frac{2\kappa}{\varepsilon^2(1-e^{-\kappa t_{i-1}})},\ W_i=c_iV_0e^{-\kappa t_{i-1}}\quad\text{and}\quad \tilde{q}=\frac{2\kappa\theta}{\varepsilon^2}.$$

Extension to simple rate of return specification

Rujivan and Zhu (2012) extend the nested expectation approach to the simple rate of return $\frac{S_{t_i}-S_{t_{i-1}}}{S_{t_{i-1}}}$ in the calculation of discrete realized variance. Correspondingly, we require evaluation of a typical term

$$E_0^Q\left[\left(\frac{S_{t_i}-S_{t_{i-1}}}{S_{t_{i-1}}}\right)^2\right]=E_0^Q\left[\frac{1}{S_{t_{i-1}}^2}\left(E_{t_{i-1}}^Q[S_{t_i}^2]-2S_{t_{i-1}}E_{t_{i-1}}^Q[S_{t_i}]\right)+1\right]. \qquad (4.30)$$

It can be shown that $E_{t_{i-1}}^Q[S_t^\gamma]$, $\gamma=1,2$, takes a similar exponential affine form with dependence on $V_{t_{i-1}}$ only. After similar calculations, we obtain

$$E_0^Q\left[\left(\frac{S_{t_i}-S_{t_{i-1}}}{S_{t_{i-1}}}\right)^2\right]=\int_0^\infty e^{r\Delta t_i}M(V_{t_{i-1}})p(V_{t_{i-1}}|V_0)\,\mathrm{d}V_{t_{i-1}}. \qquad (4.31)$$

Here, $p(V_{t_{i-1}}|V_0)$ is given in (4.28) and

$$M(V)=e^{\widetilde{C}(\Delta t_i)+\widetilde{D}(\Delta t_i)V}+e^{-r\Delta t_i}+2, \qquad (4.32)$$

where

$$\begin{aligned}
\widetilde{C}(\tau) &= r\tau+\frac{\kappa\theta}{\varepsilon^2}\left[(\widetilde{a}+\widetilde{b})\tau-2\ln\frac{1-\widetilde{g}e^{\widetilde{b}\tau}}{1-\widetilde{g}}\right],\\
\widetilde{D}(\tau) &= \frac{\widetilde{a}+\widetilde{b}}{\varepsilon^2}\frac{1-e^{\widetilde{b}\tau}}{1-\widetilde{g}e^{\widetilde{b}\tau}},\\
\widetilde{a} &= \kappa-2\rho\varepsilon,\ \widetilde{b}=\sqrt{\widetilde{a}^2-2\varepsilon^2}\ \text{and}\ \widetilde{g}=\frac{\widetilde{a}+\widetilde{b}}{\widetilde{a}-\widetilde{b}}.
\end{aligned}$$

Given that $M(V)$ is an exponential function in V plus a time-dependent function, the analytic integration of (4.31) gives the following closed-form formula:

$$E_0^Q\left[\left(\frac{S_{t_i}-S_{t_{i-1}}}{S_{t_{i-1}}}\right)^2\right] = e^{\widetilde{C}(\Delta t_i)+d_i\widetilde{D}(\Delta t_i)V_0}\left[\frac{c_i}{c_i-\widetilde{D}(\Delta t_i)}\right]^{\frac{2\kappa\theta}{\varepsilon^2}} + e^{-r\Delta t_i} - 2, \tag{4.33}$$

where

$$c_i = \frac{2\kappa}{\varepsilon^2(1-e^{-\kappa t_{i-1}})} \quad \text{and} \quad d_i = \frac{c_i e^{-\kappa t_{i-1}}}{c_i - \widetilde{D}(\Delta t_i)}.$$

The nested expectation approach can be extended to price variance swaps on discrete realized variance under stochastic volatility and interest rate, provided that the stochastic short rate follows the CIR process (Cao *et al.*, 2016; He and Zhu, 2018). With the inclusion of stochastic interest rate in the pricing formulation, one has to employ the T-forward measure in order to simplify the analytic calculations. Cao *et al.* (2016) show that an explicit closed-form solution for the forward characteristic function may not be available. Some of the coefficient functions in the exponential affine form have to be obtained via numerical integration of the Riccati system of ordinary differential equations. Alternatively, He and Zhu (2018) manage to obtain a series form solution for the fair strikes of variance swaps on discrete realized variance under the Heston-CIR hybrid model of stochastic volatility and stochastic interest rate. He and Zhu (2019) further extend the nested expectation approach to the two-factor Heston model, with one factor following the CIR process and the other factor being characterized by a Markov chain.

Besides the Heston stochastic volatility model, one may choose the mean reversion Gaussian volatility model (Zhang, 2014) or the Ornstein-Uhlenbeck process for the stochastic volatility (Cao and Fang, 2017) to price variance swaps on discrete realized variance. The transition density of the instantaneous volatility process becomes the Gaussian type under these two models. As a result, the analytic derivation of the fair strike of the swap on discrete realized variance using the nested expectation approach becomes less tedious when compared with that of the Heston stochastic volatility model.

4.2.2 Variance swaps under the 3/2-model

We assume that the dynamics of the asset price S_t and its instantaneous variance V_t under a risk neutral measure Q is governed by

$$\begin{aligned}\frac{dS_t}{S_t} &= (r-\lambda m)\,dt + \sqrt{V_t}\,dW_t^1 + (e^J-1)\,dN_t,\\ dV_t &= V_t[p(t)-qV_t]\,dt + \varepsilon V_t^{3/2}\,dW_t^2,\end{aligned} \tag{4.34}$$

where r is the risk-free interest rate, W_t^1 and W_t^2 are two correlated standard Brownian motions with $dW_t^1\,dW_t^2 = \rho\,dt$. Also, N_t is a Poisson process with constant intensity λ, assumed to be independent of W_t^1 and W_t^2. The random jump size of the log asset price is denoted by J, which has a normal distribution with mean ν and variance δ^2.

Also, J is assumed to be independent of the two Brownian motions and the Poisson process N_t. The compensator parameter m is given by

$$m = E^Q[e^J - 1] = e^{\nu+\delta^2/2} - 1.$$

The interest rate r, drift coefficient q, and correlation coefficient ρ are assumed to be constant. To allow for flexibility in model calibration, we may generalize the reversion level parameter $p(t)$ in the dynamics of V_t to be a deterministic continuous function of time.

The 3/2-dynamics of the variance process exhibits the mean reversion feature. The mean reversion rate depends on the current variance level, thus exhibiting more volatile volatility structure than that of the Heston model. We let Y_t be the reciprocal of the variance V_t. By applying Itô's lemma, Y_t is seen to follow the time-inhomogeneous CIR process [see (2.143)]:

$$\mathrm{d}\left(\frac{1}{V_t}\right) = \mathrm{d}Y_t = [q+\varepsilon^2 - p(t)Y_t]\mathrm{d}t - \varepsilon\sqrt{Y_t}\,\mathrm{d}W_t^2. \tag{4.35}$$

Certain technical conditions are required in order to avoid anomalies in the 3/2 variance process. For the CIR process (like the variance process in the Heston model and Y_t defined above), it is well known that the coefficients have to satisfy the Feller boundary condition in order to avoid hitting the zero value, causing V_t in the 3/2-process to explode with nonzero probability. In our model, this condition is expressed as $q \geq -\varepsilon^2/2$. When pricing options under the share measure, it is known that V_t has zero probability of explosion if and only if $\rho < \varepsilon/2$. By imposing the two constraints: $q > 0$ and $\rho < 0$ in our model, it becomes sufficient to avoid Y_t hitting the zero value.

Let $U_i(S,V,I_t,t)$ be the time-t price function of the contingent claim with terminal payoff $F_i(S_i,I_i)$ on maturity date t_i, where

$$U_i(S,V,I,t) = e^{-r(t_i-t)}E_t^Q[F_i(S_i,I_i)], \quad t < t_i. \tag{4.36}$$

By the Feymann-Kac theorem, the governing partial integro-differential equation (PIDE) for $U_i(S,V,I,t)$ is given by

$$\begin{aligned}
&\frac{\partial U_i}{\partial t} + \frac{VS^2}{2}\frac{\partial^2 U_i}{\partial S^2} + \rho\varepsilon V^2 S\frac{\partial^2 U_i}{\partial S\partial V} + \frac{\varepsilon^2}{2}V^3\frac{\partial^2 U_i}{\partial V^2} \\
&\quad + (r-\lambda m)S\frac{\partial U_i}{\partial S} + [p(t)V - qV^2]\frac{\partial U_i}{\partial V} \\
&\quad + \delta(t-t_{i-1})S\frac{\partial U_i}{\partial I} - rU_i + \lambda E_Q[U_i(Se^J,V,I,t) - U_i(S,V,I,t)] = 0.
\end{aligned} \tag{4.37}$$

With the inclusion of jump component in the dynamic of S_t [see (4.34)], an additional integral term is added in the governing equation (4.37). Recall that there is a jump in the value for I of amount $S_{t_{i-1}}$ across t_{i-1} while I assumes constant value across $[t_0,t_{i-1})$ and $(t_{i-1},t_i]$. However, there is no jump in the value of U_i across t_{i-1}.

Since I assumes constant value over the two subintervals (t_0,t_{i-1}) and (t_{i-1},t_i), we may omit the dependency of U_i on I in the PIDE over the respective subinterval. In summary, we have the following two-step backward procedure for solving the PIDE:

(i) For $t \in (t_{i-1}, t_i)$, the governing PIDE reduces to

$$\frac{\partial U_i}{\partial t} + \frac{VS^2}{2}\frac{\partial^2 U_i}{\partial S^2} + \rho \varepsilon V^2 S \frac{\partial^2 U_i}{\partial S \partial V} + \frac{\varepsilon^2}{2} V^3 \frac{\partial^2 U_i}{\partial V^2} + (r - \lambda m) S \frac{\partial U_i}{\partial S}$$
$$+ [p(t)V - qV^2] \frac{\partial U_i}{\partial V} - rU_i + \lambda E_Q[U_i(Se^J, V, I, t) - U_i(S, V, I, t)] = 0, \quad (4.38)$$

with terminal condition:

$$U_i(S, V, t_i) = F_i(S, S_{i-1}).$$

Here, S_{i-1} is the realized asset price at t_{i-1} as captured by I, which can be regarded as a known parameter since the value of I does not change in (t_{i-1}, t_i).

(ii) For $t \in [t_0, t_{i-1})$, we solve the same governing PIDE subject to the terminal condition at time t_{i-1} as specified by the jump condition.

$$\lim_{t \to t_{i-1}^-} U_i(S, V, S, t) = \lim_{t \to t_{i-1}^+} U_i(S, V, 0, t). \quad (4.39)$$

A close examination of (4.38) reveals that U_i is explicitly solvable over the time interval (t_{i-1}, t_i) since the terminal payoff function $F_i(S, S_{i-1})$ is independent of V and S_{i-1} is a known parameter. However, since the solution U_i at t_{i-1}^- has dependence on V, the solution of U_i at t_0 cannot be obtained in an analytic closed form. We manage to express U_i at t_0 in a quasi-closed form in terms of an integral, where the integrand is the product of the transition density and the known solution U_i at t_{i-1}^-.

For convenience, we assume uniform time interval Δt between successive monitoring dates so that $\Delta t = t_i - t_{i-1}$, $i = 1, 2, \ldots, N$. We introduce $x = \ln S$ and $\tau = t_i - t$ for $t \le t_i$. To solve (4.38) for $U_i(x, V, t)$ over (t_{i-1}, t_i), we take the Fourier transform of the governing PIDE with respect to x. Following the convention used in Yuen *et al.* (2015), the Fourier transform of $U_i(x, V, \tau)$ is defined by

$$\mathscr{F}[U_i(x, V, \tau)](k) = \int_{-\infty}^{\infty} e^{-ikx} U_i(x, V, \tau)\, dx, \quad (4.40)$$

where k is the Fourier transform variable. We define $H_i(k, V, \tau)$ to be

$$H_i(k, V, \tau) = \exp(-s(k)\tau) \mathscr{F}[U_i(x, V, \tau)](k),$$

where

$$s(k) = ik(r - \lambda m) - (r + \lambda) + \lambda \exp(ik\nu - \frac{\zeta^2}{2}k^2).$$

The symbol i denotes $\sqrt{-1}$. The impact of the jump component in S_t is captured by the terms involving λ in $s(k)$. By taking the Fourier transform on both sides, (4.38) is transformed into an one-dimensional partial differential equation in H_i as follows:

$$\frac{\partial H_i}{\partial \tau} = -c_k V H_i + [p(t) + \tilde{q}_k V^2] \frac{\partial H_i}{\partial V} + \frac{\varepsilon^2}{2} V^3 \frac{\partial^2 H_i}{\partial V^2}, \quad (4.41)$$

with initial condition:

$$H_i(k,V,0) = \mathscr{F}[F_i(e^x,S_{i-1})](k),$$

where

$$c_k = (k^2+ik)/2 \quad \text{and} \quad \tilde{q}_k = \rho\varepsilon ik - q.$$

Theorem 4.1 *Let $\widehat{H}_i(k,V,\tau)$ denote the fundamental solution to (4.41) with initial condition: $\widehat{H}_i(k,V,0)=1$, which admits the following analytic solution:*

$$\widehat{H}_i(k,V,\tau) = \frac{\Gamma(\gamma-\alpha)}{\Gamma(\gamma)}\left[\frac{2}{\varepsilon^2 y(V,t)}\right]^{\alpha} {}_1F_1\left(\alpha,\gamma,-\frac{2}{\varepsilon^2 y(V,t)}\right), \tag{4.42}$$

where $y(V,t)$, α and γ are given by

$$\begin{aligned} y(V,t) &= V\int_t^{t_i} e^{\int_t^u p(s)\,ds}\,du, \\ \alpha &= -\left(\frac{1}{2}-\frac{\tilde{q}_k}{\varepsilon^2}\right)+\sqrt{\left(\frac{1}{2}-\frac{\tilde{q}_k}{\varepsilon^2}\right)^2+\frac{2c_k}{\varepsilon^2}}, \\ \gamma &= 2\left(\alpha+1-\frac{\tilde{q}_k}{\varepsilon^2}\right),\ \tilde{q}_k=\rho\varepsilon ik-q,\ c_k=\frac{k^2+ik}{2}. \end{aligned}$$

Moreover, the solution for $U_i(x,V,t)$ over (t_{i-1},t_i) is given by

$$U_i(x,V,\tau) = \mathscr{F}^{-1}\Big[\exp(s(k)\tau)\widehat{H}_i(k,V,\tau)\mathscr{F}[F_i(e^x,S_{i-1})](k)\Big]. \tag{4.43}$$

The proof of Theorem 4.1 can be found in Appendix A in Yuen *et al.* (2015). To proceed with the solution of $U_i(x,V,\tau)$ over $[t_0,t_{i-1})$, once $U_i(x,V,t_{i-1}^+)$ has been obtained using (4.43), we apply the jump condition in (4.39) to obtain $U_i(x,V,t_{i-1}^-)$. In general, $U_i(x,V,t_{i-1}^-)$ may have functional dependence on both x and V for generalized variance swap contracts. Consequently, the joint transition density function of (x,V) is required to find the second-step solution of the PIDE. Given the joint transition density function, the solution $U_i(x,V,\tau)$ over $[t_0,t_{i-1}]$ can be obtained as a double integral by virtue of the Feynman-Kac Theorem. Note that for discrete vanilla variance swaps, $U_i(x,V,t_{i-1}^-)$ depends on V only [see (4.45a, 4.45b) later]. As a result, only the marginal density function of V is required in the solution procedure.

Vanilla variance swap

For the discrete vanilla variance swap, the generalized Fourier transform of the terminal payoff $F_i(S,S_{i-1})$ takes the following forms, depending on whether the simple rate of return or log return is used:

(i) Simple rate of return

$$\begin{aligned} \mathscr{F}\left[F_i^{(1)}(S,S_{i-1})\right] &= \mathscr{F}\left[\left(\frac{S-S_{i-1}}{S_{i-1}}\right)^2\right] = \mathscr{F}\left[\left(\frac{e^x}{S_{i-1}}-1\right)^2\right] \\ &= 2\pi\left[\frac{\delta(k+2i)}{S_{i-1}^2}-\frac{2\delta(k+i)}{S_{i-1}}+\delta(k)\right]. \end{aligned} \tag{4.44a}$$

(ii) Log return

$$\mathscr{F}\left[F_i^{(2)}(S,S_{i-1})\right] = \mathscr{F}\left[\left(\ln\frac{S}{S_{i-1}}\right)^2\right] = \mathscr{F}[(x-\ln S_{i-1})^2]$$
$$= 2\pi[-\delta''(k) - 2i\,\delta'(k)\ln S_{i-1} + \delta(k)(\ln S_{i-1})^2], \quad (4.44b)$$

In both (4.44a,b), $\delta(\cdot)$ is the Dirac function, $\delta'(\cdot)$ and $\delta''(\cdot)$ denote the first-order and second-order derivatives of $\delta(\cdot)$, respectively. The symbol i in (4.44a) denotes $\sqrt{-1}$.

We write $x = \ln S$ and let $U_i^{(n)}(x,V,\tau)$ denote the solution to (4.38) corresponding to the terminal condition: $F_i^{(n)}(e^x,S_{i-1}), n = 1,2$. Using (4.43), we derive the solution for $U_i^{(1)}$ and $U_i^{(2)}$ at $t = t_{i-1}$ ($\tau = \Delta t$) as follows:

$$\begin{aligned}
&U_i^{(1)}(x,V,\Delta t)\\
=\;& \mathscr{F}^{-1}\left[\exp(s(k)\Delta t)\widehat{H}_i(k,V,\Delta t)\mathscr{F}\left[\left(\frac{e^x}{S_{i-1}}-1\right)^2\right]\right]\\
=\;& \int_{-\infty}^{\infty} e^{ikx}\exp(s(k)\Delta t)\widehat{H}_i(k,V,\Delta t)\left[\frac{\delta(k+2i)}{S_{i-1}^2} - \frac{2\delta(k+i)}{S_{i-1}} + \delta(k)\right] dk\\
=\;& e^{s(-2i)\Delta t}\widehat{H}_i(-2i,V,\Delta t) - 2e^{s(-i)\Delta t}\widehat{H}_i(-i,V,\Delta t) + e^{-r\Delta t} \qquad (4.45a)
\end{aligned}$$

and

$$\begin{aligned}
&U_i^{(2)}(x,V,\Delta t)\\
=\;& \mathscr{F}^{-1}[\exp(s(k)\Delta t)\widehat{H}_i(k,V,\Delta t)\mathscr{F}[(x-\ln S_{i-1})^2]]\\
=\;& \int_{-\infty}^{\infty} e^{ikx}\exp(s(k)\Delta t)\widehat{H}_i(k,V,\Delta t)\left[-\delta''(k) - 2i\delta'(k)\ln S_{i-1} + \delta(k)(\ln S_{i-1})^2\right] dk\\
=\;& -f''(0) + 2if'(0)\ln S_{i-1} + f(0)(\ln S_{i-1})^2 = -g''(0), \qquad (4.45b)
\end{aligned}$$

where

$$f(k) = g(k)e^{ikx}, \quad \text{and} \quad g(k) = \exp(s(k)\Delta t)\widehat{H}_i(k,V,\Delta t).$$

The symbol i in (4.45a,b) denotes $\sqrt{-1}$. The intermediate solution $U_i^{(n)}(x,V,\tau)$, $n = 1,2$, depends on V only, which simplifies the calculation of the second-step solution since we only need to evaluate a univariate expectation using the known marginal transition density function. The required transition density of V_t can be obtained by observing that $Y_t = 1/V_t$ is a CIR process. Based on (2.113), the transition density of V_t is given by

$$\begin{aligned}
&p_{V_t}(V_t,t|V_0,0)\\
&= \tilde{p}_{Y_t}\left(\frac{1}{V_t},t\,\middle|\,\frac{1}{V_0},0\right)\frac{1}{V_t^2}\\
&= \frac{l(0,t)}{2l^*(0,t)V_t^2}\exp\left(\frac{-\frac{1}{V_0}-\frac{1}{V_t}l(0,t)}{2l^*(0,t)}\right)\left(\frac{V_0\,l(0,t)}{V_t}\right)^{\bar{\nu}/2} I_{\bar{\nu}}\left(\frac{\sqrt{l(0,t)}}{l^*(0,t)\sqrt{V_0V_t}}\right),
\end{aligned}$$
$$(4.46)$$

where

$$\bar{\nu} = \frac{2}{\varepsilon^2}(q+\varepsilon^2) - 1, \; l(s,t) = \exp\left(\int_s^t p(u)\,\mathrm{d}u\right), \; l^*(s,t) = \frac{\varepsilon^2}{2}\int_s^t l(s,u)\,\mathrm{d}u.$$

Note that $I_{\bar{\nu}}$ is the modified Bessel function of the first kind of order $\bar{\nu}$ as defined by

$$I_{\bar{\nu}}(z) = \left(\frac{z}{2}\right)^{\bar{\nu}} \sum_{k=0}^{\infty} \frac{\left(\frac{z^2}{2}\right)^k}{k!\Gamma(\bar{\nu}+k+1)}.$$

The expectation of the first term, $i = 1$ that corresponds to the discrete realized variance over $[t_0, t_1]$, can be computed by one-step backward procedure since S_0 is known. For $i \geq 2$, it is necessary to implement the two-step backward calculation from t_i to t_{i-1}, then t_{i-1} to t_0. Summarizing the above results, the fair strike of a variance swap on discrete realized variance sampling on dates: $0 = t_0 < t_1 < \cdots < t_N = T$ under the 3/2-model is given by

$$K_v^{(n)} = \frac{A}{N} e^{r\Delta t}\left[U_1^{(n)}(x_0, V_0, \Delta t) + \sum_{i=2}^{N} \int_0^{\infty} U_i^{(n)}(x, V, \Delta t) p_V(V, t_{i-1}|V_0, 0)\,\mathrm{d}V\right], \quad (4.47)$$

where $n = 1$ and $n = 2$ correspond to the simple rate of return and log return, respectively.

Gamma swap

For the gamma swap, the discrete weighted realized variance assumes the form

$$\frac{A}{N}\sum_{i=1}^{N} \frac{S_i}{S_0} \ln\left(\frac{S_i}{S_{i-1}}\right)^2$$

based on the log return. The fair strike price of the gamma swap requires the evaluation of the risk neutral expectation of the weight adjusted squared log returns, which is given by

$$E_0^Q\left[\frac{S_i}{S_0}\left(\ln\frac{S_i}{S_{i-1}}\right)^2\right] \quad i = 1, 2, \cdots, N.$$

The above weight adjusted squared log return can still be regarded as a European contingent claim with a bivariate payoff function $E_i(S_i, S_{i-1})$. As a result, the two-step PIDE approach for pricing discrete variance swaps can be applied to pricing a gamma swap.

The terminal payoff function $E_i(S_i, S_{i-1})$ is no longer homogeneous in S_i and S_{i-1}. As a result, unlike the discrete vanilla variance swaps, the time-t_{i-1} value of U_i for a weighted variance swap has dependence on both S_{i-1} and V_{i-1}. This poses the technical requirement of finding the joint transition density of $\ln S_t$ and V_t in the risk neutral expectation calculation over $[t_0, t_{i-1})$. We write $x = \ln S$, and let $p_{x,V}(x, V, t|x_s, V_s, s)$ denote the joint transition density function of x_t and V_t from time s to t. We define the log-price transformed joint density function as follows

$$\widetilde{G}(\tau; -z, V_t|x_s, V_s) = \int_{-\infty}^{\infty} e^{izy} p_{x,V}(y, V_t, t|x_s, V_s, s)\,\mathrm{d}y. \quad (4.48)$$

To achieve better analytic tractability, henceforth we take $p(t)$ to be a constant. The log-price transformed joint density admits an analytic representation as shown in Theorem 4.2.

Theorem 4.2 *The log-price transformed joint density function of the log asset price and instantaneous variance under the 3/2-model with jumps is given by*

$$\widetilde{G}(\tau;-z,V_t|x_s,V_s) = G(\tau;-z,V_t,V_s)e^{izx_s}\exp(h(z)\tau),$$

where

$$\begin{aligned} G(\tau;-z,V_t,V_s) &= \frac{e^{(1+\mu_z)p\tau}}{e^{p\tau}-1}\exp\left(-\frac{2p}{(e^{p\tau}-1)V_s\varepsilon^2}-\frac{2pe^{p\tau}}{(e^{p\tau}-1)V_t\varepsilon^2}\right) \\ & \left(\frac{2p}{\varepsilon^2 V_t}\right)\left(\frac{V_s}{V_t}\right)^{\mu_z} I_{\nu_z}\left(\frac{4pe^{p\tau/2}}{(e^{p\tau}-1)\varepsilon^2\sqrt{V_sV_t}}\right), \end{aligned}$$

and

$$\begin{aligned} h(z) &= i(r-\lambda m)z+\lambda[\exp(i\nu z-\zeta^2z^2/2)-1], \\ \mu_z &= \frac{1+\hat{\theta}_z}{2},\ \hat{\theta}_z = \frac{2(q+iz\rho\varepsilon)}{\varepsilon^2},\ c_z = \frac{z^2-iz}{2}, \\ \tilde{c}_z &= \frac{2c_z}{\varepsilon^2},\ \nu_z = 2\sqrt{\mu_z^2+\tilde{c}_z},\ \tau = t-s. \end{aligned}$$

Here, $I_{\nu_z}(\cdot)$ is the modified Bessel function of the first kind of order ν_z.

The proof of Theorem 4.2 can be found in Appendix B in Yuen *et al.* (2015). The analytic form of $\widetilde{G}$ helps reduce the dimensionality of integration by one, compared with the pricing formula obtained based on the use of the joint transition density function.

The derivation of the fair strike formula for the gamma swap in log return amounts to the following risk neutral expectation calculation:

$$E_0^Q\left[S_i\left(\ln\frac{S_i}{S_{i-1}}\right)^2\right].$$

Similar to the derivation of the fair strike formula for the vanilla variance swap, we compute the generalized Fourier transform of the terminal payoff of the gamma swap as follows:

$$\begin{aligned} \mathscr{F}[E_i(S,S_{i-1})] &= \mathscr{F}\left[S\left(\ln\frac{S}{S_{i-1}}\right)^2\right] \\ &= 2\pi[-\delta''(k+i)-2i\delta'(k+i)\ln S_{i-1}+\delta(k+i)(\ln S_{i-1})^2]. \quad (4.49) \end{aligned}$$

Let $U_i(x,V,\tau)$ denote the solution to (4.38) with initial value $E_i(e^x,S_{i-1})$. We obtain

$$\begin{aligned} & U_i(x,V,\Delta t) \\ =\ & \mathscr{F}^{-1}\left[\exp(s(k)\Delta t)\widehat{H}_i(k,V,\Delta t)\right. \\ & \left.[-\delta^{(2)}(k+i)-2i\delta^{(1)}(k+i)\ln S_{i-1}+\delta(k+i)(\ln S_{i-1})^2]\right] \\ =\ & -f''(-i)+2if'(-i)\ln S_{i-1}+f(-i)(\ln S_{i-1})^2 = e^x[-g''(-i)], \quad (4.50) \end{aligned}$$

where the functions f and g are defined in (4.45b). The symbol i denotes $\sqrt{-1}$. Note that $U_i(x,V,t_{i-1}^-)$ of the gamma swap is seen to have dependence on both x and V, plus exhibiting a separable form of $e^x f(V)$.

The risk neutral expectation calculation of the first term in the discrete realized variance does not require the two-step procedure since S_0 is known. For $i \geq 2$, it is necessary to perform the following double integration that computes the risk neutral expectation from t_{i-1} to t_0 in the second step of the backward procedure:

$$E_0^Q\left[e^{r\Delta t}U_i(x,V,\Delta t)\right] = e^{r\Delta t}\int_0^\infty \int_{-\infty}^\infty U_i(x,V,\Delta t)p_{x,V}(x,V,t_{i-1}|x_0,V_0,0)\,\mathrm{d}x\mathrm{d}V.$$

For $i \geq 2$, we denote the inner integral by $\Psi_i(V,\Delta t|x_0,V_0)$ and obtain

$$\begin{aligned}\Psi_i(V,\Delta t|x_0,V_0) &= \int_{-\infty}^\infty U_i(x,V,\Delta t)p_{x,V}(x,V,t_{i-1}|x_0,V_0,0)\,\mathrm{d}x \\ &= \int_{-\infty}^\infty \left[e^x(-g''(-i))\right]p_{x,V}(x,V,t_{i-1}|x_0,V_0,0)\,\mathrm{d}x \\ &= -g''(-i)\widetilde{G}(t_{i-1};i,V|x_0,V_0). \end{aligned} \tag{4.51}$$

Putting all results together, the fair strike of a gamma swap with discrete realized variance sampling on monitoring dates: $0 = t_0 < t_1 < \cdots < t_N = T$ is given by

$$K_v = \frac{e^{r\Delta t}}{S_0}\frac{A}{N}\left[U_1(x_0,V_0,\Delta t) + \sum_{i=2}^N \int_0^\infty \Psi_i(V,\Delta t|x_0,V_0)\,\mathrm{d}V\right]. \tag{4.52}$$

4.3 Moment generating function methods

The essence of pricing discrete variance swap products under the stochastic volatility models is the knowledge of the joint moment generating function (mgf) of the log asset price and instantaneous variance. Zheng and Kwok (2014a) propose a more succinct method that employs the tower rule in expectation calculations as an alternative to the nested expectation approach in Zhu and Lian (2012a). Based on the joint mgf of the log asset price and instantaneous variance under the SVSJ model [see (2.121), (2.123)], they derive the analytic and semi-analytic formulas of the fair strikes of various generalized variance swaps with weight feature in the discrete realized variance. This pricing framework is not restricted to the SVSJ model. It works under any affine stochastic volatility model with closed-form joint mgf of the log asset price and instantaneous variance. For example, Zhu and Lian (2015a) follow a similar approach to price the forward-start variance swaps under the Heston model with zero jumps. Pun *et al.* (2015) extend the approach to price discrete variance swaps under multifactor affine type stochastic volatility models. Dong and Wong (2017) further extend to pricing discrete variance swaps under a threshold Ornstein-Uhlenbeck model that exhibits both mean reversion and regime switching feature in the underlying asset price process. Kim and Kim (2018) derive analytic solutions for discrete variance swaps under the Heston model with stochastic mean reversion of the instantaneous variance. Later, Kim and Kim (2019) further extend the same pricing approach to the two-factor lognormal stochastic volatility model in which the log volatility is modeled by an Ornstein-Uhlenbeck process that is mean reverting and temporally homogeneous Markov Gaussian process.

4.3.1 Variance swap and gamma swap

We illustrate the derivation of the fair strike formulas of variance swap and gamma swap on discrete realized variance under the affine jump-diffusion stochastic volatility (SVSJ) model as specified in (2.121). Recall the well-known result that the kth order moment of the log asset price $E[X_t^k]$ is given by the kth order derivative of the mgf of X_t evaluated at zero (Chacko and Das, 2002). By introducing a dummy variable ϕ and applying the Fubini theorem that justifies the exchange of the order of the expectation and differentiation, we obtain the expectation of a typical term in the terminal payoff of the variance swap on discrete realized variance as follows:

$$\begin{aligned} E_0^Q\left[\left(\ln\frac{S_{t_k}}{S_{t_{k-1}}}\right)^2\right] &= E_0^Q\left[\frac{\partial^2}{\partial\phi^2}e^{\phi(X_{t_k}-X_{t_{k-1}})}\right]\bigg|_{\phi=0} \\ &= \frac{\partial^2}{\partial\phi^2}E_0^Q\left[E^Q\left[e^{\phi X_{t_k}}\Big|X_{t_{k-1}},V_{t_{k-1}}\right]e^{-\phi X_{t_{k-1}}}\right]\bigg|_{\phi=0} \\ &= \frac{\partial^2}{\partial\phi^2}E_0^Q\left[e^{B(\Theta;\Delta t_k,\mathbf{q}_1)V_{t_{k-1}}+\Gamma(\Theta;\Delta t_k,\mathbf{q}_1)+\Lambda(\Theta;\Delta t_k,\mathbf{q}_1)}\right]\bigg|_{\phi=0} \\ &= \frac{\partial^2}{\partial\phi^2}e^{B(\Theta;t_{k-1},\mathbf{q}_2)V_0+\Gamma(\Theta;t_{k-1},\mathbf{q}_2)+\Lambda(\Theta;t_{k-1},\mathbf{q}_2)}\bigg|_{\phi=0}, \end{aligned} \tag{4.53}$$

where $\mathbf{q}_1=(\phi\ 0\ 0)^T$, and

$$\mathbf{q}_2=\begin{pmatrix} 0 \\ B(\Theta;\Delta t_k,\mathbf{q}_1) \\ \Gamma(\Theta;\Delta t_k,\mathbf{q}_1)+\Lambda(\Theta;\Delta t_k,\mathbf{q}_1) \end{pmatrix}.$$

The parameter functions B, Γ, and Λ under the SVSJ model are given in (2.125). Note that when we compute the inner expectation and the outer expectation in the above procedure, we only require the marginal mgf of $\ln S_t$ and V_t, respectively. Putting the results together, the fair strike of the variance swap on discrete realized variance is given by

$$K_v(0,T;N)=\frac{A}{N}\sum_{k=1}^{N}\frac{\partial^2}{\partial\phi^2}e^{B(\Theta;t_{k-1},\mathbf{q}_2)V_0+\Gamma(\Theta;t_{k-1},\mathbf{q}_2)+\Lambda(\Theta;t_{k-1},\mathbf{q}_2)}\bigg|_{\phi=0}. \tag{4.54}$$

For numerical evaluation of the double differentiation with respect to ϕ, the use of numerical differentiation rule provides sufficient accuracy since analytic forms of B, Γ and Λ are available. It is instructive to examine the asymptotic limit of continuous sampling of realized variance on the fair strike of the variance swap. The resulting analytic pricing formula is presented in Theorem 4.3.

Theorem 4.3 *By expanding the parameter functions B, Γ, and Λ in powers of Δt_k and taking $\Delta t_k \to 0$ in formula (4.54) subsequently, we obtain the fair strike of the variance swap on continuous realized variance as follows:*

$$\begin{aligned} K_v(0,T;\infty) = \frac{1}{T}\Big\{ & \frac{1-e^{-\kappa T}}{\kappa}V_0 - \frac{\lambda\eta}{\kappa^2}(1-e^{-\kappa T}-\kappa T) \\ & + \lambda\left[\delta^2+\rho_J^2\eta^2+(\nu+\rho_J\eta)^2\right]T + \frac{\theta}{\kappa}(\kappa T - 1 + e^{-\kappa T})\Big\}, \end{aligned} \tag{4.55}$$

where we have used the convention $\frac{A}{N} = \frac{1}{T}$.

The asymptotic formula (4.55) for the fair strike of the variance swap on continuous realized variance agrees with that derived based on direct expectation of the quadratic variation of the asset price process [see (2.120) in Sec. 2.5.2]. The proof of Theorem 4.3 can be found in Zheng and Kwok (2014a).

We show how to extend this mgf approach to the gamma swap. For the gamma swap, recall that the squared returns in the discrete realized variance are weighted by the asset price ratio S_{t_k}/S_{t_0}. The expectation of a typical term in the terminal payoff of the gamma swap on discrete realized variance can be calculated as follows:

$$\begin{aligned} & E_0^Q\left[\frac{S_{t_k}}{S_{t_0}}\left(\ln\frac{S_{t_k}}{S_{t_{k-1}}}\right)^2\right] \\ = \ & e^{-X_0}E_0^Q\left[e^{X_{t_k}-X_{t_{k-1}}}(X_{t_k}-X_{t_{k-1}})^2 e^{X_{t_{k-1}}}\right] \\ = \ & e^{-X_0}E_0^Q\left[\frac{\partial^2}{\partial\phi^2}e^{\phi(X_{t_k}-X_{t_{k-1}})+X_{t_{k-1}}}\right]\Bigg|_{\phi=1} \\ = \ & e^{-X_0}\frac{\partial^2}{\partial\phi^2}E_0^Q\left[E^Q\left[e^{\phi X_{t_k}}\Big|X_{t_{k-1}},V_{t_{k-1}}\right]e^{(1-\phi)X_{t_{k-1}}}\right]\Bigg|_{\phi=1} \\ = \ & e^{-X_0}\frac{\partial^2}{\partial\phi^2}E_0^Q\left[e^{X_{t_{k-1}}+B(\Theta;\Delta t_k,\mathbf{q}_1)V_{t_{k-1}}+\Gamma(\Theta;\Delta t_k,\mathbf{q}_1)+\Lambda(\Theta;\Delta t_k,\mathbf{q}_1)}\right]\Bigg|_{\phi=1} \\ = \ & \frac{\partial^2}{\partial\phi^2}e^{B(\Theta;t_{k-1},\mathbf{q}_2)V_0+\Gamma(\Theta;t_{k-1},\mathbf{q}_2)+\Lambda(\Theta;t_{k-1},\mathbf{q}_2)}\Bigg|_{\phi=1}, \end{aligned} \tag{4.56}$$

where $\mathbf{q}_1 = (\phi\ 0\ 0)^T$ and

$$\mathbf{q}_2 = \begin{pmatrix} 1 \\ B(\Theta;\Delta t_k,\mathbf{q}_1) \\ \Gamma(\Theta;\Delta t_k,\mathbf{q}_1)+\Lambda(\Theta;\Delta t_k,\mathbf{q}_1) \end{pmatrix}.$$

The above procedure is similar to that for computing the fair strike of the variance swap except that we use the analytic expression of the joint mgf of $\ln S_t$ and V_t when we compute the outer expectation in the last but second step. Putting the results together, the fair strike of the gamma swap on discrete realized variance is given by

$$K_g(0,T;N) = \frac{A}{N}\sum_{k=1}^{N}\frac{\partial^2}{\partial\phi^2}e^{B(\Theta;t_{k-1},\mathbf{q}_2)V_0+\Gamma(\Theta;t_{k-1},\mathbf{q}_2)+\Lambda(\Theta;t_{k-1},\mathbf{q}_2)}\Big|_{\phi=1}. \tag{4.57}$$

Surprisingly, the fair strike $K_g(0,T;N)$ has no dependence on the initial asset price S_{t_0}.

Similarly, we consider the asymptotic limit of continuous sampling of realized variance on the fair strike of the gamma swap as summarized in Theorem 4.4.

Theorem 4.4 *By expanding the parameter functions B, Γ and Λ in powers of Δt_k and taking $\Delta t_k \to 0$ in formula (4.57) subsequently, we obtain the fair strike of the gamma swap on continuous realized variance as follows:*

$$\begin{aligned} K_g(0,T;\infty) = \frac{1}{T}\Bigg[&\left(V_0 - \frac{\kappa\theta}{\kappa-\rho\varepsilon} - C_2\right)\frac{e^{(r-q-\kappa+\rho\varepsilon)T}-1}{r-q-\kappa+\rho\varepsilon} \\ &+\left(\frac{\kappa\theta}{\kappa-\rho\varepsilon}+C_1+C_2\right)\frac{e^{(r-q)T}-1}{r-q}\Bigg], \end{aligned} \tag{4.58}$$

where

$$\begin{aligned} C_1 &= \frac{\lambda e^{\nu+\delta^2/2}}{1-\rho_J\eta}\left[\left(\nu+\delta^2+\frac{\rho_J\eta}{1-\rho_J\eta}\right)^2+\delta^2+\left(\frac{\rho_J\eta}{1-\rho_J\eta}\right)^2\right], \\ C_2 &= \frac{\lambda\eta e^{\nu+\delta^2/2}}{(1-\rho_J\eta)^2(\kappa-\rho\varepsilon)}. \end{aligned}$$

The proof of Theorem 4.4 can follow similar procedures as those in the proof of Theorem 4.3, the details of which can be found in Zheng and Kwok (2014a).

Recall that the moment of a random variable can be represented as the derivative of its mgf evaluated at zero. We can extend this mgf approach to price the moment swap, whose terminal payoff can be expressed as follows:

$$\frac{A}{N}\sum_{k=1}^{N}\left(\ln\frac{S_{t_k}}{S_{t_{k-1}}}\right)^m = \frac{A}{N}\sum_{k=1}^{N}\frac{\partial^m}{\partial\phi^m}e^{\phi(X_{t_k}-X_{t_{k-1}})}\Big|_{\phi=0}, \tag{4.59}$$

where m is a positive integer greater than or equal to 2. Similar pricing procedure of calculating expectation via mgf can be performed.

4.3.2 Corridor type swaps

We would like to find the fair strike price of a discretely monitored downside variance swap with an upper barrier U whose payoff at maturity T is given by

$$\frac{A}{N}\sum_{k=1}^{N}\left(\ln\frac{S_{t_k}}{S_{t_{k-1}}}\right)^2\mathbf{1}_{\{S_{t_{k-1}}\le U\}} - K.$$

The indicator function in the terminal payoff imposes extra difficulty on the evaluation of the fair strike. Fortunately, we can use the generalized inverse Fourier transform to express the indicator function as a Fourier integral of an exponential function.

We consider the expectation calculation of a typical term in the terminal payoff of the discretely monitored downside variance swap as follows:

$$
\begin{aligned}
& E_0^Q\left[\left(\ln \frac{S_{t_k}}{S_{t_{k-1}}}\right)^2 \mathbf{1}_{\{S_{t_{k-1}} \leq U\}}\right] \\
= \; & E_0^Q\left[E^Q\left[\frac{\partial^2}{\partial \phi^2} e^{\phi(X_{t_k}-X_{t_{k-1}})} \middle| X_{t_{k-1}}, V_{t_{k-1}}\right] \mathbf{1}_{\{X_{t_{k-1}} \leq \ln U\}}\right]\Bigg|_{\phi=0} \\
= \; & E_0^Q\left[\frac{\partial^2}{\partial \phi^2} e^{B(\Theta;\Delta t_k,\mathbf{q}_1)V_{t_{k-1}}+\Gamma(\Theta;\Delta t_k,\mathbf{q}_1)+\Lambda(\Theta;\Delta t_k,\mathbf{q}_1)} \mathbf{1}_{\{X_{t_{k-1}} \leq \ln U\}}\right]\Bigg|_{\phi=0} \\
= \; & \frac{\partial^2}{\partial \phi^2} E_0^Q\left[e^{B(\Theta;\Delta t_k,\mathbf{q}_1)V_{t_{k-1}}+\Gamma(\Theta;\Delta t_k,\mathbf{q}_1)+\Lambda(\Theta;\Delta t_k,\mathbf{q}_1)} \mathbf{1}_{\{X_{t_{k-1}} \leq \ln U\}}\right]\Bigg|_{\phi=0}, \qquad (4.60)
\end{aligned}
$$

where $\mathbf{q}_1 = (\phi \; 0 \; 0)^T$. For $k = 1$, the above expectation is readily seen to be

$$
\begin{aligned}
& E_0^Q\left[\left(\ln \frac{S_{t_1}}{S_{t_0}}\right)^2 \mathbf{1}_{\{S_{t_0} \leq U\}}\right] \\
= \; & \frac{\partial^2}{\partial \phi^2} e^{B(\Theta;\Delta t_1,\mathbf{q}_1)V_0+\Gamma(\Theta;\Delta t_1,\mathbf{q}_1)+\Lambda(\Theta;\Delta t_1,\mathbf{q}_1)} \mathbf{1}_{\{X_0 \leq \ln U\}}\Bigg|_{\phi=0}.
\end{aligned}
$$

For $k \geq 2$, the evaluation of expectation in (4.60) requires the representation of the indicator function $\mathbf{1}_{\{X_{t_{k-1}} \leq \ln U\}}$ in terms of an inverse Fourier transform integral. We consider the generalized Fourier transform of the indicator function $\mathbf{1}_{\{X_{t_{k-1}} \leq u\}}$ visualized as a function of u, where

$$
\int_{-\infty}^{\infty} \mathbf{1}_{\{X_{t_{k-1}} \leq \, u\}} e^{-\mathrm{i}u\omega} \, \mathrm{d}u = \int_{X_{t_{k-1}}}^{\infty} e^{-\mathrm{i}u\omega} \, \mathrm{d}u = \frac{e^{-\mathrm{i}X_{t_{k-1}}\omega}}{\mathrm{i}\omega}. \qquad (4.61\mathrm{a})
$$

Here, the Fourier transform variable ω is taken to be complex and we write $\omega = \omega_r + \mathrm{i}\omega_i$. Provided that ω_i is appropriately chosen to lie within $(-\infty, 0)$, the above generalized Fourier transform exists. By taking the corresponding generalized inverse Fourier transform, we obtain

$$
\mathbf{1}_{\{X_{t_{k-1}} \leq u\}} = \frac{1}{2\pi} \int_{-\infty}^{\infty} e^{\mathrm{i}u\omega} \frac{e^{-\mathrm{i}X_{t_{k-1}}\omega}}{\mathrm{i}\omega} \, \mathrm{d}\omega_r. \qquad (4.61\mathrm{b})
$$

This analytic representation of the indicator function expressed in terms of a generalized Fourier integral is then substituted into (4.60). By interchanging the order of performing integration with the two operations of differentiation and expectation, we obtain

$$
\begin{aligned}
& E_0^Q\left[\left(\ln\frac{S_{t_k}}{S_{t_{k-1}}}\right)^2 \mathbf{1}_{\{S_{t_{k-1}}\leq U\}}\right] \\
= & \frac{1}{2\pi}\int_{-\infty}^{\infty}\frac{\partial^2}{\partial\phi^2}E_0^Q\left[e^{-\mathrm{i}\omega X_{t_{k-1}}+B(\Theta;\Delta t_k,\mathbf{q}_1)V_{t_{k-1}}+\Gamma(\Theta;\Delta t_k,\mathbf{q}_1)+\Lambda(\Theta;\Delta t_k,\mathbf{q}_1)}\right]\Big|_{\phi=0}\frac{e^{\mathrm{i}u\omega}}{\mathrm{i}\omega}\,\mathrm{d}\omega_r \\
= & \frac{e^{\omega_i(X_0-u)}}{\pi}\int_0^{\infty}\mathrm{Re}\left\{\frac{e^{-\mathrm{i}\omega_r(X_0-u)}}{\mathrm{i}\omega_r-\omega_i}F_k(\omega_r+\mathrm{i}\omega_i)\right\}\mathrm{d}\omega_r, \quad k\geq 2, \qquad (4.62)
\end{aligned}
$$

where $\mathrm{Re}\{\cdot\}$ denotes taking the real part, $u = \ln U$ and

$$
F_k(\omega) = \frac{\partial^2}{\partial\phi^2}e^{B(\Theta;t_{k-1},\mathbf{q}_2)V_0+\Gamma(\Theta;t_{k-1},\mathbf{q}_2)+\Lambda(\Theta;t_{k-1},\mathbf{q}_2)}\Big|_{\phi=0}, \quad k\geq 2, \qquad (4.63)
$$

with $\mathbf{q}_1 = (\phi\ 0\ 0)^T$ and

$$
\mathbf{q}_2 = \begin{pmatrix} -\mathrm{i}\omega \\ B(\Theta;\Delta t_k,\mathbf{q}_1) \\ \Gamma(\Theta;\Delta t_k,\mathbf{q}_1)+\Lambda(\Theta;\Delta t_k,\mathbf{q}_1) \end{pmatrix}.
$$

Putting the results together, the fair strike of the discretely monitored downside variance swap is given by

$$
\begin{aligned}
K_d(0,T;N) = \frac{A}{N}\Bigg[& \frac{\partial^2}{\partial\phi^2}e^{B(\Theta;\Delta t_1,\mathbf{q}_1)V_0+\Gamma(\Theta;\Delta t_1,\mathbf{q}_1)+\Lambda(\Theta;\Delta t_1,\mathbf{q}_1)}\mathbf{1}_{\{X_0\leq \ln U\}}\Big|_{\phi=0} \\
& + \frac{e^{\omega_i(X_0-u)}}{\pi}\int_0^{\infty}\mathrm{Re}\left\{\frac{e^{-\mathrm{i}\omega_r(X_0-u)}}{\mathrm{i}\omega_r-\omega_i}\sum_{k=2}^{N}F_k(\omega_r+\mathrm{i}\omega_i)\right\}\mathrm{d}\omega_r\Bigg]. \qquad (4.64)
\end{aligned}
$$

The evaluation of the Fourier integral in (4.64) can be effected by adopting the fast Fourier transform (FFT) algorithm. Actually, by following a similar FFT calculation approach as in Carr and Madan (1999), one can produce the fair strike prices for all discretely monitored downside variance swaps with varying values of the upper barrier using one single FFT calculation. Similar asymptotic formula for the fair strike of the downside variance swap under the asymptotic limit of continuous sampling of realized variance can be derived [see (3.10) in Zheng and Kwok (2014a)].

Simple return specification

Though most OTC volatility derivative contracts use the log asset returns when computing the realized variance, some variance swap contracts may use the simple returns instead. The mgf approach only requires small modification to accommodate the change of the return type. As an illustration, we discuss the procedure to compute the fair strike of a variance swap on discrete realized variance with simple returns.

It suffices to consider the expectation calculation of a typical term of the terminal payoff:

$$\begin{aligned}
& E_0^Q\left[\left(\frac{S_{t_k}}{S_{t_{k-1}}}-1\right)^2\right] \\
= \;& E_0^Q\left[E^Q\left[e^{2(X_{t_k}-X_{t_{k-1}})}-2e^{(X_{t_k}-X_{t_{k-1}})}+1\middle|X_{t_{k-1}},V_{t_{k-1}}\right]\right] \\
= \;& E_0^Q\left[e^{B(\Theta;\Delta t_k,\mathbf{p}_2)V_{t_{k-1}}+\Gamma(\Theta;\Delta t_k,\mathbf{p}_2)+\Lambda(\Theta;\Delta t_k,\mathbf{p}_2)}\right. \\
& \left.\quad -2e^{B(\Theta;\Delta t_k,\mathbf{p}_1)V_{t_{k-1}}+\Gamma(\Theta;\Delta t_k,\mathbf{p}_1)+\Lambda(\Theta;\Delta t_k,\mathbf{p}_1)}\right]+1 \\
= \;& e^{B(\Theta;t_{k-1},\mathbf{q}_2)V_0+\Gamma(\Theta;t_{k-1},\mathbf{q}_2)+\Lambda(\Theta;t_{k-1},\mathbf{q}_2)} \\
& -2e^{B(\Theta;t_{k-1},\mathbf{q}_1)V_0+\Gamma(\Theta;t_{k-1},\mathbf{q}_1)+\Lambda(\Theta;t_{k-1},\mathbf{q}_1)}+1,
\end{aligned} \tag{4.65}$$

where $\mathbf{p}_j = (j\;0\;0)^T$ and

$$\mathbf{q}_j = \begin{pmatrix} 0 \\ B(\Theta;\Delta t_k,\mathbf{p}_j) \\ \Gamma(\Theta;\Delta t_k,\mathbf{p}_j)+\Lambda(\Theta;\Delta t_k,\mathbf{p}_j) \end{pmatrix}, \quad j=1,2.$$

The above formula does not involve any differentiation operation, so it is even simpler than the computation of the fair strike of the discrete variance swap with log returns. The fair strikes of other exotic variance swaps on discrete realized variance with simple return specification can be derived in a similar manner.

4.3.3 Numerical tests of the convergence for discretely monitored variance swaps

Numerical tests were performed to examine the impact of the sampling frequency of the realized variance on the fair strike prices of the discretely monitored variance swaps, gamma swaps, and corridor variance swaps to examine the convergence of the fair strikes to those of their continuously monitored counterparts. In the numerical tests, we adopted the set of parameter values shown in Table 4.1 that are calibrated to S&P 500 option prices on November 2, 1993 (Duffie *et al.*, 2000) as the basic set of parameter values. In addition, we take $r = 3.19\%$, $q = 0$, $S_0 = 1$ and assume the number of trading days in one year to be 252. Unless otherwise stated, we consider one-year swap contracts so that $A = N$ (equivalent to $T = 1$) and take $U = 1$ as the upper barrier of the downside corridor in the corridor swap.

We present numerical results that explore the convergence behavior of the fair strike prices of different types of discretely monitored generalized variance swaps under varying sampling frequencies. In Table 4.2, we report the fair strike prices of the vanilla variance swaps, gamma swaps, and downside variance swaps with varying sampling frequencies and different values of the correlation coefficient ρ. The values of the fair strike prices are all presented in variance points, which is the expected realized variance multiplied by 100^2. The fair strike prices for these discretely sampled generalized variance swaps are numerically calculated using the closed-form fair

κ	3.46	ν	−0.086
θ	$(0.0894)^2$	η	0.05
ε	0.14	λ	0.47
ρ	−0.82	ρ_J	−0.38
$\sqrt{V_0}$	0.087	δ	0.0001

TABLE 4.1: The basic set of parameter values of the SVSJ model.

Sampling frequency		$N=4$ quarterly	$N=12$ monthly	$N=26$ biweekly	$N=52$ weekly	$N=252$ daily	$N=\infty$ continuous
Variance swaps	$\rho=-1$	187.0839	183.4365	182.2551	181.7172	181.2759	181.1590
	$\rho=-0.82$	186.7823	183.3154	182.1961	181.6870	181.2695	181.1590
	$\rho=-0.3$	185.9113	182.9654	182.0257	181.5998	181.2512	181.1590
Gamma swaps	$\rho=-1$	170.1311	169.2752	169.2176	169.2203	169.2350	169.2407
	$\rho=-0.82$	171.0131	169.9908	169.8749	169.8504	169.8426	169.8423
	$\rho=-0.3$	173.6134	172.0962	171.8081	171.7036	171.6293	171.6113
Downside variance swaps	$\rho=-1$	111.5139	102.5147	101.3211	101.0009	100.8345	100.8043
	$\rho=-0.82$	110.5369	101.0294	99.6504	99.2447	99.0083	98.9599
	$\rho=-0.3$	107.8140	96.8144	94.8855	94.2254	93.7809	93.6779

TABLE 4.2: Comparison of the numerical values of the fair strike prices of variance swaps, gamma swaps, and downside variance swaps with varying sampling frequencies and different values of the correlation coefficient ρ. Here, N is the number of sampling dates within one year.

strike formulas derived earlier [see (4.54), (4.57) and (4.64)]. The fair strike prices of the continuously sampled generalized variance swaps, corresponding to $N=\infty$, are deduced by asymptotic analysis [see (4.55) and (4.58)]. Table 4.2 reveals that the fair strike prices of all variance swap products (with the exception of the continuously sampled variance swaps) have dependence on ρ. The variance swaps and gamma swaps are seen to be less sensitive to ρ for any sampling frequency specification when compared to the downside variance swaps. As $N\to\infty$ (vanishing sampling interval), all the fair strike prices of the discretely sampled generalized variance swaps converge to those of their continuously sampled counterparts. The good agreement between these numerical values of fair strike prices provides a check for accuracy of all the derived analytic formulas of fair strike.

4.3.4 Volatility swaps

The common measure of discrete realized volatility is taken to be the square root of the discrete realized variance $I(0,T;N)$ as defined in (4.1). However, finding the closed-form solution for the fair strike of the volatility swap as defined by $E_0^Q[\sqrt{I(0,T;N)}]$ is analytically intractable, except that we resort to the analytic approximation based on the convexity correction approach (Brockhaus and Long,

2000). The use of the convexity correction is strictly limited to approximation of the fair strike of a volatility swap under the assumption of continuous monitoring of realized variance. The approximation is based on the Taylor series expansion of the square root function, and it may fail badly except under well-controlled conditions that require skillful checking of convergence condition (Zhu and Lian, 2018).

As an alternative analytic approximation method, the saddlepoint approximation method is shown to be reasonably reliable and effective in pricing swaps on discrete realized variance as well as options on discrete realized variance. The saddlepoint method is based on the analytic approximation of the Laplace integral for the price function of the discrete volatility swap. Analytic valuation of the Laplace integral is simplified by deforming the integration contour to pass through a saddlepoint of the integrand function. The effective implementation of the saddlepoint method relies on availability of the analytic form of the moment generating function of the discrete realized variance. Zheng and Kwok (2014c) manage to use the small time asymptotic expansion to derive effective analytic approximation of the moment generating function of the discrete realized variance under stochastic volatility models with jumps and Lévy models. Numerical tests performed by Zheng and Kwok (2014c) reveal promising efficiency and accuracy of the saddlepoint approximation method for pricing discretely monitored variance and volatility derivatives.

Alternatively, there exists another measure of discrete realized volatility over $[0,T]$ as defined by

$$V(0,T;N) = \sqrt{\frac{\pi}{2NT}} \sum_{k=1}^{N} \left| \frac{S_{t_k} - S_{t_{k-1}}}{S_{t_k}} \right|, \tag{4.66}$$

where t_k, $k = 1,2,\ldots,N$ is the kth monitoring instant with $t_0 = 0$ and $t_N = T$. Barndorff-Nielsen and Shephard (2003) performed the actual and empirical studies on this measure of discrete realized volatility and concluded that this is a more robust measure of realized volatility. Also, closed-form solution for the fair strike of volatility swap as defined by $E_0^Q[V(0,T;N)]$ can be obtained. In the literature, volatility swap that is defined based on $V(0,T;N)$ is termed the volatility-average swap.

We follow a similar derivation procedure based on the use of the moment generating function. Assuming the current time to be zero, we let

$$y_{t,T} = \ln \frac{S_T}{S_t}, \quad t < T, \tag{4.67}$$

and define the forward moment generating function $f(\phi;t,T,V_0)$ of the future log return $y_{t,T}$ as

$$f(\phi;t,T,V_0) = E^Q[e^{\phi y_{t,T}} | y_0, V_0], \quad t < T. \tag{4.68}$$

Suppose we assume the underlying asset price process under a risk neutral measure Q follows the Heston model:

$$\begin{aligned} \frac{\mathrm{d}S_t}{S_t} &= r\,\mathrm{d}t + \sqrt{V_t}\,\mathrm{d}W_t^1 \\ \mathrm{d}V_t &= \kappa(\theta - V_t)\,\mathrm{d}t + \varepsilon\sqrt{V_t}\,\mathrm{d}W_t^2, \end{aligned} \tag{4.69}$$

where $dW_t^1 dW_t^2 = \rho\, dt$. Here, κ and θ are the mean reversion speed parameter and long term mean of the instantaneous variance process V_t, respectively, and ε is the volatility of volatility. The corresponding forward moment generating function admits the following affine representation:

$$f(\phi;t,T,V_0) = e^{C(\phi,T-t)} h(D(\phi,T-t);t,V_0) \tag{4.70}$$

where

$$C(\phi,\tau) = r\phi\tau + \frac{\kappa\theta}{\varepsilon^2}\left[(a+b)\tau - 2\ln\frac{1-ge^{b\tau}}{1-g}\right],$$

$$D(\phi,\tau) = \frac{a+b}{\varepsilon^2}\frac{1-e^{b\tau}}{1-ge^{b\tau}},$$

$$a = \kappa - \rho\varepsilon\phi, \quad b = \sqrt{a^2+\varepsilon^2(\phi-\phi^2)}, \quad g = \frac{a+b}{a-b}$$

and

$$h(\phi;t,V) = \exp\left(-\frac{2\kappa\theta}{\varepsilon^2}\ln\left(1+\frac{\varepsilon^2\phi}{2\kappa}(e^{-\kappa t}-1)\right) + \frac{2\kappa\phi}{\varepsilon^2\phi + (2\kappa-\varepsilon^2\phi)e^{\kappa t}}V\right).$$

Let $p(y_{t_{k-1},t_k})$ denote the probability density function of the stochastic variable y_{t_{k-1},t_k} and define Q_k to be the probability of the event $\{y_{t_{k-1},t_k} > 0\}$. It is known that

$$Q_k = \int_0^\infty p(y_{t_{k-1},t_k})\, dy_{t_{k-1},t_k} = \frac{1}{2} + \frac{1}{\pi}\int_0^\infty \mathrm{Re}\left\{\frac{f(i\phi;t_{k-1},t_k,V_0)}{i\phi}\right\} d\phi. \tag{4.71}$$

We define the associated density function

$$q(y_{t_{k-1},t_k}) = e^{(y_{t_{k-1},t_k} - r\Delta t_k)} p(y_{t_{k-1},t_k}), \quad \Delta t_k = t_k - t_{k-1}, \tag{4.72}$$

which is seen to satisfy

$$\int_{-\infty}^\infty q(y_{t_{k-1},t_k})\, dy_{t_{k-1},t_k} = 1 \quad \text{and} \quad q(y_{t_{k-1},t_k}) \geq 0.$$

The corresponding Fourier transform of the probability density function $q(y_{t_{k-1},t_k})$ is found to be

$$\mathcal{F}[e^{(y_{t_{k-1},t_k} - r\Delta t_k)} p(y_{t_{k-1},t_k})] = e^{-r\Delta t_k} f(i\phi+1;t_{k-1},t_k,V_0). \tag{4.73}$$

Similarly, we define

$$\begin{aligned}
\widetilde{Q}_k &= \int_0^\infty q(y_{t_{k-1},t_k})\, dy_{t_{k-1},t_k} \\
&= \int_0^\infty e^{(y_{t_{k-1},t_k} - r\Delta t_k)} p(y_{t_{k-1},t_k})\, dy_{t_{k-1},t_k} \\
&= \frac{1}{2} + \frac{1}{\pi}\int_0^\infty \mathrm{Re}\left\{\frac{e^{-r\Delta t_k} f(i\phi+1;t_{k-1},t_k,V_0)}{i\phi}\right\} d\phi.
\end{aligned} \tag{4.74}$$

Summing the results together, we deduce that

$$\begin{aligned}
&E_0^Q\left[\left|\frac{S_{t_k}}{S_{t_{k-1}}}-1\right|\right] \\
&=\int_{-\infty}^{\infty}|e^{y_{t_{k-1},t_k}}-1|\,p(y_{t_{k-1},t_k})\,\mathrm{d}y_{t_{k-1},t_k} \\
&=\int_0^{\infty}(e^{y_{t_{k-1},t_k}}-1)p(y_{t_{k-1},t_k})\,\mathrm{d}y_{t_{k-1},t_k}+\int_{-\infty}^{0}(1-e^{y_{t_{k-1},t_k}})p(y_{t_{k-1},t_k})\,\mathrm{d}y_{t_{k-1},t_k} \\
&=-\int_0^{\infty}p(y_{t_{k-1},t_k})\,\mathrm{d}y_{t_{k-1},t_k}+\int_{-\infty}^{0}p(y_{t_{k-1},t_k})\,\mathrm{d}y_{t_{k-1},t_k} \\
&\quad+e^{r\Delta t_k}\left(\int_0^{\infty}q(y_{t_{k-1},t_k})\,\mathrm{d}y_{t_{k-1},t_k}-\int_{-\infty}^{0}q(y_{t_{k-1},t_k})\,\mathrm{d}y_{t_{k-1},t_k}\right) \\
&=1-2Q_k+e^{r\Delta t_k}(2\widetilde{Q}_k-1) \\
&=\frac{2}{\pi}\int_0^{\infty}\mathrm{Re}\left\{\frac{f(i\phi+1;t_{k-1},t_k,V_0)-f(i\phi;t_{k-1},t_k,V_0)}{i\phi}\right\}\mathrm{d}\phi.
\end{aligned} \tag{4.75}$$

Lastly, the fair strike K_{vol} of the volatility-average swap is given by (Zhu and Lian, 2015b)

$$\begin{aligned}
&K_{\text{vol}} \\
&=E_0^Q[V(0,T;N)] \\
&=\sqrt{\frac{\pi}{2NT}}E_0^Q\left[\sum_{k=1}^{N}\left|\frac{S_{t_k}}{S_{t_{k-1}}}-1\right|\right] \\
&=\sqrt{\frac{2}{\pi NT}}\int_0^{\infty}\sum_{k=1}^{N}\mathrm{Re}\left\{\frac{f(i\phi+1;t_{k-1},t_k,V_0)-f(i\phi;t_{k-1},t_k,V_0)}{i\phi}\right\}\mathrm{d}\phi.
\end{aligned} \tag{4.76}$$

This pricing approach for finding the fair strike of the volatility-average swap can be extended to stochastic volatility models with regime switching (Elliott and Lian, 2013; He and Zhu, 2019), stochastic volatility, and stochastic interest rate (He and Zhu, 2018) and stochastic volatility with jump and stochastic intensity (Yang *et al.*, 2019b).

4.4 Variance swaps under time-changed Lévy processes

In this section, we consider pricing variance swaps under time-changed Lévy processes. First, we show how Carr *et al.* (2012) use a multiple of log contract for pricing swaps on continuous realized variance under jump dynamics of the underlying asset price process. We illustrate how similar analytic methods discussed in Secs. 4.1 and 4.2 can be extended to price generalized variance swaps under time changed Lévy processes. We also discuss the technical conditions on convergence of the fair strikes of discrete variance swaps to their continuous counterparts.

There are several more recent papers on pricing variance swaps under Lévy processes. Habtemicael and SenGupta (2016) study the arbitrage free pricing of continuous variance and volatility swaps under the Barndorff-Nielsen and Shephard Lévy type process. Tong and Lin (2017) consider pricing generalized variance swaps under a general time homogeneous diffusion process belonging to a symmetric pricing semigroup, time changed by a composition of a Lévy subordinator, and an absolutely continuous process. Zhu and Ruan (2019) consider pricing swaps on discrete realized higher moments under Lévy processes. Yang *et al.* (2019a) consider pricing of discrete variance swaps under stochastic volatility with Lévy jumps and stochastic interest rate in an equilibrium framework. In particular, Carr *et al.* (2021) consider pricing variance swaps on time-changed Markov processes. They show that the variance swap rate equals the price of a co-terminal European-style contract when the underlying is an exponential process, time changed by an arbitrary continuous stochastic clock when certain technical conditions are met.

4.4.1 Multiple of log contract for pricing swaps on continuous realized variance

Recall the relation between the fair strike of a variance swap on continuous realized variance under no jump of the underlying asset price process and the log contract:

$$\frac{1}{T}\int_0^T \sigma_t^2\,\mathrm{d}t = 2E_0^Q\left[-\frac{1}{T}\ln\frac{F_T}{F_0}\right], \tag{4.77}$$

where the multiplier is 2. In this subsection, we would like to show that a multiple of a log contract prices a variance swap under arbitrary exponential Lévy dynamics that is stochastically time changed by an arbitrary continuous clock having arbitrary correlation with the driving Lévy process. The multiplier depends only on the Lévy process, not on the clock; and it becomes 2 under no jump. In the presence of negatively skewed jump risk, Carr *et al.* (2012) show that the multiplier exceeds 2, which agrees with empirical studies.

Recall that $[L,L]_T$ denotes the quadratic variance of the Lévy process L_t over the time period $[0,T]$ and $E[L,L]_T$ is the expectation of $[L,L]_T$. Suppose L_t satisfies $E[e^{L_1}] < \infty$ and $E[L,L]_1 < \infty$. We define the multiplier of L_t by

$$Q_L = \frac{E[L,L]_1}{\ln E[e^{L_1}] - E[L_1]}. \tag{4.78}$$

Suppose the log contract can be properly priced (say using a replicating portfolio of vanilla options), we would like to show that pricing of the continuous variance swap under the time-changed Lévy model reduces to the calculation of the multiplier.

Proposition 4.1 *Let $\kappa_L(u)$ denote the cumulant exponent of the Lévy process L_t, where $\kappa_L(u) = \ln E[e^{uL_1}]$. The multiplier Q_L exists and satisfies*

$$0 < Q_L = \frac{\mathrm{var}(L_1)}{\ln E[e^{L_1}] - E[L_1]} = \frac{\kappa_L''(0)}{\kappa_L(1) - \kappa_L'(0)}. \tag{4.79}$$

Proof *First of all, Q_L is well-defined and positive since $E[e^{L_1}] > e^{E[L_1]}$ by convexity of the exponential function. Define $M_t = L_t - tE[L_1]$. It is seen that M_t is a martingale. By Corollary II.6.3 in Protter (2004), we have*

$$E[L,L]_1 = E[M,M]_1 = E[M_1^2] = \text{var}(L_1).$$

The second representation in (4.79), expressed in terms of $\kappa_L(u)$, follows from the definition of $\kappa_L(u)$.

Suppose the triplet of L_t is given by (μ, σ^2, Π). From Theorem 25.17 in Sato (1999), we obtain

$$\ln E[e^{L_1}] = \frac{\sigma^2}{2} + \int \left(e^x - 1 - x\mathbf{1}_{|x|\leq 1}\right) \Pi(\mathrm{d}x) + \mu.$$

Also, Example 25.12 in Sato (1999) shows that

$$-E[L_1] = -\mu - \int_{|x|>1} x\, \Pi(\mathrm{d}x).$$

It is known that

$$\text{var}(L_1) = \sigma^2 + \int x^2\, \Pi(\mathrm{d}x).$$

Combining these results together and substituting into (4.79), we obtain

$$Q_L = \frac{\sigma^2 + \int x^2\, \Pi(\mathrm{d}x)}{\frac{\sigma^2}{2} + \int (e^x - 1 - x)\, \Pi(\mathrm{d}x)}. \tag{4.80}$$

Let the log of asset price process under a risk neutral measure Q to be governed by

$$\ln \frac{S_t}{S_0} = (r-q)t + L_{Z_t} - \kappa_L(1)Z_t, \tag{4.81}$$

where Z_t is the random time-changed process and L_{Z_t} is the Lévy process generated by the time change of L_t with subordinator Z_t. The family of stopping times Z_t defines a random time change and we normalize the time change such that $E^Q[Z_t] = t$. Also, $e^{-(r-q)t}S_t$ is a Q-martingale. Let $F_t = S_t e^{(r-q)(T-t)}$ be the forward price process. We define

$$U_t = \ln \frac{F_t}{F_0} = \tilde{X}_{Z_t}, \tag{4.82}$$

where

$$\tilde{X}_t = L_t - \kappa_L(1)t = L_t - t\ln E[e^{L_1}].$$

The following proposition shows that the fair strike of the swap on continuous realized variance $\frac{1}{T}E^Q[U,U]_T$ is Q_L times the log contract's forward price $\frac{1}{T}E^Q[-U_T]$.

Proposition 4.2 (Quadratic variation and log contract) *The expectation of the quadratic variation process $[U,U]_T$ can be found to be*

$$E^Q[U,U]_T = Q_L E^Q[-U_T]. \tag{4.83}$$

In particular, the multiplier Q_L does not depend on the time-changed process Z_t.

Proof *First of all, we show that the process $[\tilde{X},\tilde{X}]_t + Q_L\tilde{X}_t$ is a martingale. Recall that $[\tilde{X},\tilde{X}]_t = [L,L]_t$. Hence, for any $s < t$, we have*

$$\begin{aligned} E^Q[[\tilde{X},\tilde{X}]_t + Q_L\tilde{X}_t|\mathcal{F}_s] &= E^Q[[L,L]_t + Q_L(L_t - t\ln E^Q[e^{L_1}])|\mathcal{F}_s] \\ &= (t-s)E^Q[L,L]_1 + (t-s)Q_L(E^Q[L_1] - \ln E^Q[e^{L_1}]) \\ &\quad + [L,L]_s + Q_L(L_s - s\ln E^Q[e^{L_1}]) \\ &= [\tilde{X},\tilde{X}]_s + Q_L\tilde{X}_s, \end{aligned}$$

where the second equality follows from the stationary increment property of a Lévy process and the third equality follows from the definition of Q_L in (4.78). Since $E^Q[Z_T] < \infty$, by Wald's first equation in continuous time, we have

$$E^Q[[\tilde{X},\tilde{X}]_{Z_T} + Q_L\tilde{X}_{Z_T}] = 0.$$

By continuity of Z_t, we have

$$[U,U]_T = [\tilde{X},\tilde{X}]_{Z_T}.$$

Therefore, we obtain

$$E^Q[U,U]_T = -Q_L E^Q[\tilde{X}_{Z_T}] = Q_L E^Q[-U_T].$$

The above result is an important extension of the replication approach discussed in Chapter 1. When the underlying asset price follows a continuous semimartingale process, the multiplier Q_L is equal to 2 [see (1.30)]. Under a general time-changed Lévy process, we have derived a similar replication result in (4.83). In particular, the quadratic variation is replicable using the log contract, except having different multiplier. Moreover, the multiplier is independent of the time-changed process.

By (4.79), Q_L is completely determined by the cumulant exponent of L_t. The following table lists the multipliers of some popular time-changed Lévy models. More examples of multipliers of various time-changed Lévy models can be found in Carr et al. (2012).

4.4.2 Swaps on discrete realized variance

Let L_t be a general Lévy process under a risk neutral measure Q, whose characteristic function $\phi_{L_t}(u)$ and characteristic exponent $\psi_L(u)$ are related by [see (2.52)]

$$\phi_{L_t}(u) = E^Q[e^{iuL_t}] = e^{-t\psi_L(u)}. \tag{4.84}$$

Let the log return of the asset price process S_t under Q be governed by (4.81) and expressed as

$$\ln\frac{S_t}{S_0} = (r-q)t + Y_t. \tag{4.85}$$

We write $X_t = \ln S_t$. The forward characteristic function of X_t is defined by

$$\phi_{t,T}(u) = E_0^Q[\exp(iu(X_T - X_t))] = e^{iu(r-q)t}E_0^Q[\exp(iu(Y_T - T_t))], \tag{4.86}$$

Process	Variance	Lévy density	Multiplier
Brownian	σ^2	0	2
Kou[a]	σ^2	$p\eta_+ e^{-\eta_+ x}\mathbf{1}_{\{x\geq 0\}}+$ $(1-p)\eta_- e^{\eta_- x}\mathbf{1}_{\{x<0\}}$	$\dfrac{\sigma^2+\frac{2p}{\eta_+^2}+\frac{2(1-p)}{\eta_-^2}}{\frac{\sigma^2}{2}+\frac{p}{\eta_+(\eta_+-1)}+\frac{(1-p)}{\eta_-(\eta_-+1)}}$
CGMY[b]	0	$\frac{C}{\lvert x\rvert^{1+Y}}\left(e^{Gx}\mathbf{1}_{\{x<0\}}+e^{-Mx}\mathbf{1}_{\{x\geq 0\}}\right)$	$\dfrac{G^{-2}+M^{-2}}{G^{-1}-\ln(1+G^{-1})-M^{-1}-\ln(1-M^{-1})}$
NIG[c]	0	$\frac{\delta\alpha}{\pi}\frac{\exp(\beta x)I_1(\alpha\lvert x\rvert)}{\lvert x\rvert}$	$\dfrac{\alpha^2/(\alpha^2-\beta^2)}{\alpha^2-\beta^2-\beta-\sqrt{(\alpha^2-\beta^2)[\alpha^2-(\beta+1)^2]}}$

TABLE 4.3: Examples of multipliers for some popular time-changed Lévy processes.
a. $\eta_+ \geq 1$ and $\eta_- > 0$, $p > 0$.
b. $C > 0, G > 0, M > 1$ and $Y < 2$.
c. $\delta > 0$, $\alpha > 0$ and $-\alpha < \beta < \alpha - 1$. Here, I_1 denote the modified Bessel function of the second kind and order 1.

which will be shown to play a vital role in pricing discrete variance swaps under the time-changed Lévy models. To compute $\phi_{t,T}(u)$, Itkin and Carr (2010) introduce a new complex-valued measure $\widetilde{Q}$ that is absolutely continuous with respect to Q and is defined by a complex-valued exponential martingale

$$\left.\frac{\mathrm{d}\tilde{Q}}{\mathrm{d}Q}\right|_T = D_T(u) = \exp\left(iuY_T + \psi_L(u)Z_T\right). \tag{4.87}$$

The optimal stopping theorem ensures that

$$D_t(u) = E_t^Q[D_T(u)] = \exp(iuY_t + \psi_L(u)Z_t) \tag{4.88}$$

is a Q-martingale and

$$E_t^{\tilde{Q}}[Z_T] = E_t^Q\left[\frac{\widetilde{Q}_T}{\widetilde{Q}_t}Z_T\right]. \tag{4.89}$$

As a result, the forward characteristic function of X_t can be expressed as

$$\begin{aligned}
\phi_{t,T}(u) &= e^{iu(r-q)\tau}E_0^Q\left[e^{iu(Y_T-Y_t)}\right] \\
&= e^{iu(r-q)\tau}E_0^Q\left[E_t^Q\left[\frac{D_T(u)}{D_t(u)}e^{-\psi_L(u)(Z_T-Z_t)}\right]\right] \\
&= e^{iu(r-q)\tau}E_0^Q\left[E_t^{\tilde{Q}}\left[e^{-\psi_L(u)(Z_T-Z_t)}\right]\right] \\
&= e^{iu(r-q)\tau}E_0^Q\left[E_t^{\tilde{Q}}\left[e^{-\psi_L(u)\int_t^T v_s\,\mathrm{d}s}\right]\right],
\end{aligned} \tag{4.90}$$

where $\tau = T - t$ and v_t is the activity rate. Suppose v_t is affine under the new measure $\tilde{Q}$, then the Laplace transform of $\int_t^T v_s\,\mathrm{d}s$ under the measure $\tilde{Q}$ is also an exponential affine function in v_t and takes the form

$$E_t^{\tilde{Q}}\left[e^{\psi_L(u)\int_t^T v_s\,\mathrm{d}s}\right] = \exp(\alpha(\tau,\psi_L(u)) + \beta(\tau,\psi_L(u))v_t), \tag{4.91}$$

where $\alpha(\cdot,\cdot)$ and $\beta(\cdot,\cdot)$ are some coefficient functions that can be determined by a Riccati system of ordinary differential equations. This gives (Itkin and Carr, 2010)

$$\phi_{t,T}(u) = e^{iu(r-q)\tau+\alpha(\tau,\psi_L(u))}\phi_{v_t}(-i\beta(\tau,\psi_L(u))), \tag{4.92}$$

where $\phi_{v_t}(\cdot)$ is the characteristic function of v_t under the measure Q.

A typical term in the discrete realized variance can be represented in terms of the forward characteristic function $\phi_{t,T}(u)$ as follows:

$$\begin{aligned} E_0^Q\left[\left(\ln\frac{S_{t_j}}{S_{t_{j-1}}}\right)^2\right] &= \frac{\partial^2}{\partial u^2}E_0^Q\left[e^{u(X_{t_j}-X_{t_{j-1}})}\right]\Big|_{u=0} \\ &= \frac{\partial^2}{\partial u^2}\phi_{t_{j-1},t_j}(-iu)\Big|_{u=0} = -\phi''_{t_{j-1},t_j}(0). \end{aligned} \tag{4.93}$$

According to (4.4), the fair strike of the discrete variance swap is given by

$$K_v(0,T;N) = -\frac{A}{N}\sum_{j=1}^{N}\phi''_{t_{j-1},t_j}(0). \tag{4.94}$$

For the discrete variance swaps with simple returns, a typical term can be expressed in terms of the forward characteristic function $\phi_{t,T}(u)$ as follows:

$$\begin{aligned} E_0^Q\left[\left(\frac{S_{t_j}}{S_{t_{j-1}}}-1\right)^2\right] &= E_0^Q\left[e^{2(X_{t_j}-X_{t_{j-1}})}-2e^{(X_{t_j}-X_{t_{j-1}})}+1\right] \\ &= \phi_{t_{j-1},t_j}(-2i)-2\phi_{t_{j-1},t_j}(-i)+1. \end{aligned} \tag{4.95}$$

The pricing of generalized discrete vanilla swaps will be discussed in Sec.4.4.3.

Like the decomposition approach proposed by Broadie and Jain (2008) and later generalized by Bernard and Cui (2011) under stochastic volatility models (see Sec. 4.1), Carr *et al.* (2012) derive a similar result that links the fair strike of the discrete variance swap to that of its continuous counterpart under the time-changed Lévy models. The decomposition formula is summarized in Proposition 4.3.

Proposition 4.3 (Variance swaps) *Let the asset price process S_t be defined by* (4.81). *Suppose the base Lévy process L_t and the continuous random time process Z_t are independent, then we have*

$$\begin{aligned} &K_v(0,T;N) = K_v(0,T;\infty) \\ &\quad +\frac{1}{T}\sum_{k=1}^{N}\left(E_0^Q\left[\ln\frac{S_{t_k}}{S_{t_{k-1}}}\right]\right)^2+\frac{1}{T}\sum_{k=1}^{N}\mathrm{var}\left(E_0^Q\left[\ln\frac{S_{t_k}}{S_{t_{k-1}}}\Big|\mathcal{F}_T^Z\right]\right), \end{aligned} \tag{4.96}$$

where $\mathcal{F}_T^Z$ is the σ-algebra generated by $\{Z_t : 0 \le t \le T\}$. Moreover, the last term has the explicit form

$$\begin{aligned} \sum_{k=1}^{N}\mathrm{var}\left(E_0^Q\left[\ln\frac{S_{t_k}}{S_{t_{k-1}}}\Big|\mathcal{F}_T^Z\right]\right) &= [E_0^Q[L_1]-\kappa_L(1)]^2\sum_{k=1}^{N}E_0^Q\left[(Z_{t_k}-Z_{t_{k-1}})^2\right] \\ &\quad -\sum_{k=1}^{N}\left(E_0^Q\left[\ln\frac{S_{t_k}}{S_{t_{k-1}}}\right]-(r-q)\Delta t_k\right)^2. \end{aligned} \tag{4.97}$$

As deduced from (4.96), the fair strike of the discrete variance swap is always greater than that of its continuous counterpart since the correction term is seen to be always positive.

Proof *Recall that $M_t = L_t - tE_0^Q[L_1]$ is a Q-martingale and set $Y_t = L_{Z_t}$. For any $t \in [0,T]$, Wald's first equation implies that*

$$\tilde{M}_t = M_{Z_t} = Y_t - Z_t E_0^Q[L_1]$$

is also a Q-martingale. Consider a sequence of monitoring times: $0 = t_0 < t_1 < \cdots < t_N = T$, and for any stochastic process U_t, we write $\Delta U_{t_k} = U_{t_{k+1}} - U_{t_k}$, $k = 0,1,\ldots,N-1$. We then obtain

$$\begin{aligned} E_0^Q[\Delta[\ln S, \ln S]_{t_k}] &= E_0^Q[\Delta[Y,Y]_{t_k}] = E_0^Q[\Delta[\tilde{M},\tilde{M}]_{t_k}] \\ &= E_0^Q[(\Delta\tilde{M}_{t_k})^2] = E_0^Q[(\Delta Y_{t_k} - \Delta Z_{t_k} E[L_1])^2]. \end{aligned}$$

The second equality follows from the fact that $\tilde{M}_t - Y_t = Z_t E[L_1]$ has finite variation, as Z_t is assumed to be a nondecreasing process. Therefore, the quadratic variations of $\tilde{M}$ and Y_t are equal. The third equality follows from Corollary 27.3 in Protter (2004). By the independence condition, we have

$$E_0^Q[\Delta Y_{t_k} | \mathcal{F}_T^Z] = \Delta Z_{t_k} E_0^Q[L_1].$$

As a result, we obtain

$$\begin{aligned} E^Q[\Delta[\ln S, \ln S]_{t_k}] &= E_0^Q[(\Delta Y_{t_k} - E^Q[\Delta Y_{t_k} | \mathcal{F}_T^Z])^2] \\ &= E_0^Q[\text{var}(\Delta Y_{t_k} | \mathcal{F}_T^Z)] \\ &= E_0^Q[\text{var}((r-q)\Delta t_k + \Delta Y_{t_k} - \kappa_L(1)\Delta Z_{t_k} | \mathcal{F}_T^Z)] \\ &= \text{var}((r-q)\Delta t_k + \Delta Y_{t_k} - \kappa_L(1)\Delta Z_{t_k}) \\ &\quad - \text{var}(E_0^Q[(r-q)\Delta t_k + \Delta Y_{t_k} - \kappa_L(1)\Delta Z_{t_k} | \mathcal{F}_T^Z]) \\ &= E_0^Q\left[(\Delta \ln S_{t_k})^2\right] - \left(E_0^Q\left[\Delta \ln S_{t_k}\right]\right)^2 - \text{var}\left(E_0^Q\left[\Delta \ln S_{t_k} \middle| \mathcal{F}_T^Z\right]\right). \end{aligned}$$

Finally, summing from $k = 1$ to N, rearranging the terms and multiplying the annualizing factor, we obtain (4.96). *To prove (4.97), we first obtain*

$$E_0^Q[\Delta Y_{t_k} - \kappa_L(1)\Delta Z_{t_k} | \mathcal{F}_T^Z] = (E_0^Q[L_1] - \kappa_L(1))\Delta Z_{t_k},$$

which implies that

$$\begin{aligned} &\text{var}(E^Q[\Delta \ln S_{t_k} | \mathcal{F}_T^Z]) \\ = &\ [E_0^Q[L_1] - \kappa_L(1)]^2 \text{var}(\Delta Z_{t_k}) \\ = &\ [E_0^Q[L_1] - \kappa_L(1)]^2 E^Q[\Delta Z_{t_k}^2] - [E_0^Q[E^Q[L_1]\Delta Z_{t_k} - \kappa_L(1)\Delta Z_{t_k}]]^2 \\ = &\ [E_0^Q[L_1] - \kappa_L(1)]^2 E^Q[\Delta Z_{t_k}^2] - [E_0^Q[\Delta \ln S_{t_k}] - (r-q)\Delta t_k]^2. \end{aligned}$$

By summing the above equation from $k = 1$ to N, we obtain (4.97).

4.4.3 Generalized variance swaps

We would like to extend the decomposition approach to price generalized variance swaps (Crosby and Davis, 2011). We define the generalized joint forward characteristic function as follows:

$$\Psi_k(u_1,u_2,u_3) = E_0^Q\left[\exp\left(iu_1\ln\frac{S_{t_{k-1}}}{S_0}+iu_2\ln\frac{S_{t_k}}{S_{t_{k-1}}}+iu_3\ln\frac{S_T}{S_{t_k}}\right)\right],\ k=1,2,\ldots,N. \tag{4.98}$$

By differentiating $\Psi_k(u_1,u_2,u_3)$ with respect to u_2 twice, we obtain

$$-\frac{\partial^2\Psi_k(u_1,u_2,u_3)}{\partial u_2^2} = E_0^Q\left[\left(\ln\frac{S_{t_k}}{S_{t_{k-1}}}\right)^2\exp\left(iu_1\ln\frac{S_{t_{k-1}}}{S_0}+iu_2\ln\frac{S_{t_k}}{S_{t_{k-1}}}+iu_3\ln\frac{S_T}{S_{t_k}}\right)\right]. \tag{4.99}$$

By letting (u_1,u_2,u_3) be $(0,0,0)$, $(-i,-i,0)$ and $(-i,-i,-i)$, respectively, we recover the respective individual term in evaluating the expected payoffs of the variance swap, gamma swap, and self-quantoed variance swap. The joint characteristic function is closely related to the expected payoffs of the generalized variance swaps through its partial derivatives.

Recall that the asset price process under Q is governed by

$$\ln\frac{S_t}{S_0} = (r-q)t + L_{Z_t} - \psi_L(-i)Z_t \tag{4.100}$$

and

$$D_t = \exp(iuL_{Z_t}+\psi_L(u)Z_t),$$

we obtain

$$\exp\left(iu_2\ln\frac{S_{t_k}}{S_{t_{k-1}}}\right) = \frac{D_{t_k}(u_2)}{D_{t_{k-1}}(u_2)}\exp\left(iu_2(r-q)\Delta t_k+[\psi_L(u_2)-iu_2\psi_L(-i)](Z_{t_k}-Z_{t_{k-1}})\right),$$

where $D_t(u)$ is defined in (4.88). Also, we write

$$\left(\ln\frac{S_{t_k}}{S_{t_{k-1}}}\right)^2 = a_k(u_2)+b_k(u_2)+c_k(u_2)+d_k(u_2), \tag{4.101}$$

where

$$\begin{aligned}
a_k(u_2) &= [p_k(u_2)]^2,\\
b_k(u_2) &= 2p_k(u_2)q_k(u_2),\\
c_k(u_2) &= [q_k(u_2)]^2+\psi_L''(u_2)(Z_{t_k}-Z_{t_{k-1}}),\\
d_k(u_2) &= -\psi_L''(u_2)(Z_{t_k}-Z_{t_{k-1}}),\\
p_k(u_2) &= (r-q)(t_k-t_{k-1})-[i\psi_L'(u_2)+\psi_L(-i)](Z_{t_k}-Z_{t_{k-1}}),\\
q_k(u_2) &= Y_{t_k}-Y_{t_{k-1}}+i\psi_L'(u_2)(Z_{t_k}-Z_{t_{k-1}}).
\end{aligned}$$

The motivation of the decomposition in (4.101) is related to the martingale property of $D_t(u)$, where

$$E^Q_{t_{k-1}}\left[\frac{D_{t_k}(u)}{D_{t_{k-1}}(u)}\right] = 1. \tag{4.102}$$

By differentiating (4.102) once and twice, respectively, we obtain

$$E^Q_{t_{k-1}}\left[\frac{D_{t_k}(u)}{D_{t_{k-1}}(u)}q_k(u)\right] = 0,$$
$$E^Q_{t_{k-1}}\left[\frac{D_{t_k}(u)}{D_{t_{k-1}}(u)}c_k(u)\right] = 0.$$

As a result, summing (4.99) from $k=1$ to N gives

$$\begin{aligned}-\sum_{k=1}^{N}\frac{\partial^2\Psi_k(u_1,u_2,u_3)}{\partial u_2^2} &= \Omega_1(iu_1,iu_2,iu_3;N)+\Omega_2(iu_1,iu_2,iu_3;N)\\ &\quad +\Omega_3(iu_1,iu_2,iu_3;N)+\Omega_4(iu_1,iu_2,iu_3;N),\end{aligned} \tag{4.103}$$

where the four terms are given by (rewriting ϕ_k as iu_k, $k=1,2,3$)

$$\begin{aligned}\Omega_1(\phi_1,\phi_2,\phi_3;N) &= \sum_{k=1}^{N}\exp((r-q)(t_k-t_{k-1})\phi_2)E^Q_0\Bigg[\exp\left(\phi_1\ln\frac{S_{t_{k-1}}}{S_0}\right)\\ &\quad E^Q_{t_{k-1}}\left[\frac{D_{t_k}(-i\phi_2)}{D_{t_{k-1}}(-i\phi_2)}\exp\left((\psi_L(-i\phi_2)-\phi_2\psi_L(-i))(Z_{t_k}-Z_{t_{k-1}})\right)\right.\\ &\quad \left.\left. a_k(-i\phi_2)E^Q_{t_k}\left[\exp\left(\phi_3\ln\frac{S_T}{S_{t_k}}\right)\right]\right]\right],\\ \Omega_2(\phi_1,\phi_2,\phi_3;N) &= \sum_{k=1}^{N}\exp((r-q)(t_k-t_{k-1})\phi_2)E^Q_0\Bigg[\exp\left(\phi_1\ln\frac{S_{t_{k-1}}}{S_0}\right)\\ &\quad E^Q_{t_{k-1}}\left[\frac{D_{t_k}(-i\phi_2)}{D_{t_{k-1}}(-i\phi_2)}\exp\left((\psi_L(-i\phi_2)-\phi_2\psi_L(-i))(Z_{t_k}-Z_{t_{k-1}})\right)\right.\\ &\quad \left.\left. b_k(-i\phi_2)E^Q_{t_k}\left[\exp\left(\phi_3\ln\frac{S_T}{S_{t_k}}\right)\right]\right]\right],\\ \Omega_3(\phi_1,\phi_2,\phi_3;N) &= \sum_{k=1}^{N}\exp((r-q)(t_k-t_{k-1})\phi_2)E^Q_0\Bigg[\exp\left(\phi_1\ln\frac{S_{t_{k-1}}}{S_0}\right)\\ &\quad E^Q_{t_{k-1}}\left[\frac{D_{t_k}(-i\phi_2)}{D_{t_{k-1}}(-i\phi_2)}\exp\left((\psi_L(-i\phi_2)-\phi_2\psi_L(-i))(Z_{t_k}-Z_{t_{k-1}})\right)\right.\\ &\quad \left.\left. c_k(-i\phi_2)E^Q_{t_k}\left[\exp\left(\phi_3\ln\frac{S_T}{S_{t_k}}\right)\right]\right]\right],\end{aligned}$$

$$\begin{aligned}\Omega_4(\phi_1,\phi_2,\phi_3;N) &= \sum_{k=1}^{N}\exp((r-q)(t_k-t_{k-1})\phi_2)E_0^Q\Bigg[\exp\left(\phi_1\ln\frac{S_{t_{k-1}}}{S_0}\right)\\ &\quad E_{t_{k-1}}^Q\left[\frac{D_{t_k}(-i\phi_2)}{D_{t_{k-1}}(-i\phi_2)}\exp\left((\psi_L(-i\phi_2)-\phi_2\psi_L(-i))(Z_{t_k}-Z_{t_{k-1}})\right)\right.\\ &\quad \left.d_k(-i\phi_2)E_{t_k}^Q\left[\exp\left(\phi_3\ln\frac{S_T}{S_{t_k}}\right)\right]\right]\Bigg].\end{aligned}$$

The auxiliary functions $\Omega_i(\phi_1,\phi_2,\phi_3;N)$, $i=1,\ldots,4$, defined in the above observe the following properties:

1. Adding the kth summands of the four functions together gives $-\frac{\partial^2\Psi_k(u_1,u_2,u_3)}{\partial u_2^2}$ evaluated at $(-i\phi_1,-i\phi_2,-i\phi_3)$.
2. For any $\phi_2\in\mathbb{R}$, $a_k(-i\phi_2)\geq 0$, due to the fact that $p_k(-i\phi_2)$ is real. As a result, $\Omega_1(\phi_1,\phi_2,\phi_3;N)$ is larger than or equal to zero, with equality only in the case $p_k(-i\phi)=0$.
3. $\Omega_2(\phi_1,\phi_2,\phi_3;N)$ and $\Omega_3(\phi_1,\phi_2,\phi_3;N)$ vanish if the random time process Z_t is independent of L_t. To see this, we may evaluate the expectations in Ω_2 and Ω_3 by conditioning on the path of Z_t, $0\leq t\leq T$ and then using the identities in (4.103). Furthermore, $\Omega_3(\phi_1,\phi_2,\phi_3;N)$ also vanishes when $\phi_2=0,1$ and $\phi_3=0,1$. Apparently, $\psi_L(-i\phi_2)-\phi_2\psi_L(-i)$ vanishes when $\phi_2=0,1$ and $E_{t_k}^Q\left[\exp\left(\phi_3\ln\frac{S_T}{S_{t_k}}\right)\right]$ is constant when $\phi_3=0,1$. Then, the conclusion follows by using (4.103).
4. $\Omega_4(\phi_1,\phi_2,\phi_3;N)$ is strictly positive. To see this, it suffices to show that

   $$-\psi_L''(-i\phi_2)>0$$

   for any $\phi_2\in\mathbb{R}$. By definition, we obtain

   $$-\psi_L''(-i\phi_2)=E^Q\left(\frac{e^{\phi_2L_1}}{E^Q[e^{\phi_2L_1}]}L_1^2\right)-\left(E^Q\left[\frac{e^{\phi_2L_1}}{E^Q[e^{\phi_2L_1}]}L_1\right]\right)^2. \quad (4.104)$$

   Apparently, the RHS of the above equation can be regarded as the variance of L_1 under the new probability measure $\hat{Q}$ for which the Radon-Nikodym derivative is given by

   $$\frac{d\hat{Q}}{dQ}=\frac{e^{\phi_2L_1}}{E^Q[e^{\phi_2L_1}]}.$$

The fair strikes of various generalized variance swaps can now be represented in terms of these auxiliary functions.

Proposition 4.4 (Variance swaps) *The fair strike of the variance swap is given by*

$$\begin{aligned}K_v(0,T;N) &= \frac{1}{T}(\Omega_1(0,0,0;N)+\Omega_2(0,0,0;N)+\Omega_4(0,0,0;N))\\ &= \frac{1}{T}E_0^Q\left[\sum_{k=1}^{N}E_{t_{k-1}}^Q[a_k(0)+b_k(0)]\right]-\frac{1}{T}\psi_L''(0)E_0^Q[Z_T-Z_0]. \quad (4.105)\end{aligned}$$

Proof *The proof follows directly by letting* $(u_1, u_2, u_3) = (0,0,0)$ *in (4.103). After simplification, we have*

$$\begin{aligned}
\Omega_1(0,0,0;N) &= E_0^Q\left[\sum_{k=1}^{N} E_{t_{k-1}}^Q[a_k(0)]\right],\\
\Omega_2(0,0,0;N) &= E_0^Q\left[\sum_{k=1}^{N} E_{t_{k-1}}^Q[b_k(0)]\right],\\
\Omega_4(0,0,0;N) &= -\psi_L''(0)E^Q[Z_T - Z_0].
\end{aligned}$$

By the third remark above, we obtain $\Omega_3(0,0,0;N) = 0$.

It will be shown later that the second term in (4.105) is indeed the continuous fair strike, that is,

$$K_v(0,T;\infty) = -\frac{1}{T}\psi_L''(0)E^Q[Z_T - Z_0].$$

By direct calculations, it can be shown that (4.96) is a special case of (4.105) under the independence assumption. In fact, under the assumption of independence between L_t and Z_t, we have

$$\begin{aligned}
E_0^Q[E_{t_{k-1}}^Q[q_k(0)]] &= E_0^Q[E_{t_{k-1}}^Q[L_{Z_{t_k}} - L_{Z_{t_{k-1}}} + i\psi_L'(0)(Z_{t_k} - Z_{t_{k-1}})|\mathcal{F}_T^Z]]\\
&= E_0^Q[-i\psi_L'(0)(Z_{t_k} - Z_{t_{k-1}}) + i\psi_L'(0)(Z_{t_k} - Z_{t_{k-1}})] = 0.
\end{aligned}$$

Therefore, we obtain

$$\begin{aligned}
E_0^Q[E_{t_{k-1}}^Q[b_k(0)]] &= E_0^Q[E_{t_{k-1}}^Q[2p_k(0)q_k(0)|\mathcal{F}_T^Z]]\\
&= E_0^Q[2p_k(0)E_{t_{k-1}}^Q[q_k(0)|\mathcal{F}_T^Z]] = 0.
\end{aligned}$$

It suffices to show that

$$E_0^Q[a_k(0)] = \left(E_0^Q\left[\ln\frac{S_{t_k}}{S_{t_{k-1}}}\right]\right)^2 + \mathrm{var}\left(E_0^Q\left[\ln\frac{S_{t_k}}{S_{t_{k-1}}}\middle|\mathcal{F}_T^Z\right]\right). \tag{4.106}$$

This is valid since the RHS of (4.106) can be rewritten as

$$\begin{aligned}
RHS &= E_0^Q\left[\left(E_0^Q\left[\ln\frac{S_{t_k}}{S_{t_{k-1}}}\middle|\mathcal{F}_T^Z\right]\right)^2\right]\\
&= E_0^Q\left[\left\{(r-q)\Delta t_k - \phi_L(-i)(Z_{t_k} - Z_{t_{k-1}})E^Q\left[Y_{t_k} - Y_{t_{k-1}}\middle|\mathcal{F}_T^Z\right]\right\}^2\right]\\
&= E^Q\left[\left\{(r-q)\Delta t_k - [i\phi_L'(0) + \phi_L(-i)](Z_{t_k} - Z_{t_{k-1}})\right\}^2\right]\\
&= E_0^Q[a_k(0)].
\end{aligned}$$

From the above argument, one can see that the correlation between L_t and Z_t affects the fair strike through the term involving $E_0^Q[b_k(0)]$. Further analysis of $E_0^Q[b_k(0)]$ implies that $E_0^Q[b_k(0)]$ becomes larger when the correlation becomes more negative. To see this, by Jensen's inequality, we have

$$E^Q[e^{L_1}] > e^{E^Q[L_1]} \Rightarrow e^{\psi_L(-i)} > e^{-i\psi_L'(0)} \Rightarrow i\psi_L'(0) + \psi_L(-i) > 0.$$

As a result, by omitting all other terms that are independent of the correlation, we conclude that

$$E_0^Q[b_k(0)] = -2[i\psi_L'(0)+\psi_L(-i)]E_0^Q[(Z_{t_k}-Z_{t_{k-1}})(Y_{t_k}-Y_{t_{k-1}})]+\cdots,$$

is a decreasing function of the correlation. Hence, $E_0^Q[b_k(0)] \geq 0$ under negative correlation between L_t and Z_t. Furthermore, by the second remark of the auxiliary functions, $a_k(0) \geq 0$. Therefore, (4.105) indicates that the correction term is always positive and is larger under negative correlation than under zero correlation.

Proposition 4.5 (Self-quantoed variance swaps and gamma swaps) *The respective fair strike of the self-quantoed variance swap and gamma swap are given by*

$$K_{sq}(0,T;N) = \frac{1}{T}[\Omega_1(1,1,1;N)+\Omega_2(1,1,1;N)+\Omega_4(1,1,1;N)], \quad (4.107)$$

$$K_g(0,T;N) = \frac{1}{T}[\Omega_1(0,1,0;N)+\Omega_2(0,1,0;N)+\Omega_4(0,1,0;N)]. \quad (4.108)$$

Note that for discrete variance swaps, self-quantoed variance swaps and gamma swaps, $\Omega_4(\phi_1,\phi_2,\phi_3;N)$ corresponds to the fair strike of their respective continuous counterpart; $\Omega_3(\phi_1,\phi_2,\phi_3;N)$ is identically equal to zero; $\Omega_1(\phi_1,\phi_2,\phi_3;N)$ is nonnegative and related to the discretization effect, while $\Omega_2(\phi_1,\phi_2,\phi_3;N)$ vanishes under the independence assumption. Therefore, it can be interpreted as the term that captures the correlation between the base Lévy process and the time changing process. For the variance swap, we have shown that $\Omega_2(\phi_1,\phi_2,\phi_3;N)$ is a decreasing function of the correlation due to a negative coefficient in front of the cross term. In a similar manner, one can observe that for the self-quantoed variance swaps and gamma swaps, the coefficient becomes $-[i\psi_L'(-i)+\psi_L(-i)]$ and is positive. To see this, we have

$$-i\psi_L'(-i)e^{\psi_L(-i)} = E^Q[e^{L_1}L_1] > E^Q[e^{L_1}]\ln E^Q[e^{L_1}] = e^{\psi_L(-i)}\psi_L(-i),$$

where we have applied Jensen's inequality to the convex function $f(x)=x\ln x$ in the second step. Hence, for the self-quantoed variance swaps and gamma swaps, their fair strikes are increasing functions of the correlation.

The fair strikes of other exotic variance swaps can also be represented in terms of the extended joint characteristic function. Their formulas are summarized in the following propositions.

Proposition 4.6 (Proportional variance swaps) *The fair strike of the proportional variance swap (variance swap with simple returns) is given by*

$$K_p(0,T;N) = \frac{1}{T}\sum_{k=1}^{N}(\Psi_k(0,-2i,0)-2\Psi_k(0,-i,0)+1). \quad (4.109)$$

Proposition 4.7 (Skewness swaps) *The fair strike of the skewness swap is given by*

$$K_s(0,T;N) = \frac{i}{T}\sum_{k=1}^{N}\left.\frac{\partial^3\Psi_k(u_1,u_2,u_3)}{\partial u_2^3}\right|_{(u_1,u_2,u_3)=(0,0,0)}. \quad (4.110)$$

4.4.4 Convergence of fair strikes

We explore the convergence of the fair strikes of various discretely sampled generalized variance swaps under the time-changed Lévy models in the limit $N \to \infty$, where N is the number of monitoring instants. Without loss of generality, we assume

$$E^Q_{t_{k-1}}[Z_{t_k} - Z_{t_{k-1}}] \sim O(N^{-1}),$$
$$E^Q_{t_{k-1}}[(Z_{t_k} - Z_{t_{k-1}})^2] \sim O(N^{-2}),$$
$$E^Q[(E^Q_{t_{k-1}}[q_k(0)|\mathcal{F}^Z_T])^2] \sim O(N^{-2}).$$

Proposition 4.8 *The fair strike of the continuously sampled variance swap under the time-changed Lévy model is given by*

$$K_v(0,T;\infty) = -\frac{1}{T}\phi_L^{(2)}(0)E[Z_T - Z_0]. \tag{4.111}$$

In other words, as $N \to \infty$, we have

$$K_v(0,T;N) - K_v(0,T;\infty) = \frac{1}{T}E^Q\left[\sum_{k=1}^{N} E^Q_{t_{k-1}}\left[a_k(0) + b_k(0)\right]\right] \sim O(N^{-1}).$$

Proof *It suffices to consider the convergence of $\Omega_1(0,0,0;N)$, $\Omega_2(0,0,0;N)$ and $\Omega_4(0,0,0;N))$ step by step. We consider*

$$\begin{aligned}
\Omega_1(0,0,0;N) &= \sum_{k=1}^{N} E^Q[E^Q_{t_{k-1}}[a_k(0)]] \\
&= \sum_{k=1}^{N}(r-q)^2\Delta t_k^2 + (i\psi_L'(0) + \psi_L(-i))^2 \sum_{k=1}^{N} E^Q[E^Q_{t_{k-1}}[(Z_{t_k} - Z_{t_{k-1}})^2]] \\
&\quad -(r-q)(i\psi_L'(0) + \psi_L(-i)) \sum_{k=1}^{N} \Delta t_k E^Q[E^Q_{t_{k-1}}[Z_{t_k} - Z_{t_{k-1}}]] \\
&\sim \sum_{k=1}^{N} O(N^{-2}) \sim O(N^{-1}), \\
\Omega_2(0,0,0;N) &= \sum_{k=1}^{N} E^Q[E^Q_{t_{k-1}}[b_k(0)|\mathcal{F}^Z_T]] \\
&= 2\sum_{k=1}^{N} E^Q[p_k(0)E^Q_{t_{k-1}}[q_k(0)|\mathcal{F}^Z_T]] \\
&\leq 2\sum_{k=1}^{N} \left\{E^Q[p_k^2(0)]E^Q[(E^Q_{t_{k-1}}[q_k(0)|\mathcal{F}^Z_T])^2]\right\}^{1/2} \\
&\sim O(N^{-1}),
\end{aligned}$$

where the inequality used in the last but second step is a result of the Cauchy-Schwarz inequality. So far, we have shown that

$$\Omega_1(0,0,0;N) + \Omega_2(0,0,0;N) \sim O(N^{-1}).$$

Apparently, $\Omega_4(0,0,0;N)$, which is seen to be independent of N, corresponds to the continuous limit of $K_v(0,T;N)$ as $N \to \infty$.

From Proposition 4.8, the continuous fair strike is seen to be dependent on the Lévy process via $\phi_L^{(2)}(0)$ and on the time-changed process via $E^Q[Z_T - Z_0]$.

Proposition 4.9 (Self-quantoed variance swaps and gamma swaps) *The fair strikes of the continuously sampled self-quantoed variance swap and gamma swap under the time-changed Lévy models are given by*

$$K_{sq}(0,T;\infty) = -\frac{1}{T}\psi_L''(-i)E^Q\left[\frac{S_T}{S_0}(Z_T - Z_0)\right], \tag{4.112}$$

$$K_g(0,T;\infty) = -\frac{1}{T}\psi_L''(-i)E^Q[Z_T - Z_0]. \tag{4.113}$$

Furthermore, as $N \to \infty$,

$$K_{sq}(0,T;N) - K_{sq}(0,T;\infty) \sim O(N^{-1}),$$
$$K_g(0,T;N) - K_g(0,T;\infty) \sim O(N^{-1}).$$

Proof *We consider the self-quantoed variance swap case. Like the variance swap, it suffices to consider the convergence of $\Omega_1(1,1,1;N)$, $\Omega_2(1,1,1;N)$ and $\Omega_4(0,0,0;N)$ as $N \to \infty$. Since $D_t(-i)$ is a Q-martingale, we may change the probability measure to $\tilde{Q}$ for which the Radon-Nykodym derivative is defined by [see (4.87)]*

$$\left.\frac{d\tilde{Q}}{dQ}\right|_{\mathcal{F}_t} = D_t(-i).$$

We then have

$$\Omega_1(1,1,1;N) = \sum_{k=1}^{N} e^{(r-q)(T-t_{k-1})}E^Q\left[\frac{S_{t_{k-1}}}{S_0}E_{t_{k-1}}^{\tilde{Q}}[a_k(-i)]\right],$$

where we have used the martingale property of $S_t e^{-(r-q)t}$ as follows:

$$E_{t_k}^Q\left[\exp\left(\ln\frac{S_T}{S_{t_k}}\right)\right] = \frac{E_{t_k}^Q[S_T]}{S_{t_k}} = e^{(r-q)(T-t_k)}.$$

By making similar assumptions as in (4.111) under the new measure $\tilde{Q}$, one can show that $\Omega_1(1,1,1;N) \sim O(N^{-1})$. Similarly, $\Omega_2(1,1,1;N)$ can be shown to be bounded by a term of order $O(N^{-1})$ under some proper technical assumptions. As for the last term, we have

$$\begin{aligned}\Omega_4(1,1,1;N) &= \sum_{k=1}^{N} E_0^Q\left[\frac{S_T}{S_0}[-\psi_L''(-i)(Z_{t_k} - Z_{t_{k-1}})]\right] \\ &= -\psi_L''(-i)E_0^Q\left[\frac{S_T}{S_0}(Z_T - Z_0)\right].\end{aligned}$$

The proof of the result for the gamma swap in the proposition can be done in a similar manner.

Proposition 4.10 (Proportional variance swaps and skewness swaps) *The fair strikes of the continuously sampled proportional variance swap and skewness swap are given by*

$$K_p(0,T;\infty) = \frac{1}{T}[\psi_L(-2i) - 2\psi_L(-i)]E^Q[Z_T - Z_0], \quad (4.114)$$

$$K_s(0,T;\infty) = \frac{i}{T}\psi_L'''(0)E^Q[Z_T - Z_0]. \quad (4.115)$$

Furthermore, as $N \to \infty$,

$$K_p(0,T;N) - K_p(0,T;\infty) \sim O(N^{-1}),$$
$$K_s(0,T;N) - K_s(0,T;\infty) \sim O(N^{-1}).$$

The proof of this proposition requires some additional tedious technical assumptions, the details of which can be found in Crosby and Davis (2011).

4.4.5 Conditions on convergence in expectation

The discrete realized variance converges to its continuous counterpart in probability, but this does not guarantee convergence in expectation. To ensure the convergence of the discrete fair strikes to the continuous counterparty, we explore the technical condition that allows the exchange of the limit operator (the discrete sampling period goes to zero) with an expectation operator.

Consider a semimartingale M_t such that the asset price process is given by

$$S_t = S_0 Y_t,$$

where Y_t satisfies

$$\mathrm{d}Y_t = Y_t \, \mathrm{d}M_t.$$

For simplicity, suppose S_t is a positive local martingale under the risk neutral measure Q, then it can be shown that M_t is also a local martingale. We write $M_t = M_t^c + M_t^d$, where M_t^c is the continuous local martingale part and M_t^d is the jump part of M_t. Let $\mu = \mu(\omega; \mathrm{d}t, \mathrm{d}x)$ be the random measure associated with the jumps of M_t and ν be the predictable compensator of μ. We write

$$M_t^d = x * (\mu - \nu) = \int_0^t \int x(\mu(\mathrm{d}t, \mathrm{d}x) - \nu(\mathrm{d}t, \mathrm{d}x)). \quad (4.116)$$

The asset price process can be represented as

$$S_t = S_0 \exp\left(M_t - \frac{1}{2}\langle M^c, M^c\rangle_t - (x - \ln(1+x)) * \mu\right), \quad (4.117)$$

where $\langle M^c, M^c\rangle_t$ is the unique predictable increasing process such that $\langle M^c, M^c\rangle_0 = 0$ and $(M_t^c)^2 - \langle M^c, M^c\rangle_t$ is a local martingale. Under this semimartingale model setting, we consider the convergence of the realized variance and realized volatility toward their continuous counterparts in L^1. For convenience, we denote

$$\begin{aligned} P_n(T) &= \sum_{k=1}^{n}\left(\ln\frac{S_{t_k}}{S_{t_{k-1}}}\right)^2, \quad P(T)=[\ln S,\ln S]_T, \\ V_n(T) &= \sqrt{P_n(T)}, \quad V(T)=\sqrt{P(T)}. \end{aligned} \tag{4.118}$$

Finiteness of expectation

One preliminary condition for the realized variance to converge to its continuous counterpart in L^1 is that both have finite expectation. Given that $E[P(T)]$ is known to be finite, the necessary and sufficient conditions for $E[P_n(T)]$ to be finite are summarized in Theorems 4.1, 4.2, and 4.3. The proof of these theorems can be found in Jarrow *et al.* (2013).

Theorem 1 *Assume that $P(T)\in L^1$. The following statements are equivalent:*

(i) $P_n(T)\in L^1$ *for at least one* $n\geq 1$;

(ii) $\begin{cases} \langle M^c,M^c\rangle_T \in L^2, \\ (x-\ln(1+x))*\nu_T \in L^2. \end{cases}$

It can be shown that the above theorem also implies that if $P_n(T)\in L^1$ for some n then $P_n(T)\in L^1$ for all $n\geq 1$.

Theorem 2 *Assume that $V(T)\in L^1$. The following statements are equivalent:*

(i) $V_n(T)\in L^1$ *for at least one* $n\geq 1$;

(ii) $\begin{cases} \langle M^c,M^c\rangle_T \in L^2, \\ (x-\ln(1+x))*\nu_T \in L^2. \end{cases}$

When S_t is continuous, we have $[\ln S,\ln S]=\langle M^c,M^c\rangle=\langle M,M\rangle$. Moreover, we have the following modified version of Theorem 1.

Theorem 3 *Assume that S_t is continuous. The following statements are equivalent:*

1. $M_t\in L^1$ *and* $P_n(T)\in L^1$;
2. $P(T)\in L^2$.

From the above theorem, we observe that $E[P_n(T)]<\infty$ is a stronger requirement than $E[P(T)]<\infty$.

Convergence in expectation

We assume that the probability space supports an m-dimensional Brownian motion $\mathbf{W} = (W^1, \cdots, W^m)$ and a Poisson random measure $\mu = \mu(\mathrm{d}t, \mathrm{d}z)$ on $\mathbb{R}_+ \times \mathbb{R}$ with intensity measure $\mathrm{d}t \times F(\mathrm{d}z)$, where $\in (z^2 \wedge 1)F(\mathrm{d}z) < \infty$. Moreover, we assume that M^c is a stochastic integral with respect to $\mathbf{W}$ and M^d is a stochastic integral with respect to $\mu - \mathrm{d}t \times F(\mathrm{d}z)$. That is, we assume that there are predictable processes $\sigma^1, \cdots, \sigma^m$ and a predictable function $\psi > -1$ such that

$$M_t^c = \sum_{k=1}^{m} \int_0^t \sigma_s^k \, \mathrm{d}W_s^k \quad \text{and} \quad M_t^d = \psi * (\mu - \mathrm{d}t \times F(\mathrm{d}z))_t.$$

It is known from a classical result that there exists a one-dimensional Brownian motion B and a nonnegative predictable process σ such that

$$M_t^c = \int_0^t \sigma_s \, \mathrm{d}B_s. \tag{4.119}$$

The following theorem gives the conditions on σ and ψ such that the expectation of $P_n(T)$ converges to the expectation of $P(T)$ as $n \to \infty$.

Theorem 4 *Assume that the following conditions hold:*

$$\begin{cases} E\left[\int_0^T \sigma_s^4 \, \mathrm{d}s\right] < \infty, \\ E\left[\int_0^T \int (1 \vee |z|^{-p})[\psi(s,z)^2 + (\ln(1+\psi(s,z)))^2]F(\mathrm{d}z) \, \mathrm{d}s\right] < \infty, \end{cases}$$

for some $p \geq 0$ such that $\int (1 \wedge |z|^p)F(\mathrm{d}z) < \infty$. We then have

$$\lim_{n\to\infty} |E[P(T)] - E[P_n(T)]| = 0.$$

More specifically, there are constants C and D such that

$$|E[P(T)] - E[P_n(T)]| \leq Cn^{-1} + Dn^{-1/2}.$$

As a remark, the above result applies to most popular asset price models, including the Black-Scholes model, Heston stochastic volatility model, Merton jump-diffusion model, and most stochastic volatility models with jumps.

Convergence under the 3/2 stochastic volatility model

As an example, we consider the convergence under the 3/2 stochastic volatility model that is governed by the following dynamic equations:

$$\mathrm{d}S_t = S_t\sqrt{V_t}\, \mathrm{d}W_t^S, \tag{4.120a}$$

$$\mathrm{d}V_t = \kappa V_t(\theta - V_t)\mathrm{d}t + \varepsilon V_t^{3/2}\, \mathrm{d}W_t^V, \tag{4.120b}$$

where W_t^S and W_t^V are two correlated Brownian motions, and κ, θ and ε are constants such that $\kappa > -\varepsilon^2/2$ and $\varepsilon > 0$. The first condition is imposed to avoid explosion of

V_t in finite time, or equivalently, to prevent the reciprocal $\tilde{V}_t = 1/V_t$ from reaching zero. It can be shown that $\tilde{V}_t$ becomes a square root process that is governed by

$$d\tilde{V}_t = (\varepsilon^2 + \kappa - \kappa\theta\tilde{V}_t)dt - \varepsilon\, dW_t^V.$$

It is well known that for $\tilde{V}_t$ to stay positive, the coefficients must satisfy the Feller condition: $\kappa > -\varepsilon^2/2$. We state without proof the following proposition on the finiteness of moments of the square root process (Jarrow *et al.*, 2013).

Proposition 4.11 *Let $\bar{V} = \frac{2(\kappa+\varepsilon^2)}{\varepsilon^2}$. We then have*

$$\begin{cases} E[\tilde{V}_t^{-p}] < \infty, & \text{if } p < \bar{V}, \\ E[\tilde{V}_t^{-p}] = \infty, & \text{if } p \geq \bar{V}, \end{cases}$$

and for all $p \geq -\bar{V}$,

$$E[\tilde{V}_t^p] = \mu_t^p e^{-\lambda_t} \frac{\Gamma(\bar{V}+p)}{\Gamma(\bar{V})} M(\bar{V}+p, p, \lambda_t), \tag{4.121}$$

where $\mu_t = \frac{\varepsilon^2}{2}\frac{1-e^{-pt}}{p}$, $\lambda_t = \frac{2pv_0}{\varepsilon^2(e^{-pt}-1)}$, Γ is the Gamma function, and M is the confluent hypergeometric function

$$M(\alpha,\gamma,z) = \sum_{n=0}^{\infty} \frac{(\alpha)_n z^n}{(\gamma)_n n!},$$

with the notation $(x)_n = \prod_{k=0}^{n-1}(x+k)$.

The conditions for convergence of $P_n(T)$ to $P(T)$ as $n \to \infty$ under the 3/2 stochastic volatility model can now be summarized in the following proposition.

Proposition 4.12 *For the 3/2 stochastic volatility model, we observe the following properties.*

1. *Both $P_n(T)$ and $P(T)$ have finite expectation.*
2. *If $\kappa > 0$, so that the instantaneous variance process is mean reverting, $P_n(T)$ converges to $P(T)$ in L^1 as $n \to \infty$.*
3. *If $\kappa \leq 0$, we can no longer guarantee that $P_n(T)$ converges to $P(T)$ in L^1 as $n \to \infty$.*

Proof *Let $M_t = \int_0^t \sqrt{V_s}\, dB_s$, we then have*

$$\langle M,M\rangle_t = \int_0^t V_s\, ds.$$

Carr and Sun (2007) gives the explicit form of the Laplace transform of the continuous realized variance $\int_0^t V_s\, ds$, which implies that all moments of $\int_0^t V_s\, ds$ exist. Therefore, $\langle M,M\rangle_T \in L^2$. By Theorem 4.3, both $P_n(T)$ and $P(T)$ have finite expectation.

(i) When $\kappa > 0$, V_t is mean-reverting with a rate of mean reversion proportional to V_t. Moreover, $\bar{V} > 2$, hence by Proposition 4.11, we observe that $E[V_t^2] = E[\tilde{V}_t^{-2}]$ is finite and it is integrable over $[0,T]$ as a continuous function on this compact interval. As a result, the condition of Theorem 4 is satisfied and the expectation of $P_n(T)$ converges to that of $P(T)$ as $n \to \infty$.

(ii) When $-\frac{\varepsilon^2}{2} < \kappa \leq 0$, it follows that $1 < \bar{V} \leq 2$ and Proposition 4.11 implies that $E[V_t^2] = \infty$. By Fubini's Theorem, we observe $E\left[\int_0^T V_t^2 \, dt\right] = \infty$. The sufficient condition in Theorem 4 fails, so convergence is not guaranteed.

Appendix

Proof of (4.22)

We let $x = \ln S$ and $\tau = t_i - t$, then (4.20a) becomes

$$\begin{aligned}\frac{\partial U_i}{\partial \tau} &= \frac{V}{2}\frac{\partial^2 U_i}{\partial x^2} + \rho\varepsilon V \frac{\partial^2 U_i}{\partial x \partial V} + \frac{\varepsilon^2}{2} V \frac{\partial^2 U_i}{\partial V^2} \\ &\quad + \left(r - \frac{V}{2}\right)\frac{\partial U_i}{\partial x} + \kappa(\theta - V)\frac{\partial U_i}{\partial V} - rU_i = 0, \ 0 < \tau < t_i - t_{i-1},\end{aligned}$$

with initial condition: $U_i(x,V,0;I) = (x - \ln I)^2$. We consider the partial Fourier transform of $U_i(x,V,I,t)$ with respect to x, where

$$\widetilde{U}_i(\omega, V, \tau) = \mathcal{F}\{U_i(x,V,I,\tau)\} = \int_{-\infty}^{\infty} e^{i\omega x} U_i(x,V,\tau) \, dx.$$

The resulting one-dimensional partial differential equation for $\widetilde{U}_i$ is given by

$$\begin{aligned}\frac{\partial \widetilde{U}_i}{\partial \tau} &= \frac{\varepsilon^2}{2} V \frac{\partial^2 \widetilde{U}_i}{\partial V^2} + [\kappa\theta + (\rho\varepsilon\omega i - \kappa)V]\frac{\partial \widetilde{U}_i}{\partial V} \\ &\quad + \left[r(\omega i - 1) - \frac{\omega i + \omega^2}{2} V\right]\widetilde{U}_i, \ 0 < \tau < t_i - t_{i-1},\end{aligned}$$

with initial condition: $\widetilde{U}_i = \mathcal{F}\{(x - \ln I)^2\}$.

Following the standard procedure in solving the affine type model, the solution of the above partial differential equation takes the form:

$$\widetilde{U}_i(\omega, V, \tau) = e^{C(\omega,\tau) + D(\omega,\tau)V}\, \widetilde{U}_i(\omega, V, 0).$$

Substituting the assumed exponential affine form of the solution into the one-dimensional partial differential equation for $\widetilde{U}_i$, we obtain the following Riccati system of ordinary differential equations:

$$\begin{aligned} \frac{dD}{d\tau} &= \frac{\varepsilon^2}{2}D^2 + (\rho\varepsilon\omega i - \kappa)D - \frac{\omega i + \omega^2}{2}, \\ \frac{dC}{d\tau} &= \kappa\theta D + r(\omega i - 1), \end{aligned}$$

with initial conditions: $D(\omega, 0) = C(\omega, 0) = 0$. The solution of the above Riccati system gives

$$C(\omega, \tau) = r(\omega i - 1)\tau + \frac{\kappa\theta}{\varepsilon^2}[(a+b)\tau - 2\ln\frac{1 - ge^{b\tau}}{1-g}],$$

$$D(\omega, \tau) = \frac{a+b}{\varepsilon^2}\frac{1 - e^{b\tau}}{1 - ge^{b\tau}},$$

$$a = \kappa - \rho\varepsilon\omega i, \quad b = \sqrt{a^2 + \varepsilon^2(\omega^2 + \omega i)}, \quad g = \frac{a+b}{a-b}.$$

Finally, the solution of the original differential equation (4.20a) is obtained by taking the Fourier inversion of $\widetilde{U_i}(\omega, V, \tau)$ and this leads to (4.22).

Chapter 5

Options on Discrete Realized Variance

In this chapter, we consider pricing of options on discrete realized variance $I(0,T;N)$ of the asset price S_t under stochastic volatility models and Lévy models. Here, N is the number of monitoring instants over the time period $[0,T]$. Mathematically, $I(0,T;N)$ converges to the annualized quadratic variation $\frac{1}{T}[\ln S_t, \ln S_t]_0^T$ of the log price $\ln S_t$ when the number of monitoring dates N tends to infinity. Letting $\Delta t = \max_{1\leq k\leq N} \Delta t_k \to 0$, where Δt_k is the time interval between successive monitoring dates t_{k-1} and t_k, the discrete realized variance converges to the continuous realized variance as shown below:

$$\begin{aligned}\lim_{\Delta t\to 0} I(0,T;N) &= \frac{A}{N}\sum_{k=1}^{N}\left(\ln\frac{S_{t_k}}{S_{t_{k-1}}}\right)^2 = \frac{1}{T}[\ln S_t, \ln S_t]_0^T \\ &= \frac{1}{T}\left[\int_0^T V_t\,\mathrm{d}t + \sum_{n=1}^{N_T}(J_n^S)^2\right]. \end{aligned} \tag{5.1}$$

Here, A is the annualization factor, where $\frac{A}{N} = \frac{1}{T}$, V_t is the instantaneous variance process, J_n^S is the nth jump in the asset price S_t, and N_T is the number of random jumps over $[0,T]$. Taking advantage of analytic tractability of the moment generating function (mgf) of the continuous realized variance, analytic integral price formulas of options on continuous realized variance have been obtained under various Lévy models (Carr *et al.*, 2005), the Heston model (Sepp, 2008), general affine stochastic volatility model (Kallsen *et al.*, 2011) and 3/2-model (Drimus, 2012). Carr and Lee (2007) propose approximate pricing and hedging strategies for options on continuous realized variance and volatility using continuously monitored variance and volatility swaps.

However, convergence in probability does not guarantee convergence in expectation. In other words, even though the quadratic variation provides good approximation of the discrete realized variance, it is not certain whether the expectation of the quadratic variation remains to be a good approximation of the expectation of the discrete realized variance when we price options on discrete realized variable. Bühler (2006) documents that while the approximation of realized variance via quadratic variation works very well for variance swaps, it may not be sufficient for non-linear payoffs with short maturities. This observation is common to most stochastic volatility models. In particular, he presents some numerical examples for call options on realized variance in the Heston model, where the approximation by quadratic variation diverges quite significantly from the true value for short maturities. As a result,

DOI: 10.1201/9781003263524-5

it is necessary to consider the discrete realized variance directly instead of using the continuous counterpart.

Various analytic approximation methods, semi-analytic methods, and numerical algorithms have been developed to price options on discrete realized variance. Sepp (2012) proposes an analytic approximation method that involves mixing the discrete variance in a lognormal model and the quadratic variance in the Heston model. This serves to find approximation of the discrete variance in the Heston model by adding an adjustment term for the discretization effect via the lognormal distribution. However, Drimus *et al.* (2016) argue that Sepp's mixing method works well only for near-the-money options and medium to long maturities. They present a comprehensive treatment of the discretization effect under general stochastic volatility models and do not restrict to particular strike ranges or maturity ranges. They show that conditional on the realization of the stochastic variance, the discrete realized variance is asymptotically normal as the number of monitoring instants N goes to infinity. By using this limiting distribution, they are able to derive the conditional Black-Scholes method for pricing options on discrete realized variance based on a simulation path of the stochastic variance process. They also develop several other Fourier-based analytic pricing formulas using additional asymptotic approximation of the conditional mean and variance. As expected, their approach works well for options with long maturity, under which a sufficiently large value of N is ensured. However, for short maturity options on discrete realized variance, corresponding to a relatively small value of N, the approximation based on the central limit theorem would not perform so well and would affect the performance of their approximation method. Zheng and Kwok (2015) derive analytic approximation formulas for pricing options on discrete realized variance under affine stochastic volatility models with jumps using the partially exact and bounded (PEB) approximation method. The PEB method relies on an enhanced conditioning variable approach. By adopting either the conditional normal or gamma distribution approximation derived based on the asymptotic behavior of the discrete realized variance of the underlying asset price process, they manage to obtain the PEB approximation formulas that achieve high level of numerical accuracy even for short maturity options. As an additional analytic approximation method to explore the adjustment required due to discrete sampling of realized variance, Keller-Ressel and Muhle-Karbe (2013) use the small time limit close to maturity of the price difference between prices of options on discrete realized variance and continuous realized variance as the adjustment term. As $T \to 0^+$, it can be shown that $I(0,T;N)$ and $I(0,T;\infty)$ can be approximated by the gamma distribution under the Lévy models and general semimartingale models. As expected, this approach works surprisingly well for options on discrete realized variance at small time limit. In addition, Zheng *et al.* (2016) propose recursive Laplace transform algorithms for pricing options on discrete realized variance under time changed Lévy processes by adopting the randomization of the Laplace transform of the discrete log return with a standard normal random variable.

Besides analytic approximation methods, one can price options on discrete realized variance using the direct numerical methods, like the lattice tree algorithms and Fourier transform algorithms. Windcliff *et al.* (2006) pioneer the use of finite difference schemes for pricing discretely monitored variance and volatility derivatives under the local volatility function. Their algorithms face with the curse of dimensionality since additional state variables are required to capture the accumulation of discrete realized variance over successive monitoring dates. Zheng and Kwok (2014) develop fast Fourier transform algorithms for pricing and hedging discretely sampled variance and volatility derivatives under additive processes (time-inhomogeneous Lévy processes). Cui *et al.* (2017) develop efficient numerical algorithms that blend the regime switching models, Markov chain approximation techniques, and Fourier method for pricing swaps and options on discrete realized variance under general stochastic volatility models with jumps.

This chapter focuses on analytic approximation methods for pricing options on discrete realized variance. Sec. 5.1 discusses the adjustment for the discretization effect via the lognormal approximation as suggested by Sepp (2012). In Sec. 5.2, we discuss the conditional Black-Scholes method and Fourier-based discretization adjustment proposed by Drimus *et al.* (2016). Sec. 5.3 presents the partially exact and bound approximation for pricing options on discrete realized variance under the Heston model using the conditional normal distribution approximation and conditional gamma distribution approximation. In Sec. 5.4, we discuss the small time asymptotic approximation of $I(0,T;N)$ and $I(0,T;\infty)$ by the gamma distribution under the Lévy models and semimartingale models. The theoretical results are used to derive the correction term added to adjust the impact of discrete sampling on prices of options on realized variance.

5.1 Adjustment for discretization effect via lognormal approximation

Given availability of analytic pricing formulas of derivatives on continuous realized variance, we investigate the effect of the discrete sampling of realized variance by deriving the approximate price formulas of the derivatives on the discrete realized variance via finding an approximation of the discrete sampling adjustment. Suppose the discrete realized variance $\widetilde{I}(0,T;N)$ and discrete realized variance $\widetilde{I}(0,T;\infty)$ of some chosen auxiliary asset price process can be found readily, Sepp (2012) proposes to use the difference $\widetilde{I}(0,T;N)-\widetilde{I}(0,T;\infty)$ as control and use it as an approximation to the adjustment term required for the Heston model.

Under a risk neutral pricing measure Q, the dynamics of the asset price S_t and instantaneous variance V_t under the Heston stochastic volatility model are governed by

$$\frac{dS_t}{S_t} = (r-q)dt + \sqrt{V_t}\, dW_t \tag{5.2a}$$

$$dV_t = \kappa(\theta - V_t)dt + \varepsilon\sqrt{V_t}\, dW_t^V, \tag{5.2b}$$

where $dW_t\, dW_t^V = \rho\, dt$, r and q are constant interest rate and dividend yield, respectively, κ is the mean reversion speed, and θ is the long-term mean reversion level. For nice analytical tractability, we choose the lognormal model process $\tilde{S}_t$ with deterministic variance $\tilde{V}_t$ as the auxiliary process, whose dynamics under Q is governed by

$$\frac{d\tilde{S}_t}{\tilde{S}_t} = (r-q)\,dt + \sqrt{\tilde{V}_t}\, d\tilde{W}_t, \tag{5.3}$$

where $\tilde{W}_t$ is a standard Brownian motion independent of the Heston asset price process S_t. The deterministic evolution of the instantaneous variance $\tilde{V}_t$ is assumed to be

$$d\tilde{V}_t = \kappa(\theta - \tilde{V}_t)dt \tag{5.4a}$$

so that

$$\tilde{V}_t = e^{-\kappa t}\tilde{V}_0 + \left(1 - e^{-\kappa t}\right)\theta. \tag{5.4b}$$

Based on the conjecture that the discrete sampling effect is not sensitive to stochastic variance, at least in terms of distributional property, Sepp proposes the following distributional approximation of the discrete realized variance:

$$I(0,T;N) \overset{d}{\approx} I(0,T;\infty) + \tilde{I}(0,T;N) - \tilde{I}(0,T;\infty). \tag{5.5}$$

The auxiliary lognormal process $\tilde{S}_t$ and the Heston asset price process S_t have different terminal distributions. Let V_t denote the instantaneous variance process of S_t, with $V_0 = \tilde{V}_0$. These two processes have the same expected quadratic variation, where

$$E^Q\left[\int_0^T V_t\, dt\right] = \theta T + (\tilde{V}_0 - \theta)\frac{1-e^{-\kappa T}}{\kappa} = \int_0^T \tilde{V}_t\, dt. \tag{5.6}$$

One may query whether the lognormal model suffices to capture the discrete sampling effect due to the diffusion component of the discrete realized variance. We argue that the discretization effect between the discrete realized variance and its continuous counterpart is mainly driven by the randomness arising from the asset price dynamics, not much from the instantaneous variance process. One may ask what would be the impact of including a jump component that resembles the return jumps in the SVSJ model. The numerical experiments performed by Sepp (2012) show that discrete sampling does not give noticeable impact on the distribution of the jump component of the realized variance. As a result, one may take the jump component of the continuous realized variance as a proxy of that of the discrete counterpart.

5.1.1 Discrete realized variance under the lognormal model

Let $\tilde{y}_k = \ln \frac{\tilde{S}_{t_k}}{\tilde{S}_{t_{k-1}}}$ be the log return of the auxiliary lognormal process $\tilde{S}_t$ over $[t_{k-1}, t_k]$. In terms of distribution, $\tilde{y}_k$ can be represented as

$$\tilde{y}_k \overset{d}{=} \mu_k + \sigma_k \varepsilon_k, \tag{5.7}$$

where ε_k is a standard normal random variable. The mean μ_k and variance σ_k^2 are given by

$$\mu_k = (r-q)\Delta t_k - \frac{\sigma_k^2}{2} \quad \text{and} \quad \sigma_k^2 = \int_{t_{k-1}}^{t_k} \tilde{V}_t \, dt = \theta \Delta t_k + (\tilde{V}_{t_{k-1}} - \theta)\frac{1-e^{-\kappa \Delta t_k}}{\kappa},$$

where $\Delta t_k = t_k - t_{k-1}$. By using the independent increment property of the lognormal process, the discrete realized variance over the sampling period $[0,T]$ is then given by

$$\tilde{I}(0,T;N) \stackrel{d}{=} \frac{1}{T}\sum_{k=1}^{N}(\mu_k + \sigma_k \varepsilon_k)^2. \tag{5.8}$$

The moment generating function (mgf) of $\tilde{I}(0,T;N)$ under the lognormal dynamics is given by

$$G_{\tilde{I}}(u) = E^Q\left[e^{u\tilde{I}(0,T;N)}\right] \quad = \quad \prod_{k=1}^{N} E^Q\left[e^{(\mu_k^2 + 2\mu_k\sigma_k\varepsilon_k + \sigma_k^2\varepsilon_k^2)u/T}\right].$$

For each individual term in the above expression, we obtain

$$E^Q\left[e^{(\mu_k^2 + 2\mu_k\sigma_k\varepsilon_k + \sigma_k^2\varepsilon_k^2)u/T}\right] = \exp\left(\frac{\mu_k^2 u/T}{1-2\sigma_k^2 u/T}\right)\frac{1}{\sqrt{1-2\sigma_k^2 u/T}},$$

for any u satisfying $\text{Re}\{u\} < \frac{T}{2\sigma_k^2}$. We then have

$$G_{\tilde{I}}(u) = E^Q\left[e^{u\tilde{I}(0,T;N)}\right] = \exp\left(\sum_{k=1}^{N}\frac{\mu_k^2 u/T}{1-2\sigma_k^2 u/T}\right)\prod_{k=1}^{N}\frac{1}{\sqrt{1-2\sigma_k^2 u/T}}, \tag{5.9}$$

for any u satisfying $\text{Re}\{u\} < \min_{1 \le k \le N} \frac{T}{2\sigma_k^2}$.

In particular, we consider the Black-Scholes model, whose variance assumes the constant value σ^2. The mgf of $\tilde{I}(0,T;N)$ of the Black-Scholes model with uniform monitoring interval Δt is given by

$$G_{\tilde{I}}^{BS}(u) = \exp\left(\frac{\left(r-q-\frac{\sigma^2}{2}\right)^2 uT}{N-2\sigma^2 u}\right)\frac{1}{\left(1-\frac{2\sigma^2 u}{N}\right)^{N/2}}, \tag{5.10}$$

for any u satisfying $\text{Re}\{u\} < \frac{N}{2\sigma^2}$.

Inclusion of jump component

We also consider the mgf of $\tilde{I}(0,T;N)$ under the lognormal model with jumps. Assuming constant parameters, the dynamic equation in (5.3) is modified to become

$$\frac{d\tilde{S}_t}{\tilde{S}_t} = (r-q-\lambda m)dt + \sqrt{V_0}\, d\tilde{W}_t + (e^{J^S}-1)d\tilde{N}_t, \tag{5.11}$$

where $\tilde{J^S}$ and $\tilde{N}_t$ are independent copies of J^S and N_t in the SVSJ model. Also, λ is the constant intensity of $\tilde{N}_t$ and $m = E[e^{\tilde{J^S}} - 1]$. By conditioning on the number of jumps in the Poisson process during the monitoring interval and assuming the Gaussian jump distribution with $\tilde{J^S} \sim N(\nu, \delta^2)$, one can obtain

$$G_{\tilde{I}}(u) = \left[\sum_{n=0}^{\infty} \frac{(\lambda T/N)^n}{n!} \frac{\exp\left(\frac{[(\mu-\lambda m)T/N+\nu n]^2 u/T}{1-2(TV_0/N+n\delta^2)u/T} - \lambda T/N\right)}{\sqrt{1-2(V_0T/N+n\delta^2)u/T}} \right]^N. \tag{5.12a}$$

Numerical tests show that the expression of $G_{\tilde{I}}$ in (5.12a) can be well approximated by the following simplified formula:

$$\hat{G}_{\tilde{I}}(u) = \frac{\exp\left(\frac{\mu^2 uT}{N(1-2V_0u/N)}\right)}{(1-2V_0u/N)^{N/2}} \exp\left(\frac{\lambda T \exp\left(\frac{\nu^2 u}{T(1-2\delta^2 u/T)}\right)}{\sqrt{1-2\delta^2 u/T}} - \lambda T \right). \tag{5.12b}$$

As seen from (5.12b), the first factor of $\hat{G}_{\tilde{I}}$ is the mgf of the discrete realized variance neglecting the jump contribution. The second factor is the mgf of $\frac{1}{T}\sum_{n=1}^{N_T}\left(J_n^S\right)^2$, which is the jump part of the continuous realized variance. Hence, $\hat{G}_{\tilde{I}}$ indicates a decoupled approximation to the distribution of the discrete realized variance under the jump-diffusion model. In particular, the jump part of the continuous realized variance can be taken to be the proxy of that of the discrete realized variance.

As a remark, similar observations can be made for the SVSJ model. That is, the contribution from the jump component to the realized variance is only slightly affected by discrete or continuous sampling. For this reason, an auxiliary jump-free lognormal process is chosen in (5.5) that is sufficient to capture the discrete sampling effect of the diffusion part of the realized variance under the SVSJ model.

5.1.2 Approximation formulas for moment generating function

It is known from Sec. 4.3.3 that the continuous realized variance has no dependence on ρ while the discrete realized variance is also insensitive to ρ (Zheng and Kwok, 2014a). When considering the adjustment due to discrete sampling, it suffices to assume $\rho = 0$ and no jumps in the subsequent derivation.

We define

$$y_k = \ln \frac{S_{t_k}}{S_{t_{k-1}}}$$

to be log return of S_t under the Heston model over $[t_{k-1}, t_k]$. In distribution, we observe

$$y_k \overset{\text{d}}{=} \left[(r-q)\Delta t_k - \frac{w_k}{2}\right] + \sqrt{w_k}\,\varepsilon_k \quad \text{and} \quad w_k = \int_{t_{k-1}}^{t_k} V_t \, dt,$$

where ε_k is a standard normal random variable. The discrete realized variance can be written as

$$I(0,T;N) = \frac{1}{T}\sum_{k=1}^{N} y_k^2 = \frac{1}{T}\sum_{k=1}^{N} w_k\varepsilon_k^2 + \frac{2}{T}\sum_{k=1}^{N}\left[(r-q)\Delta t_k - \frac{w_k}{2}\right]\sqrt{w_k}\,\varepsilon_k + \frac{1}{T}\sum_{k=1}^{N}\left[(r-q)\Delta t_k - \frac{w_k}{2}\right]^2.$$

Since w_k can be regarded as order Δt_k, the first term in the above expression is the leading linear order Δt_k, the second term is of order $\Delta t_k^{3/2}$ while the last term is of order Δt_k^2. Suppose we replace w_k in the last two terms by its mean $\bar{w}_k = E[w_k]$, which is given in (5.6), the error introduced is approximately the same order as the second moment of w_k. This imposes little impact on the overall distribution. Therefore, we obtain the approximation:

$$I(0,T;N) \overset{d}{\approx} \frac{1}{T}\sum_{k=1}^{N} w_k\varepsilon_k^2 + \frac{2}{T}\sum_{k=1}^{N}\mu_k\sqrt{\bar{w}_k}\,\varepsilon_k + \frac{1}{T}\sum_{k=1}^{N}\mu_k^2,$$

where $\mu_k = (r-q)\Delta t_k - \bar{w}_k/2$. We propose another approximation to $I(0,T;N)$:

$$I(0,T;N) \overset{d}{\approx} \hat{I}(0,T;N) = \frac{1}{T}\sum_{k=1}^{N} w_k\varepsilon_k^2 + \frac{2}{T}\sum_{k=1}^{N}\mu_k\sqrt{\bar{w}_k}\,v_k + \frac{1}{T}\sum_{k=1}^{N}\mu_k^2, \tag{5.13}$$

where v_k is a standard normal random variable that is independent of ε_k. Apparently, replacing ε_k in the second term by v_k preserves the mean and variance. In our later calculations, we use the identity: $E^Q[\exp(\alpha v_k)] = \frac{\alpha^2}{2}$, $k = 1,2,\ldots,N$.

Recall that $\rho = 0$ is assumed, so the three random vectors

$$\mathbf{e} = (\varepsilon_1^2,\cdots,\varepsilon_N^2)^T,\ \mathbf{v} = (v_1,\cdots,v_N)^T \text{ and } \mathbf{w} = (w_1,\cdots,w_N)^T$$

are mutually independent. For the sum of the last two terms in (5.13), we consider

$$\begin{aligned} H(u) &= E^Q\left[\exp\left(2\frac{u}{T}\sum_{k=1}^{N}\mu_k\sqrt{\bar{w}_k}v_k + \frac{u}{T}\sum_{k=1}^{N}\mu_k^2\right)\right] \\ &= \exp\left(2\frac{u^2}{T^2}\sum_{k=1}^{N}\mu_k^2\bar{w}_k + \frac{u}{T}\sum_{k=1}^{N}\mu_k^2\right). \end{aligned} \tag{5.14}$$

For the first term in (5.13), we consider

$$G(\mathbf{e}) = E^Q\left[\exp\left(\frac{u}{T}\sum_{k=1}^{N} w_k\varepsilon_k^2\right)\middle|\,\mathbf{e}\right] = E^Q\left[\exp\left(\frac{u}{T}\mathbf{e}^T\mathbf{w}\right)\middle|\,\mathbf{e}\right].$$

Since G is a continuous function with continuous Jacobian, we deduce the following approximation derived from the Taylor expansion:

$$\begin{aligned} E^Q[G(\mathbf{e})] &\approx G(\bar{\mathbf{e}}) + \frac{1}{2}\sum_{k=1}^{N}\frac{\partial^2 G(\bar{\mathbf{e}})}{\partial e_k^2}\mathrm{var}(e_k) \\ &= E^Q\left[\exp\left(\frac{u}{T}\sum_{k=1}^{N} w_k\right)\right] + E^Q\left[\sum_{k=1}^{N}\frac{u^2}{T^2}w_k^2\exp\left(\frac{u}{T}\sum_{k=1}^{N} w_k\right)\right], \end{aligned}$$

where we have used the results: $\bar{\mathbf{e}} = \mathbf{1}$ and $\text{var}(e_k) = 2$. Note that the first term in the above expression is identified as the mgf of the continuous realized variance

$$\begin{aligned} G_I^\infty(u) = E^Q\left[e^{uI(0,T;\infty)}\right] &= E^Q\left[\exp\frac{u}{T}\sum_{k=1}^N \int_{t_{k-1}}^{t_k} V_t \, dt\right] \\ &= E^Q\left[\exp\left(\frac{u}{T}\sum_{k=1}^N w_k\right)\right]. \end{aligned} \tag{5.15}$$

The second term can be further approximated by freezing w_k^2 by $\bar{w}_k^2$:

$$E^Q\left[\sum_{k=1}^N \frac{u^2}{T^2} w_k^2 \exp\left(\frac{u}{T}\sum_{k=1}^N w_k\right)\right] \approx \left[\frac{u^2}{T^2}\sum_{k=1}^N \bar{w}_k^2\right] G_I^\infty(u).$$

Putting all the above results together, we obtain the following approximation to the mgf of the discrete realized variance $I(0,T;N)$:

$$\begin{aligned} G_I(u) &\approx E^Q\left[e^{u\hat{I}(0,T;N)}\right] \\ &= H(u)E^Q[G(\mathbf{e})] \approx H(u)\left(1 + \frac{u^2}{T^2}\sum_{k=1}^N \bar{w}_k^2\right) G_I^\infty(u), \end{aligned} \tag{5.16}$$

where $H(u)$ is given by (5.14).

For the subsequent analysis, we assume $|2\bar{w}_k u| \leq 1$, $k = 1, \cdots, N$. We have obtained the mgf $G_{\bar{I}}(u)$ of the discrete realized variance under the lognormal dynamics [see (5.9)]. By performing the Taylor expansion of $G_{\bar{I}}(u)$ up to the second-order terms, we obtain

$$\begin{aligned} G_{\bar{I}}(u) &= \left[\exp\left(\sum_{k=1}^N \mu_k^2 \frac{u}{T}\left(1 + 2\bar{w}_k \frac{u}{T} + \cdots\right)\right)\right] \prod_{k=1}^N \left(1 + \bar{w}_k \frac{u}{T} + \frac{3}{2}\bar{w}_k^2 \frac{u^2}{T^2} + \cdots\right) \\ &= \left[\exp\left(\frac{u}{T}\sum_{k=1}^N \mu_k^2 + 2\frac{u^2}{T^2}\sum_{k=1}^N \mu_k^2 \bar{w}_k + \cdots\right)\right] \\ &\quad \left(1 + \frac{u}{T}\sum_{k=1}^N \bar{w}_k + \frac{3}{2}\frac{u^2}{T^2}\sum_{k=1}^N \bar{w}_k^2 + \frac{u^2}{T^2}\sum_{k=1}^N\sum_{l>k}^N \bar{w}_k \bar{w}_l + \cdots\right) \\ &\approx H(u)\left[1 + \frac{u}{T}\sum_{k=1}^N \bar{w}_k + \frac{u^2}{T^2}\sum_{k=1}^N \bar{w}_k^2 + \frac{1}{2}\frac{u^2}{T^2}\left(\sum_{k=1}^N \bar{w}_k\right)^2\right]. \end{aligned}$$

Combining the above results, we further obtain the following approximation:

$$\begin{aligned}
\exp\left(-\frac{u}{T}\sum_{k=1}^{N}\bar{w}_k\right)G_{\tilde{I}}(u) \approx\ & \left[1-\frac{u}{T}\sum_{k=1}^{N}\bar{w}_k+\frac{1}{2}\frac{u^2}{T^2}\left(\sum_{k=1}^{N}\bar{w}_k\right)^2+\cdots\right]H(u)\\
& \left[1+\frac{u}{T}\sum_{k=1}^{N}\bar{w}_k+\frac{u^2}{T^2}\sum_{k=1}^{N}\bar{w}_k^2+\frac{1}{2}\frac{u^2}{T^2}\left(\sum_{k=1}^{N}\bar{w}_k\right)^2\right]\\
\approx\ & H(u)\left(1+\frac{u^2}{T^2}\sum_{k=1}^{N}\bar{w}_k^2\right).
\end{aligned}$$

Consequently, the approximation formula (5.16) can be expressed as

$$G_I(u) \approx G_I^{\infty}(u)G_{\tilde{I}}(u)\exp\left(-\frac{u}{T}\sum_{k=1}^{N}\bar{w}_k\right), \tag{5.17}$$

where $G_{\tilde{I}}(u)$ and $G_I^{\infty}(u)$ are defined in (5.9) and (5.15), respectively. By observing that $\tilde{I}(0,T;N)$ is deterministic and $\tilde{I}(0,T;\infty) = \frac{1}{T}\sum_{k=1}^{N}\bar{w}_k$, the above approximation formula (5.17) confirms the distributional approximation formula (5.5). Once the approximate formula of the mgf of the discrete realized variance is available, we can derive the corresponding approximate price formulas of swaps and options on discrete realized variance.

Remark

While the approximation is fairly good under most circumstances, there are cases that approximation formula (5.17) may not give satisfactory approximation results. Drimus *et al.* (2016) argue that the distributional approximation formula (5.5) overlooks the phenomenon that the discretization effect itself is dependent on the continuous realized variance, not just only on the deterministic part but also on the uncertainty associated with it. More precisely, the discrete sampling effect increases (decreases) as $I(0,T;\infty)$ increases (decreases). In other words, the approximation formula (5.17) tends to underprice out-of-the-money calls and overprice out-of-the-money puts.

5.2 Normal approximation to conditional distribution of discrete realized variance

Due to the presence of the stochastic volatility, the log returns are no longer independent nor identically distributed under the stochastic volatility models. Nevertheless, the log returns are independent conditioning on a realization path of the instantaneous variance process. Using a generalized version of the Central Limit Theorem,

Drimus *et al.* (2016) propose a normal approximation to the conditional distribution of the discrete realized variance. By performing further asymptotic approximation to the conditional mean and variance of the discrete realized variance as $N \to \infty$, where N is the number of monitoring intervals, they derive an approximation of the conditional distribution of the discrete realized variance that depends on the stochastic volatility process only through the continuous realized variance. Consequently, they manage to express the price of an option on discrete realized variance in terms of the price of its continuous counterpart plus an adjustment term due to discrete sampling.

5.2.1 Conditional normal approximation pricing scheme

Consider a jump-free simplification of the dynamics of asset price S_t and instantaneous variance V_t of the general stochastic volatility model under a risk neutral measure Q [by letting $\lambda = 0$ in (4.5)], where

$$\begin{aligned} \frac{\mathrm{d}S_t}{S_t} &= (r-q)\mathrm{d}t + \sqrt{V_t}\left(\rho\, \mathrm{d}W^1 + \sqrt{1-\rho^2}\, \mathrm{d}W_t^2\right), \\ \mathrm{d}V_t &= \alpha(V_t)\mathrm{d}t + \beta(V_t)\mathrm{d}W_t^1. \end{aligned} \tag{5.18}$$

The two processes S_t and V_t are correlated with constant correlation coefficient ρ. The other parameters and parameter functions have the same interpretation as those in (4.5). Let N be the number of monitoring instants and $[0,T]$ is the sampling period. We assume that the monitoring interval Δt is uniform, where $N\Delta t = T$. Recall the two auxiliary function $f(x)$ and $g(x)$ as defined by [see (4.6)]

$$f(x) = \int_0^x \frac{\sqrt{z}}{\beta(z)}\,\mathrm{d}z \quad \text{and} \quad g(x) = \alpha(x)f'(x) + \frac{\beta^2(x)}{2}f''(x).$$

The kth log return $\ln \frac{S_{t_k}}{S_{t_{k-1}}}$ over $[t_{k-1}, t_k]$ is normally distributed whose conditional mean and variance are given by

$$\mu_{k,N} = (r-q)\frac{T}{N} - \int_{t_{k-1}}^{t_k} \frac{V_t}{2}\,\mathrm{d}t + \rho\left[f(V_{t_k}) - f(V_{t_{k-1}}) - \int_{t_{k-1}}^{t_k} g(V_t)\,\mathrm{d}t\right] \tag{5.19a}$$

$$\sigma_{k,N}^2 = (1-\rho^2)\int_{t_{k-1}}^{t_k} V_t\,\mathrm{d}t, \tag{5.19b}$$

respectively. Based on the Lindeberg-Feller generalized Central Limit Theorem, Drimus *et al.* (2016) establish the following result.

Theorem 5 *Let $\mathcal{F}_T^V$ be the σ-algebra generated by the instantaneous variance process $\{V_t,\ 0 \le t \le T\}$. Conditional on $\mathcal{F}_T^V$, the discrete realized variance converges in distribution to a normal random variable. That is, as $N \to \infty$, we have*

$$\frac{T}{s_N}\left(I(0,T;N) - \frac{\sum_{k=1}^N \mu_{k,N}^2 + \sigma_{k,N}^2}{T}\right) \xrightarrow{d} N(0,1), \tag{5.20}$$

where

$$s_N^2 = \sum_{k=1}^{N} 2\sigma_{k,N}^4 + 4\mu_{k,N}^2\sigma_{k,N}^2.$$

The above theorem states that conditioning on a realization path of the instantaneous variance, the discrete realized variance is asymptotically normal with mean and variance given by

$$M_N = \frac{1}{T}\sum_{k=1}^{N} \mu_{k,N}^2 + \sigma_{k,N}^2 \quad \text{and} \quad \Sigma_N^2 = \frac{s_N^2}{T^2}, \tag{5.21}$$

respectively. The undiscounted price of a call option on discrete realized variance with strike price K conditional on the realization path of the instantaneous variance is given by

$$\tilde{c}_0 = \Sigma_N \phi\left(\frac{M_N - K}{\Sigma_N}\right) + (M_N - K)\Phi\left(\frac{M_N - K}{\Sigma_N}\right), \tag{5.22}$$

where $\phi(\cdot)$ and $\Phi(\cdot)$ represent the density and distribution function of the standard normal distribution, respectively.

Based on the conditional property of the discrete realized variance, the conditional normal approximation pricing scheme for a call option on discrete realized variance can be summarized as follows:

1. Simulate an instantaneous variance process path $\{V_t, 0 \le t \le T\}$.
2. Compute the conditional mean and variance μ_k and σ_k for each k; compute the option price using (5.22).
3. Repeat the previous two steps sufficiently many simulation runs and take the sample average.

As a remark, the normal approximation also works well in the traditional Black-Scholes model (absence of stochastic volatility). Since the variance assumes the constant value σ^2 in the Black-Scholes model, we have independent and identically distributed normal log asset returns:

$$\ln\frac{S_{t_k}}{S_{t_{k-1}}} = N\left((r - q - \frac{\sigma^2}{2})\frac{T}{N}, \sigma^2\frac{T}{N}\right).$$

Here, $N(\mu, \sigma^2)$ denotes a normal distribution with mean μ and variance σ^2. The discrete realized variance is known to be

$$I(0,T;N) \stackrel{d}{=} \frac{\sigma^2}{N}\chi'(N,\gamma), \tag{5.23}$$

where $\chi'(N,\gamma)$ denotes the noncentral chi-squared distribution with N degrees of freedom and non-centrality parameter γ as given by

$$\gamma = \frac{(r - q - \frac{\sigma^2}{2})^2 T}{\sigma^2}. \tag{5.24}$$

It is well known that as the number of degrees of freedom $N \to \infty$, we observe

$$\frac{\chi'(N,\gamma)-(N+\gamma)}{\sqrt{2(N+2\gamma)}} \xrightarrow{d} N(0,1). \tag{5.25}$$

Correspondingly, we deduce that $I(0,T;N)$ converges asymptotically in distribution to a normal random variable as given by

$$N\left(\sigma^2+\frac{(r-q-\frac{\sigma^2}{2})^2T}{N}, \frac{2\sigma^4}{N}+\frac{4(r-q-\frac{\sigma^2}{2})^2\sigma^2T}{N^2}\right).$$

5.2.2 Simplified conditional pricing schemes

Although the conditional normal approximation pricing scheme has reduced the dimension of the simulation procedure, it involves lengthy calculation of the conditional mean and variance $\mu_{k,N}$ and $\sigma_{k,N}^2$ for each k. By using the asymptotic property of M_n and Σ_N^2, one can simplify the procedure by making further analytic approximation. The simplification procedure relies on the following lemma.

Lemma 5.1 *Let $(V_t)_{t\geq 0}$ be a process which is almost surely locally bounded and has at most a countable number of discontinuity points on every finite interval. For any fixed $T>0$ and positive integer k, we have*

$$(\Delta t)^{k-1}\sum_{j=1}^{N}\left(\int_{t_{j-1}}^{t_j} V_t\,\mathrm{d}t\right)^k \to \int_0^T V_t^k\,\mathrm{d}t \quad a.s.$$

as $N \to \infty$.

The proof of Lemma 5.1 can be found in Drimus *et al.* (2016).

As explained earlier, the impact of ρ on the value of options on realized variance is insignificant. It suffices to assume $\rho = 0$. We obtain the following reduced forms of $\mu_{k,N}$ and $\sigma_{k,N}^2$:

$$\begin{aligned}
\mu_{k,N} &= (r-q)\Delta t - \frac{1}{2}\int_{t_{k-1}}^{t_k} V_t\,\mathrm{d}t,\\
\sigma_{k,N}^2 &= \int_{t_{k-1}}^{t_k} V_t\,\mathrm{d}t.
\end{aligned}$$

Let $I_k = \frac{1}{T}\int_0^T V_t^k\,\mathrm{d}t$, $k=1,2,3$. By Lemma 5.1, we have

$$\begin{aligned}
\frac{1}{T}\sum_{k=1}^{N}\sigma_{k,N}^4 &= \frac{T}{N}I_2+o(N^{-1}).\\
\frac{1}{T}\sum_{k=1}^{N}\sigma_{k,N}^6 &= \frac{T^2}{N^2}I_3+o(N^{-2}).
\end{aligned}$$

Consequently, we obtain

$$\begin{aligned}
\frac{1}{T}\sum_{k=1}^{N}\mu_{k,N}^2 &= \frac{1}{T}\sum_{k=1}^{N}\left[(r-q)^2\Delta t^2-(r-q)\Delta t\sigma_{k,N}^2+\frac{\sigma_{k,N}^4}{4}\right]\\
&= (r-q)^2\Delta t-(r-q)\Delta t I_1+\frac{\Delta t}{4}I_2+o(N^{-1}),\\
\frac{1}{T}\sum_{k=1}^{N}\mu_{k,N}^2\sigma_{k,N}^2 &= \frac{1}{T}\sum_{k=1}^{N}\left[(r-q)^2\Delta t^2\sigma_{k,N}^2-(r-q)\Delta t\sigma_{k,N}^4+\frac{\sigma_{k,N}^6}{4}\right]\\
&= \frac{T^2}{N^2}\left[(r-q)^2 I_1-(r-q)I_2+\frac{I_3}{4}\right]+o(N^{-2}).
\end{aligned}$$

Recall the definition of M_N and Σ_N^2 in (5.21), so

$$\begin{aligned}
M_N &= (r-q)^2\Delta t-(r-q)\Delta t I_1+\frac{\Delta t}{4}I_2+I_1+o(N^{-1})=I_1+o(N^{-1}),\\
\Sigma_N^2 &= \frac{2}{N}I_2+o(N^{-1})+\frac{4\Delta t}{N}\left[(r-q)^2 I_1-(r-q)I_2+\frac{I_3}{4}\right]\\
&= \frac{2}{N}I_2+o(N^{-1}).
\end{aligned}$$

As a result, the conditional distribution of $I(0,T;N)$ can be further approximated by

$$I(0,T;N)\overset{d}{\approx}N\left(I_1,\ \frac{2}{N}I_2\right)=N\left(\frac{1}{T}\int_0^T V_t\,\mathrm{d}t,\ \frac{2}{N}\frac{1}{T}\int_0^T V_t^2\,\mathrm{d}t\right). \tag{5.26}$$

Using the approximation in (5.26), the conditional pricing scheme is simplified as follows:

1. Simulate an instantaneous variance process path $\{V_t,\, 0\le t\le T\}$.
2. Compute $M_N=I_1$ and $\Sigma_N=\sqrt{\frac{2}{N}I_2}$; compute the option price using (5.22).
3. Repeat the previous two steps sufficiently many simulation runs and take the sample average.

In the above approach, though it no longer requires to compute the quantities $\mu_{k,N}$ and $\sigma_{k,N}^2$, one still needs to perform the simulation of the entire path of the instantaneous variance process over $[0,T]$. To further simplify the conditional pricing scheme, by virtue of (5.26), it is seen that the conditional distribution of $I(0,T;N)$ depends on the simulation path of V_t through the two quantities I_1 and I_2. Note that I_1 is the continuous realized variance, which is known to have good tractability, while I_2 does not. Therefore, it is natural to find some method to approximate I_2 by I_1. By Jensen's inequality, we have

$$I_2\ge I_1^2,$$

which dictates that one may approximate I_2 using this lower bound I_1^2. As a result, the conditional distribution of $I(0,T;N)$ is approximated by

$$I(0,T;N) \overset{d}{\approx} N\left(I_1, \frac{2}{N}I_1^2\right) = N\left(\frac{1}{T}\int_0^T V_t \, dt, \frac{2}{N}\left(\frac{1}{T}\int_0^T V_t \, dt\right)^2\right). \tag{5.27}$$

Under the new approximation in (5.27), the simulation of the path of V_t can be replaced by sampling directly from the distribution of I_1. Though (5.27) is obtained by an asymptotic expansion, it can be regarded as a simple extension of the approximation formula under the Black-Scholes model. The approximation (5.27) models the discrete sampling effect by randomizing the discrete realized variance around the continuous realized variance. Moreover, the extra randomness as captured by the conditional variance depends on the continuous realized variance. It is increasing as a function of the continuous realized variance, which means that the discretization effect would be greater when the level of continuous realized variance is higher. Therefore, compared to the independency assumption of Sepp's approximation that uses an auxiliary lognormal model, the conditional normal approximation in (5.27) represents better approximation. However, one needs to be aware that approximation (5.27) is indeed biased since it implies that

$$E^Q[I(0,T;N)] = E^Q[E^Q[I(0,T;N)|I(0,T;\infty)]] = E^Q[I(0,T;\infty)]. \tag{5.28}$$

This contradicts to the empirical finding that the fair strike of the discrete variance swap is in general larger than that of its continuous counterpart.

5.2.3 Non-simulation asymptotic approximation pricing scheme

Using the approximation in (5.27), we can obtain the characteristic function of $I(0,T;N)$ as follows:

$$\phi_N(u) = E^Q\left[e^{iuI(0,T;N)}\right] = E^Q\left[E^Q[e^{iuI(0,T;N)}|I_1]\right] = E^Q\left[e^{iuI_1 - \frac{u^2 I_1^2}{N}}\right]. \tag{5.29}$$

By using the Taylor expansion, the above expectation can be approximated by

$$\phi_N(u) \approx E^Q\left[e^{iuI_1}\left(1 - \frac{u^2 I_1^2}{N}\right)\right] = \phi(u) + \frac{u^2}{N}\phi''(u), \tag{5.30}$$

where $\phi(u) = E^Q[e^{iuI_1}]$ is the characteristic function of the continuous realized variance.

Given the characteristic function of $I(0,T;N)$, the undiscounted call option price c_0 can be conveniently calculated by the Fourier inversion formula:

$$\begin{aligned} c_0 &= E^Q[(I(0,T;N)-K)^+] = \frac{1}{2\pi}\int_{-\infty}^{\infty} \frac{e^{-(\alpha+i\beta)K}\phi_N(\beta - i\alpha)}{(\alpha+i\beta)^2} \, d\beta, \\ &\approx \frac{1}{2\pi}\int_{-\infty}^{\infty} \frac{e^{-(\alpha+i\beta)K}\phi(\beta - i\alpha)}{(\alpha+i\beta)^2} \, d\beta \\ &\quad - \frac{1}{2\pi N}\int_{-\infty}^{\infty} e^{-(\alpha+i\beta)K}\phi''(\beta - i\alpha) \, d\beta, \end{aligned} \tag{5.31}$$

where the damping factor $\alpha > 0$ is some chosen fixed constant to ensure existence of the Fourier integral. Note that the first integral is essentially the undiscounted price of a call option on the continuous realized variance, namely, $E^Q[(I(0,T;\infty)-K)^+]$. The second integral, denoted by Δ_N, is recognized as the correction term due to discrete sampling. By performing integration by parts, Δ_N can be simplified as follows:

$$\begin{aligned}
\Delta_N &= -\frac{1}{2\pi N}\left[e^{-(\alpha+i\beta)K}\phi'(\beta-i\alpha)\Big|_{-\infty}^{\infty} + \int_{-\infty}^{\infty} iKe^{-(\alpha+i\beta)K}\phi'(\beta-i\alpha)\,\mathrm{d}\beta\right] \\
&= -\frac{1}{2\pi N}\Big[e^{-(\alpha+i\beta)K}\phi'(\beta-i\alpha)\Big|_{-\infty}^{\infty} + iKe^{-(\alpha+i\beta)K}\phi(\beta-i\alpha)\Big|_{-\infty}^{\infty} \\
&\qquad - \int_{-\infty}^{\infty} K^2 e^{-(\alpha+i\beta)K}\phi(\beta-i\alpha)\,\mathrm{d}\beta\Big] \\
&= \frac{K^2}{N}\frac{1}{2\pi}\int_{-\infty}^{\infty} e^{-(\alpha+i\beta)K}\phi(\beta-i\alpha)\,\mathrm{d}\beta, \qquad (5.32)
\end{aligned}$$

where the boundary terms resulting from integration by parts vanish. In conclusion, we have

$$c_0 \approx \frac{1}{2\pi}\int_{-\infty}^{\infty} e^{-(\alpha+i\beta)K}\phi(\beta-i\alpha)\left[\frac{1}{(\alpha+i\beta)^2}+\frac{K^2}{N}\right]\mathrm{d}\beta, \qquad (5.33)$$

which can be efficiently calculated using the FFT algorithm, provided that analytic formula of $\phi(u)$ is available. By the put-call parity, the price of the put option on discrete realized variance can be obtained by adding the same correction term Δ_N as given by (5.32) to the price of its continuous counterpart.

As a remark, it is well known that the probability density function of $I(0,T;\infty)$ has the following representation:

$$f_{I(0,T;\infty)}(K) = \frac{1}{2\pi}\int_{-\infty}^{\infty} e^{-(\alpha+i\beta)K}\phi(\beta-i\alpha)\,\mathrm{d}\beta. \qquad (5.34)$$

Therefore, we have

$$\Delta_N = \frac{K^2}{N} f_{I(0,T;N)}(K) \geq 0, \qquad (5.35)$$

which implies that options on discrete realized variance are always more expensive than their continuous counterparts.

Comments

Both the conditional normal approximation pricing schemes and the analytic approximation pricing scheme are easy to be implemented. For the conditional pricing schemes, provided that there exists an efficient simulation method for the instantaneous variance process, the pricing problem would be straightforward. Moreover, the option greeks like delta, gamma, and vega can be computed in a similar manner. The analytic approximation pricing scheme has very compact pricing formulas and can be computed efficiently using FFT algorithms. While these approaches are presented in the context of jump-free stochastic volatility model, it is possible to include Poisson jumps in the asset returns to the pricing model. Despite the simple formulations of these approaches, one also needs to be aware of their limitations. The underpinning of these approximation approaches is the generalized Central Limit Theorem and various asymptotic approximation techniques, whose validity is subject to the condition that the number of monitoring instants N is sufficiently large. While this is acceptable for options with long maturities (3 months or longer), it is probably not quite acceptable for short-maturity options.

Alternative analytic approximation to moment generating function

Instead of considering the normal approximation of the conditional discrete realized variance, one may compute analytic approximation of the mgf of the discrete realized variance up to quadratic power in Δt, where $\Delta t = T/N$ is the uniform time interval between successive monitoring times. We consider the same jump-free stochastic volatility model as defined in (5.18). Recall that the kth order return $y_k = \ln \frac{S_{t_k}}{S_{t_{k-1}}}$ conditional on $\mathcal{F}_T^V$ is normally distributed with mean μ_k and variance σ_k^2 as given by (5.19a) and (5.19b), respectively. According to (5.9), the mgf of the conditional discrete realized variance is given by

$$E^Q\left[e^{uI(0,T;N)}\middle|\mathcal{F}_T^V\right] = E^Q\left[e^{u\sum_{k=1}^N y_k^2}\middle|\mathcal{F}_T^V\right]$$
$$= \exp\left(\sum_{k=1}^N \left[\frac{u\mu_k^2}{T-2\sigma_k^2 u} - \frac{1}{2}\ln\left(1-\frac{2\sigma_k^2 u}{T}\right)\right]\right). \tag{5.36}$$

Lian *et al.* (2014) perform the tedious procedure of expanding

$$\frac{u\mu_k^2}{T-2\sigma_k^2 u} - \frac{1}{2}\ln\left(1-\frac{2\sigma_k^2 u}{T}\right)$$

in powers of Δt and taking $\rho = 0$ [using the same justification as the assumption made in the approximations proposed by Sepp (2012) and Drimus *et al.* (2016)]. By taking the tower rule in iterated expectation, they obtain the following analytic approximation to the mgf of the discrete realized variance $I(0,T;N)$:

$$E^Q\left[e^{uI(0,T;N)}\right] = E^Q\left[E^Q e^{uI(0,T;N)}\middle|\mathcal{F}_T^V\right]$$
$$\approx \exp\left(u\mu + uc_0 + c_2\sum_{k=1}^N E^Q[I_k^2]\right)E^Q\left[\exp\left(c_1\int_0^T V_t\,dt\right)\right], \tag{5.37}$$

where $\mu = E^Q[I(0,T;N)]$, $I_k^2 = \left(\int_{t_{k-1}}^{t_k} V_t \, dt\right)^2$,

$$c_0 = [(r-q)\Delta t - 1]E^Q\left[\frac{1}{T}\int_0^T V_t \, dt\right],$$

$$c_1 = \frac{2u^2(r-q)^2\Delta t^2}{T^2} - \frac{u(r-q)\Delta t}{T} + \frac{u}{T},$$

$$c_2 = \frac{u^2}{T^2} - \frac{2u^2(r-q)\Delta t}{T^2} + \frac{4u^3(r-q)^2\Delta t^2}{T^3},$$

$$c_3 = \frac{u^2(3T+8u)}{6T^3} - \frac{4u^3(r-q)\Delta t}{T^3} + \frac{8u^4(r-q)^2\Delta t^2}{T^4}.$$

For the Heston model, the expansion of $E^Q[(\int_{t_{k-1}}^{t_k} V_t \, dt)^2]$ in powers of Δk can be readily found (Lian *et al.*, 2014).

5.3 Partially exact and bounded approximation for options on discrete realized variance

The partial exact and bounded (PEB) approximation procedure consists of two steps. The first step is to derive the lower bound of the undiscounted price of the option on discrete realized variance based on the conditioning variable approach. In the second step, we derive the adjustment terms that approximate the residual component. To derive the lower bound, Zheng and Kwok (2015) choose the conditioning variable to be $I_T = [\ln S, \ln S]_T$, the time-T continuous realized variance of the underlying asset price process S_t. The lower bound can be easily computable, provided that the characteristic function of I_T is available in an analytic form. They then use the conditional normal distribution approximation and conditional gamma distribution approximation to derive the corresponding adjustment terms.

Lower bound of the call price based on conditioning variable

Let $\{\mathcal{F}_t\}_{0 \le t \le T}$ be the natural filtration generated by the asset price process S_t, $I_T = [\ln S, \ln S]_T$ and A be the event $\{I_T > c\}$ with $c > 0$ such that $A \in \mathcal{F}_T$. We write A^c as the complement of the event A and $I_T^{(N)}$ as the (unannualized) discrete realized variance of asset price S_t on monitoring dates: $0 = t_0 < t_1 < \cdots < t_N = T$ over $[0,T]$, which is defined to be

$$I_T^{(N)} = \sum_{k=1}^{N}\left(\ln\frac{S_{t_k}}{S_{t_{k-1}}}\right)^2 = \sum_{k=1}^{N}(X_{t_k} - X_{t_{k-1}})^2, \tag{5.38}$$

where $X_t = \ln S_t$ is the log asset price process. For the undiscounted price of the call option on the discrete realized variance $I_T^{(N)}$ with strike price K, we have

$$
\begin{aligned}
E^Q[(I_T^{(N)} - K)^+] &= E^Q[(I_T^{(N)} - K)^+ 1_{A^c}] + E^Q[(I_T^{(N)} - K)^+ 1_A] \\
&= E^Q[(I_T^{(N)} - K)^+ 1_{A^c}] + E^Q[(K - I_T^{(N)})^+ 1_A] + E^Q[(I_T^{(N)} - K) 1_A] \\
&\geq E^Q[(I_T^{(N)} - K) 1_{\{I_T > c\}}]. \qquad (5.39)
\end{aligned}
$$

The last term gives a lower bound for the undiscounted call option price and the corresponding maximum value among all choices of nonnegative values of c providing the best lower bound. Note that I_T is chosen to be the conditioning variable since $I_T^{(N)}$ and I_T are highly correlated. We take the advantage that I_T is analytically tractable. We define

$$
g(c) = E^Q[(I_T^{(N)} - K) 1_{\{I_T > c\}}] = \int_c^\infty E^Q[I_T^{(N)} - K | I_T = y] f_I(y) \, dy, \qquad (5.40a)
$$

where f_I is the density function of I_T. The conditional expectation $g(c)$ increases in value with increasing c when c is small and eventually drops to zero as c becomes sufficiently large due to the rapid decay of f_I. Therefore, we expect that $g(c)$ achieves its maximum value at some finite value c^* that is close to K. The critical value c^* satisfies the first-order condition:

$$
g'(c) = -E^Q[I_T^{(N)} - K | I_T = c] f_I(c) = 0. \qquad (5.40b)
$$

5.3.1 Lower bound with known characteristic function

We consider the implementation of the first step in the PEB approximation of deriving the lower bound $E^Q\left[(I_T^{(N)} - K)\mathbf{1}_{\{I_T > c\}}\right]$ when the analytic form of the joint characteristic function of the squared increment $(X_{t_k} - X_{t_{k-1}})^2$ and continuous realized variance I_T is known. Though our argument works even when the joint characteristic function does not have a closed form, such analytic tractability indeed facilitates the efficient computation of the lower bound. For a typical quadratic term $(X_{t_k} - X_{t_{k-1}})^2$ in $I_T^{(N)}$, $k = 1, 2, \ldots, N$, evaluation of the conditional expectation can be performed via the following transformation:

$$
E^Q[(X_{t_k} - X_{t_{k-1}})^2 | I_T = c] = -\frac{\partial^2}{\partial \phi^2} E^Q[e^{i\phi(X_{t_k} - X_{t_{k-1}})} | I_T = c]\Big|_{\phi=0}.
$$

We write $\Delta_k = X_{t_k} - X_{t_{k-1}}$ and let $f_{\Delta_k, I}$ and $\Phi_{\Delta_k, I}$ denote the joint density function and joint characteristic function of Δ_k and I_T, respectively. By virtue of the Parseval identity and interchanging order of integration, we obtain

$$\begin{aligned}
& E^Q[e^{\mathrm{i}\phi\Delta_k}|I_T = c] \\
= & \frac{1}{f_I(c)} \int_{-\infty}^{\infty} e^{\mathrm{i}\phi x} f_{\Delta_k,I}(x,c)\ \mathrm{d}x \\
= & \frac{1}{f_I(c)} \int_{-\infty}^{\infty} e^{\mathrm{i}\phi x} \frac{1}{4\pi^2} \int_{\mathrm{i}\beta_i-\infty}^{\mathrm{i}\beta_i+\infty} \int_{\mathrm{i}\alpha_i-\infty}^{\mathrm{i}\alpha_i+\infty} \Phi_{\Delta_k,I}(\alpha,\beta) e^{-\mathrm{i}\alpha x - \mathrm{i}\beta c}\ \mathrm{d}\alpha\ \mathrm{d}\beta\ \mathrm{d}x \\
= & \frac{1}{f_I(c)} \int_{\mathrm{i}\beta_i-\infty}^{\mathrm{i}\beta_i+\infty} \frac{1}{2\pi} \Phi_{\Delta_k,I}(\phi,\beta) e^{-\mathrm{i}\beta c}\ \mathrm{d}\beta, \qquad (5.41)
\end{aligned}$$

where $\alpha = \alpha_r + \mathrm{i}\alpha_i$ and $\beta = \beta_r + \mathrm{i}\beta_i$ are the complex Fourier transform variables. The respective imaginary parts α_i and β_i of the pair of transform variables α and β are chosen to be some appropriate fixed constants to ensure the existence of the generalized Fourier transform integral. Summing all the individual expectations of $e^{\mathrm{i}\phi\Delta_k}$ conditional on $I_T = c$, where $k = 1, 2, \cdots, N$, and substituting into (5.40b), the first-order condition can be expressed as

$$g'(c) = \int_0^{\infty} \frac{1}{\pi} \sum_{k=1}^{N} \mathrm{Re}\left\{ \frac{\partial^2}{\partial\phi^2} \Phi_{\Delta_k,I}(\phi,\beta_r + \mathrm{i}\beta_i)\Big|_{\phi=0} e^{-\mathrm{i}(\beta_r+\mathrm{i}\beta_i)c} \right\} \mathrm{d}\beta_r + K f_I(c) = 0,$$

where $\mathrm{Re}\{\cdot\}$ represents taking the real part. It is known that the characteristic function Φ_I of I exists in closed form for most affine jump-diffusion models, like the Heston stochastic volatility model with jumps. It is convenient to express $f_I(c)$ as a Fourier inversion integral of Φ_I such that

$$f_I(c) = \frac{1}{\pi} \int_0^{\infty} \mathrm{Re}\left\{ \Phi_I(\beta_r + \mathrm{i}\beta_i) e^{-\mathrm{i}(\beta_r+\mathrm{i}\beta_i)c} \right\} \mathrm{d}\beta_r.$$

As a result, the first-order condition can be expressed in the following compact form:

$$g'(c) = \frac{1}{\pi} \int_0^{\infty} \mathrm{Re}\left\{ \Psi(\beta_r + \mathrm{i}\beta_i) e^{-\mathrm{i}(\beta_r+\mathrm{i}\beta_i)c} \right\} \mathrm{d}\beta_r = 0, \qquad (5.42)$$

where

$$\Psi(\beta) = \sum_{k=1}^{N} \frac{\partial^2}{\partial\phi^2} \Phi_{\Delta_k,I}(\phi,\beta)\big|_{\phi=0} + K\Phi_I(\beta).$$

The integral representation of $g''(c)$ is readily available and it takes the form

$$g''(c) = \frac{1}{\pi} \int_0^{\infty} \mathrm{Re}\left\{ -\mathrm{i}(\beta_r + \mathrm{i}\beta_i) \Psi(\beta_r + \mathrm{i}\beta_i) e^{-\mathrm{i}(\beta_r+\mathrm{i}\beta_i)c} \right\} \mathrm{d}\beta_r.$$

One may solve (5.42) for the critical value c^* via Newton's iteration method with the initial guess $c = K$ and check for the second-order condition $g''(c) < 0$ to ensure that c^* is a maximizer of $g(c)$.

Finally, we can calculate the optimal lower bound for the undiscounted option price as follows:

$$\begin{aligned} g(c^*) &= \sum_{k=1}^{N} E^Q[\,\Delta_k^2 1_{\{I_T>c^*\}}] - KQ[I_T > c^*] \\ &= -\frac{1}{\pi}\int_0^\infty \mathrm{Re}\left\{ \sum_{k=1}^{N} \frac{\partial^2}{\partial\phi^2}\Phi_{\Delta_k,I}(\phi,\beta_r+\mathrm{i}\beta_i)\Big|_{\phi=0} \frac{e^{-\mathrm{i}c^*(\beta_r+\mathrm{i}\beta_i)}}{\mathrm{i}(\beta_r+\mathrm{i}\beta_i)} \right\} \mathrm{d}\beta_r \\ &\quad -K\frac{1}{\pi}\int_0^\infty \mathrm{Re}\left\{ \Phi_I(\beta_r+\mathrm{i}\beta_i)\frac{e^{-\mathrm{i}c^*(\beta_r+\mathrm{i}\beta_i)}}{\mathrm{i}(\beta_r+\mathrm{i}\beta_i)} \right\} \mathrm{d}\beta_r \\ &= -\frac{1}{\pi}\int_0^\infty \mathrm{Re}\left\{ \Psi(\beta_r+\mathrm{i}\beta_i)\frac{e^{-\mathrm{i}c^*(\beta_r+\mathrm{i}\beta_i)}}{\mathrm{i}(\beta_r+\mathrm{i}\beta_i)} \right\} \mathrm{d}\beta_r. \end{aligned} \tag{5.43}$$

Note that it is necessary to restrict β_i to be negative for the existence of the Fourier integral.

It is relatively easy to establish a lower bound for the undiscounted call option price via an approximation of $I_T^{(N)}$ by the conditional variable $E[I_T^{(N)}|I_T]$. Indeed, by the Jensen inequality, we deduce that

$$E^Q[(I_T^{(N)} - K)^+] = E^Q[E^Q[(I_T^{(N)} - K)^+|I_T]] \geq E[(E[I_T^{(N)}|I_T] - K)^+].$$

The above observation relies on the assumption that $g'(c)$ defined by (5.42) changes sign only once at some interior point. Though a rigorous proof of this assumed property of $g'(c)$ may not be straightforward, one can visualize this analytic property intuitively. Since $I_T^{(N)}$ and I_T are strongly correlated, when I_T is set at some value of c that is significantly smaller than K, the conditional density of $I_T^{(N)}$ is supposed to be concentrated around c. As a result, $-E[I_T^{(N)} - K|I_T = c]$ is positive. This gives $g'(c) > 0$ if $f_I(c)$ is refrained from hitting zero. As c increases in value and gets closer to K, the above negative of the conditional expectation decreases in value and eventually $g'(c)$ hits the zero value at some critical value c^*. When c increases beyond c^* further, $E[I_T^{(N)} - K|I_T = c]$ remains positive, so $g'(c)$ stays negative. Finally, $|g'(c)|$ decreases in magnitude and $g'(c)$ approaches the zero value from below at some asymptotically large value of c since $f_I(c)$ converges to zero as c tends to infinity. Based on the above assumed analytic property of $g'(c)$, one then obtain

$$\begin{aligned} E^Q[(E^Q[I_T^{(N)}|I_T] - K)^+] &= \int_0^\infty (E^Q[I_T^{(N)}|I_T = c] - K)^+ f_I(c)\,\mathrm{d}c \\ &= \int_0^\infty \left(E^Q[I_T^{(N)}|I_T = c]f_I(c) - Kf_I(c)\right)^+ \mathrm{d}c \\ &= \int_0^\infty [-g'(c)]^+\,\mathrm{d}c = \int_{c^*}^\infty -g'(c)\,\mathrm{d}c = g(c^*) \end{aligned} \tag{5.44}$$

Interestingly, the lower bound on the right-hand side is simply the optimal lower bound $g(c^*)$ as given by (5.43).

5.3.2 Partially exact and bounded approximation

The lower bound derived from the conditioning variable approach works quite well for arithmetic Asian options based on conditioning on the geometric average counterpart, whereas the lower bound $g(c^*)$ defined by (5.43) is seen to fail to provide sufficiently accurate approximation formulas for short-maturity options on discrete realized variance. One major difference is that while we observe dominance of arithmetic average over geometric average, there is lack of strict dominance of the discrete realized variance over the continuous counterpart or vice versa. Due to lack of dominance, optionality on the continuous realized variance may not be carried over to optionality on the discrete counterpart. This explains the significant gap between the lower bound and the exact price of an option on discrete realized variance. Indeed, the discrepancy between the discrete realized variance and continuous realized variance becomes more profound when maturity or sampling period becomes shorter. Therefore, the lower bound approximation becomes more unreliable for short-maturity options on discrete realized variance. As a remark, the crude approximation of $I_T^{(N)}$ by I_T in the option valuation provides an even worse approximation than the lower bound $g(c^*)$ derived by conditioning.

Henceforth, we drop the subscript T in both $I_T^{(N)}$ and I_T for notational convenience in our later exposition when no ambiguity arises. To provide better approximation, it is natural to consider an analytic approximation to the sum of the residual terms

$$E^Q[(I^{(N)}-K)^+ 1_{\{I \leq c^*\}}] + E^Q[(K-I^{(N)})^+ 1_{\{I > c^*\}}]$$

in the decomposition of the option price shown in (5.39). In the literature on pricing arithmetic Asian options, this approach is termed the *partially exact and bounded* (PEB) approximation. The essence of the PEB approximation is to consider an approximation to the conditional distribution of $I^{(N)}|I$ so that evaluation of the two residual terms can be performed efficiently. In the implementation of the second step of the PEB approximation for the call option on discrete realized variance, we propose two analytic approximation methods based on the normal distribution and gamma distribution, both are derived from some asymptotic behavior of the realized variance of the underlying asset price process.

Conditional normal distribution approximation

Based on the generalized Central Limit Theorem and asymptotic analysis of the discrete realized variance of an asset price process under stochastic volatility, Drimus *et al.* (2016) show that one may approximate $I^{(N)}|I$ by $\hat{I}^{(N)}|I$ for a sufficiently large value of number of monitoring instants N, where

$$\hat{I}^{(N)}|I \sim N\left(I, \frac{2}{N}I^2\right). \tag{5.45}$$

Here, $N(\mu, \sigma^2)$ denotes a normal distribution with mean μ and variance σ^2. Though their result is derived under the stochastic volatility framework, it can be seen that it remains to work well under stochastic volatility with jumps.

In the derivation of the PEB approximation for the call option on discrete realized variance, it is important to introduce another more efficient approximation that has the same order as that proposed by Drimus *et al.* (2016). Let $\Phi_{\hat{I}^{(N)},I}(\alpha,\beta)$ denote the joint characteristic function of $\hat{I}^{(N)}$ and I. Suppose we adopt the approximation in (5.45), by introducing the approximation: $e^{-\alpha^2 I^2/N} \approx 1 - \frac{\alpha^2 I^2}{N}$ under $O(N^{-2})$ approximation, we have

$$\begin{aligned}\Phi_{\hat{I}^{(N)},I}(\alpha,\beta) &= E^Q[e^{\mathrm{i}\alpha\hat{I}^{(N)}+\mathrm{i}\beta I}] = E^Q[E^Q[e^{\mathrm{i}\alpha\hat{I}^{(N)}+\mathrm{i}\beta I}|I]] = E^Q[e^{\mathrm{i}\alpha I-\frac{\alpha^2 I^2}{N}}e^{\mathrm{i}\beta I}]\\ &\approx E^Q\left[e^{\mathrm{i}(\alpha+\beta)I}\left(1-\frac{\alpha^2 I^2}{N}\right)\right] = \Phi_I(\alpha+\beta)+\frac{\alpha^2}{N}\Phi_I''(\alpha+\beta),\end{aligned} \tag{5.46}$$

where Φ_I denotes the characteristic function of I and Φ_I'' refers to the second-order derivative of Φ_I.

We derive an analytic approximation of the two residual terms by expressing them as Fourier integrals via the Parseval Theorem. For the first residual term, we propose the approximation

$$\begin{aligned}E^Q[(I^{(N)}-K)^+\mathbf{1}_{\{I\le c^*\}}] &\approx E^Q[(\hat{I}^{(N)}-K)^+\mathbf{1}_{\{I\le c^*\}}]\\ &= \frac{1}{4\pi^2}\int_{\mathrm{i}b-\infty}^{\mathrm{i}b+\infty}\int_{\mathrm{i}a-\infty}^{\mathrm{i}a+\infty} e^{-\mathrm{i}\alpha K-\mathrm{i}\beta c^*}\frac{\Phi_{\hat{I}^{(N)},I}(\alpha,\beta)}{\mathrm{i}\beta\alpha^2}\,d\alpha\,d\beta,\end{aligned}$$

where $a<0$ and $b>0$ are chosen so that the integration contours are within the domain of analyticity of the two-dimensional generalized Fourier transform. In the next step, we apply an analytic approximation of the joint characteristic function $\Phi_{\hat{I}^{(N)},I}(\alpha,\beta)$ given by (5.46). For convenience, we write $z=\alpha+\beta$ so that

$$\begin{aligned}&E^Q[(I^{(N)}-K)^+\mathbf{1}_{\{I\le c^*\}}]\\ \approx\ &\frac{1}{2\pi}\int_{\mathrm{i}u-\infty}^{\mathrm{i}u+\infty} e^{-\mathrm{i}zc^*}\Phi_I(z)\frac{1}{2\pi}\int_{\mathrm{i}a-\infty}^{\mathrm{i}a+\infty}\frac{e^{-\mathrm{i}\alpha(K-c^*)}}{\mathrm{i}(z-\alpha)\alpha^2}\,d\alpha\,dz\\ &+\frac{1}{2\pi}\int_{\mathrm{i}u-\infty}^{\mathrm{i}u+\infty} e^{-\mathrm{i}zc^*}\frac{\Phi_I''(z)}{N}\frac{1}{2\pi}\int_{\mathrm{i}a-\infty}^{\mathrm{i}a+\infty}\frac{e^{-\mathrm{i}\alpha(K-c^*)}}{\mathrm{i}(z-\alpha)}\,d\alpha\,dz,\end{aligned} \tag{5.47}$$

where $u=a+b>a$ specifies the horizontal contour in the complex integral with respect to z. In a similar manner, we may approximate the second residual term by

$$\begin{aligned}E^Q[(K-I^{(N)})^+\mathbf{1}_{\{I>c^*\}}] &\approx E^Q[(K-\hat{I}^{(N)})^+\mathbf{1}_{\{I>c^*\}}]\\ &= \frac{1}{4\pi^2}\int_{\mathrm{i}\hat{b}-\infty}^{\mathrm{i}\hat{b}+\infty}\int_{\mathrm{i}\hat{a}-\infty}^{\mathrm{i}\hat{a}+\infty} e^{-\mathrm{i}\alpha K-\mathrm{i}\beta c^*}\frac{\Phi_{\hat{I}^{(N)},I}(\alpha,\beta)}{-\mathrm{i}\beta\alpha^2}\,d\alpha\,d\beta,\end{aligned}$$

where $\hat{a}>0$ and $\hat{b}<0$ are chosen to ensure that the integration contours are within the domain of analyticity of the two-dimensional generalized Fourier transform. Again, by applying the approximation in (5.46) and letting $z = \alpha + \beta$, we obtain

$$\begin{aligned}
& E^Q[(K - I^{(N)})^+ 1_{\{I>c^*\}}] \\
\approx & -\frac{1}{2\pi}\int_{i\hat{u}-\infty}^{i\hat{u}+\infty} e^{-izc^*}\Phi_I(z)\frac{1}{2\pi}\int_{i\hat{a}-\infty}^{i\hat{a}+\infty}\frac{e^{-i\alpha(K-c^*)}}{i(z-\alpha)\alpha^2}\,d\alpha\,dz \\
& -\frac{1}{2\pi}\int_{i\hat{u}-\infty}^{i\hat{u}+\infty} e^{-izc^*}\frac{\Phi_I''(z)}{N}\frac{1}{2\pi}\int_{i\hat{a}-\infty}^{i\hat{a}+\infty}\frac{e^{-i\alpha(K-c^*)}}{i(z-\alpha)}\,d\alpha\,dz, \qquad (5.48)
\end{aligned}$$

where $\hat{u} = \hat{a} + \hat{b} < \hat{a}$ specifies the horizontal contour in the complex integral with respect to z.

Interestingly, the corresponding integrands in the Fourier integrals in (5.47) and (5.48) are identical. The two Fourier integrals differ only in the choices of the contours, where one is along a horizontal contour below the real axis oriented in the positive direction while the other is along a horizontal contour above the real axis oriented in the negative direction. This is not surprising since the two quantities in the two residual terms have the same analytic form but differ in sign. We include the vertical contours at the two extreme ends on the right and left sides of the complex plane that join the two horizontal contours to form a closed contour C. The values of the contour integrals along the two vertical contours at the positive and negative far-end side of the complex plane are seen to tend to zero value in the asymptotic limit.

We can combine the Fourier integrals in (5.47) and (5.48) that approximate the two residual terms. We choose a common contour for the integral with respect to z. That is, we choose the horizontal contour to be from $i\tilde{u} - \infty$ to $i\tilde{u} + \infty$, where $a < \tilde{u} < \hat{a}$. Also, we use the Cauchy Residue Theorem to evaluate the inner contour integral with respect to the closed contour C. Since we have chosen $a < \tilde{u} < \hat{a}$, where $a < 0$ and $\hat{a} > 0$, the poles are included inside the closed contour C. By combining the respective first terms in (5.47) and (5.48), we obtain

$$\begin{aligned}
A &= \frac{1}{2\pi}\int_{i\tilde{u}-\infty}^{i\tilde{u}+\infty} e^{-izc^*}\Phi_I(z)\frac{1}{2\pi}\oint_C\frac{e^{-i\alpha(K-c^*)}}{i(z-\alpha)\alpha^2}\,d\alpha\,dz \\
&= \frac{1}{2\pi}\int_{i\tilde{u}-\infty}^{i\tilde{u}+\infty} e^{-izc^*}\Phi_I(z)\frac{1+iz(c^*-K)-e^{iz(c^*-K)}}{z^2}\,dz. \qquad (5.49)
\end{aligned}$$

In a similar manner, by combining the respective second terms in (5.47) and (5.48), we obtain

$$\begin{aligned}
B &= \frac{1}{2\pi}\int_{i\tilde{u}-\infty}^{i\tilde{u}+\infty} e^{-izc^*}\frac{\Phi_I''(z)}{N}\frac{1}{2\pi}\oint_C\frac{e^{-i\alpha(K-c^*)}}{i(z-\alpha)}\,d\alpha\,dz \\
&= \frac{1}{2\pi}\int_{i\tilde{u}-\infty}^{i\tilde{u}+\infty} e^{-izc^*}\frac{\Phi_I''(z)}{N}[-e^{-iz(K-c^*)}]\,dz, \\
&= \frac{K^2}{2\pi N}\int_{i\tilde{u}-\infty}^{i\tilde{u}+\infty} e^{-izK}\Phi_I(z)\,dz \quad \text{(applying integration by parts twice)} \\
&= \frac{K^2}{N}f_I(K). \qquad (5.50)
\end{aligned}$$

We manage to express the approximation of the two residual terms as the sum of a one-dimensional integral and an explicitly known term.

One can provide the financial interpretation of the above two terms. It is easily visualized that the term A is simply equal to the following quantity:

$$E^Q[(I-K)^+ 1_{\{I \leq c^*\}}] + E^Q[(K-I)^+ 1_{\{I > c^*\}}].$$

In other words, keeping the single term A alone in the analytic approximation would be equivalent to approximating the two residual terms by simply replacing $I^{(N)}$ by I. Since the optimal solution $c^* \approx K$, we expect that both $\{K < I \leq c^*\}$ and $\{K \geq I > c^*\}$ are small probability events. Therefore, the correction contributed by A would be small and secondary. The second term B is seen to be identical to the discretization adjustment term [see (5.35)] in Drimus *et al.* (2016). This discretization adjustment arises when Drimus *et al.* attempt to account for the discrete sampling effect of discrete realized variance in the approximation of $E^Q[(I^{(N)} - K)^+]$ by $E^Q[(I-K)^+]$. It is interesting to observe that B has dependence on N but no dependence on c^* while A has the reverse properties of functional dependence. The term B provides the discretization gap between $I^{(N)}$ and I that is not captured by the optimal lower bound. In general, the contribution of B as an adjustment term added to the optimal lower bound is more significant compared to that of A.

Conditional gamma distribution approximation

The conditional normal distribution is based on the asymptotic behavior of $I^{(N)}$ as $N \to \infty$. When we consider pricing of short-maturity options on discrete realized variance, the asymptotic behavior of the discrete realized variance as $T \to 0$ is more relevant. In this regard, Keller-Ressel and Muhle-Karbe (2013) propose the asymptotic gamma distribution of the discrete realized variance as $T \to 0$. More specifically, it can be shown that the annualized continuous realized variance tends to the initial value of the instantaneous variance V_0 as $T \to 0$ while the discrete realized variance converges in distribution to a gamma distribution with shape parameter $N/2$ and scale parameter $2V_0/N$ (see Theorem 5.3 in Sec. 5.5.1). Motivated by this theoretical result, we propose to approximate $I^{(N)}$ by $\hat{I}^{(N)}$, which is a gamma distribution with shape parameter $N/2$ and scale parameter $2I/N$ conditional on I. We write

$$\hat{I}^{(N)}|I \sim \text{gamma}(N/2, 2I/N). \tag{5.51}$$

The above gamma distribution approximation has the same conditional mean and variance as the normal distribution approximation discussed earlier. Specifically, the gamma distribution approximation is advantageous over the normal distribution approximation in the following two aspects. First, it becomes exact in asymptotic limit as $T \to 0$. Second, the gamma distribution approximation retains nonnegativity of $I^{(N)}|I$.

As the first step in deriving the analytic approximation of the residual terms using the conditional gamma distribution approximation, we express the residual terms in terms of nested conditional expectation:

$$E^Q[E^Q[(K - I^{(N)})^+|I] 1_{\{I > c^*\}}] + E[E[(I^{(N)} - K)^+|I] 1_{\{I \leq c^*\}}].$$

Substituting the explicit form of the gamma density function and applying the put-call parity relation, the inner expectation can be evaluated as follows:

$$
\begin{aligned}
& E^Q[(K-I^{(N)})^+|I] \\
\approx\ & \int_0^K (K-y)\frac{y^{N/2-1}e^{-\frac{Ny}{2I}}}{\Gamma(N/2)(2I/N)^{N/2}}\,\mathrm{d}y \\
=\ & \frac{1}{\Gamma(\frac{N}{2})}\left[(K-I)\gamma\left(\frac{N}{2},\frac{KN}{2I}\right)+K\exp\left(\left(\frac{N}{2}-1\right)\ln\frac{KN}{2I}-\frac{KN}{2I}\right)\right] \\
& E^Q[(I^{(N)}-K)^+|I] \\
=\ & E[I^{(N)}|I]-K+E[(K-I^{(N)})^+|I] \\
\approx\ & I-K+\frac{1}{\Gamma(\frac{N}{2})}\left[(K-I)\gamma\left(\frac{N}{2},\frac{KN}{2I}\right)\right. \\
& \left.\qquad +K\exp\left(\left(\frac{N}{2}-1\right)\ln\frac{KN}{2I}-\frac{KN}{2I}\right)\right],
\end{aligned}
$$

where $\Gamma(s)=\int_0^\infty z^{s-1}e^{-z}\,\mathrm{d}z$ is the gamma function and $\gamma(s,x)=\int_0^x z^{s-1}e^{-z}\,\mathrm{d}z$ is the lower incomplete gamma function. Putting the above results together, the correction term C_g that is added to the optimal lower bound based on the conditional gamma distribution approximation is given by

$$C_g=\int_0^\infty G(y)f_I(y)\,\mathrm{d}y+\int_0^{c^*}(y-K)f_I(y)\,\mathrm{d}y, \tag{5.52}$$

where

$$G(y)=\frac{1}{\Gamma(\frac{N}{2})}\left[(K-y)\gamma\left(\frac{N}{2},\frac{KN}{2y}\right)+K\exp\left(\left(\frac{N}{2}-1\right)\ln\frac{KN}{2y}-\frac{KN}{2y}\right)\right].$$

Unlike the earlier derivation of the conditional normal distribution approximation, it is not necessary to use the method of double Fourier transform in the above derivation of the conditional gamma distribution approximation.

The above correction formula also conforms well with financial intuition. The first integral in C_g is seen to be $E^Q[(K-\hat{I}^{(N)})^+]$ under the conditional gamma distribution approximation. The second term can be interpreted as

$$E^Q[\hat{I}^{(N)}-K]-E^Q[(\hat{I}^{(N)}-K)\mathbf{1}_{\{I>c^*\}}]$$

under the same approximate distribution. The sum gives

$$E^Q[(\hat{I}^{(N)}-K)^+]-E^Q[(\hat{I}^{(N)}-K)\mathbf{1}_{\{I>c^*\}}],$$

which is exactly the residual with $I^{(N)}$ being replaced by $\hat{I}^{(N)}$ under the conditional gamma distribution. The small time asymptotic approximation approach by Keller-Ressel and Muhle-Karbe (2013) attempts to approximate the "discretization gap" between the price of an option on continuous realized variance and that of the continuous counterpart (see Sec. 5.4). The PEB approximation considers approximating

$I^{(N)}$ by $\hat{I}^{(N)}$ under the approximate gamma distribution in the residual terms. As a result, while the small time asymptotic approximation is only guaranteed to perform well for small T, the PEB approximation would provide high level of accuracy over a wider range of T.

Connection between the normal distribution approximation and gamma distribution approximation

We would like to connect the above conditional normal distribution approximation and conditional gamma distribution approximation through the well-known normal approximation to the gamma distribution. By virtue of the Central Limit Theorem, it is well known that the gamma distribution with shape parameter k and scale parameter θ converges to the normal distribution with mean $k\theta$ and variance $k\theta^2$ when k is sufficiently large. We would like to show that when N is sufficiently large, the gamma distribution given by (5.51) converges to the normal distribution given by (5.45). We consider the Taylor expansion in powers of $1/N$ of the moment generating function $M_g(z)$ of the gamma distribution:

$$\begin{aligned} M_g(z) &= \left(1-\frac{2I}{N}z\right)^{-N/2} = \exp\left(-\frac{N}{2}\ln\left(1-\frac{2I}{N}z\right)\right) \\ &= \exp\left(-\frac{N}{2}\left[-\frac{2I}{N}z-\frac{1}{2}\left(\frac{2I}{N}z\right)^2+O(N^{-3})\right]\right) \\ &= \exp\left(Iz+\frac{I^2}{N}z^2+O(N^{-2})\right). \end{aligned} \tag{5.53}$$

Suppose we ignore the higher order terms $O(N^{-2})$ in the above Taylor expansion, it becomes identical to the moment generating function of the normal distribution in (5.45). This connection helps explain why the performances of the two approximations for long-maturity options on discrete realized variance are almost indistinguishable (see Table 5.2). Finally, we remark that since we have made the simplification shown in (5.46), simply replacing the gamma density in (5.51) with its normal approximation would not lead to the same formulas as shown in (5.49) and (5.50).

Though the PEB approximation procedure does not depend on any specific stochastic volatility model of the asset price process, the success of the implementation of the procedure relies on availability of the joint characteristic function of Δ_k and I in analytic form. Thanks to the affine structure of the Heston stochastic volatility model with jumps, we are able to express $\Phi_{\Delta_k,I}(\alpha,\beta)$ in an exponential affine form.

5.3.3 Numerical calculations of partially exact and bounded approximation

In this subsection, we present the numerical calculations that were performed to examine accuracy of the proposed PEB approximations as reported by Zheng and Kwok (2015). Though the PEB method can be applied for pricing options on discrete realized variance under a general model assumption, it is particularly effective for the affine stochastic volatility models with jumps since they admit closed-form characteristic functions. For illustrative purpose, the numerical examples are confined to the Heston stochastic volatility model with compound Poisson jumps [see (2.121)].

The model parameter values for the Heston model with jumps in the numerical calculations are adopted from Duffie *et al.* (2000) and they are shown in Table 5.1. Furthermore, we choose $r = 0.0319$, $q = 0$ and $S_0 = 1$.

κ	3.46	ν	−0.086
θ	$(0.0894)^2$	λ	0.47
ε	0.14	Δ	0.0001
ρ	−0.82	$\sqrt{V_0}$	0.087

TABLE 5.1: The basic set of parameter values of the Heston model with jumps in asset price.

Numerical accuracy

Zheng and Kwok (2015) calculated the prices of the call options on daily sampled realized variance with varying sampling periods and strike prices. They made three choices of maturities, $N = 20$, $N = 126$ and $N = 252$, representing one month (short), half a year (intermediate) and a year (long), respectively. For each maturity, they choose three representative strike prices that correspond to deep in-the-money (ITM), at-the-money (ATM) and deep out-of-the-money (OTM) call options. They also list the prices of the call options on the continuous realized variance to compare with the prices of the discrete counterparts. The benchmark Monte Carlo simulation results are generated by simulating 8×10^5 paths with step size $\Delta t = \frac{1}{252} \times \frac{1}{16}$ according to the following Monte Carlo simulation scheme:

$$\ln S_{t+\Delta t} = \ln S_t + \left(r - q - \frac{V_t^+}{2}\right)\Delta t + \sqrt{V_t^+ \Delta t}\left(\rho Z_2 + \sqrt{1-\rho^2} Z_1\right) + \sum_{i=1}^{N_{\Delta t}} J_i$$

$$V_{t+\Delta t} = V_t + \kappa \Delta t(\theta - V_t^+) + \varepsilon \sqrt{V_t^+ \Delta t} Z_2, \tag{5.54}$$

where $V_t^+ = \max(V_t, 0)$, Z_1 and Z_2 are two independent standard normal random variables, and J_i are independent copies of the random jump size. To hasten the rate of convergence of the simulation, they use the discrete realized variance as a control variate [see Broadie and Jain (2008)].

The numerical results in Table 5.2 reveal that the performance of the lower bound approximation is quite similar to the crude approximation using the price of the call option on the continuous realized variance. For short-maturity options, though numerical accuracy is not quite satisfactory in general, the lower bound approximation slightly outperforms the "Cont" approximation. Both the PEB approximation methods with the normal or gamma distribution approximation have shown significant improvement over the lower bound approximation in terms of numerical accuracy.

Maturity	Strike	Cont (RE)	LB (RE)	PEBn (RE)	PEBg (RE)	MC (SE)
	7.049	2.938(10.4%)	2.956(9.84%)	3.423(4.41%)	3.309(0.93%)	3.278(0.002)
N = 20	8.812	2.685(7.01%)	2.703(6.39%)	2.879(0.27%)	2.908(0.74%)	2.887(0.002)
	10.574	2.595(3.23%)	2.595(3.23%)	2.624(2.16%)	2.679(0.11%)	2.682(0.002)
	45.087	18.817(1.25%)	18.773(1.48%)	19.041(0.08%)	19.033(0.12%)	19.055(0.008)
N = 126	56.358	14.721(1.29%)	14.698(1.45%)	14.903(0.07%)	14.898(0.10%)	14.914(0.010)
	67.630	11.696(0.88%)	11.671(1.10%)	11.788(0.10%)	11.791(0.09%)	11.801(0.010)
	90.836	34.210(0.62%)	34.160(0.76%)	34.382(0.12%)	34.379(0.13%)	34.423(0.013)
N = 252	113.545	23.131(0.89%)	23.088(1.07%)	23.328(0.04%)	23.341(0.01%)	23.338(0.017)
	136.254	14.652(2.28%)	14.642(2.35%)	15.077(0.55%)	15.059(0.43%)	14.994(0.019)

TABLE 5.2: All the option prices are interpreted as basis points, where the calculated results are multiplied by 10^4. "Cont" refers to the prices of the call options on the continuous realized variance, "LB" means the lower bound approximation given by (5.43), "PEBn" means the PEB approximation with normal distribution, "PEBg" means the PEB approximation with gamma distribution, and "MC" refers to the Monte Carlo simulation results using the scheme (5.54). The numbers in brackets after numerical option prices represent the relative errors (RE) with the Monte Carlo simulation results as the benchmark for comparison. The numbers in brackets after the Monte Carlo simulation values represent the standard error (SE) in the Monte Carlo simulation calculations.

However, the PEB method with the normal distribution approximation fails to deliver consistent accurate approximation for the one-month call options. On the other hand, the PEB method with the gamma distribution approximation provides very accurate results even for the short-maturity options. This is expected since the gamma distribution approximation is exact in the asymptotic limit when $T \to 0^+$. The gamma distribution approximation remains to perform equally well for relatively long maturities, which supports the theoretical result that the gamma distribution approximation converges to the normal distribution when N is sufficiently large. The numerical experiment once again confirms the significant discrepancy between the discrete realized variance and continuous realized variance when the time to maturity is small. The two PEB approximation methods, especially the gamma distribution approximation, have been shown to be an efficient and accurate analytic approximation method for pricing options on discrete realized variance under all ranges of maturities and strikes.

As a summary, the numerical tests demonstrate that the PEB approximation formulas provide very good numerical accuracy, for pricing options on discrete realized variance under the Heston stochastic volatility model with jumps, without the shortcoming exhibited in other analytic approximation methods where accuracy may deteriorate substantially in pricing options with short maturities. The high level of numerical accuracy is attributed to the adoption of either the normal or gamma approximation of the distribution of discrete realized variance conditional on continuous realized variance. Since the gamma distribution approximation is exact in asymptotic limit as maturity tends to zero, the PEB method with the gamma distribution approximation is more reliable when pricing short-maturity options on discrete realized variance. For options with medium to long maturities, the gamma distribution is closely connected to the normal distribution. The approximation using either distribution is seen to be highly reliable and provide numerical accuracy within 1% error for most reasonable ranges of model parameter values.

5.4 Small time asymptotic approximation

To explore the impact of discrete sampling on the prices of options on discrete realized variance, Keller-Ressel and Muhle-Karbe (2013) consider the small time asymptotic limit close to maturity of the continuous and discrete realized variance under general semimartingale framework. It is found that as $T \to 0^+$, $I(0,T;N)$ and $I(0,T;\infty)$ can be approximated by the gamma distribution, which then enables the quantification of the discrete sampling effect in small time limit. It follows that the price of an option on discrete realized variance can be approximated by the price of its continuous counterpart plus a correction term, which is taken to be the small time limit of the price difference. It is shown that this approach works surprisingly well for short-maturity options on discrete realized variance. Also, the discrete sampling effect does not change much with different values of T. In what follows, we first present the method under the setting of Lévy models and generalization to the semimartingale framework is discussed subsequently.

5.4.1 Small time asymptotics under Lévy models

Consider a square-integrable Lévy process L_t which is characterized by its triplet (b, σ^2, F) with respect to the truncation function $h(x) = x$. By Lévy-Khintchine's Theorem, its Lévy exponent is given by

$$\psi_L(u) = ub + \frac{\sigma^2}{2}u^2 + \int (e^{ux} - 1 - ux)F(\mathrm{d}x), \tag{5.55}$$

for which $E[e^{uL_t}] = \exp(t\psi_L(u))$ and $u \in \mathcal{D}_L \equiv \{u \in \mathbb{C} : E[e^{uL_t}] < \infty\}$. Moreover, L_t admits the following decomposition:

$$L_t = bt + \sigma W_t + J_t, \tag{5.56}$$

where W_t is a standard Brownian motion and J_t is an independent centered pure jump process. Let the log asset price be driven by L_t under a risk neutral measure Q as follows:

$$X_t = \ln S_t = X_0 + (r-q)t + L_t - t\psi_L(1). \tag{5.57}$$

We establish the following two theorems on the asymptotic small time limit of the continuous realized variance and its discrete counterpart.

Theorem 6 (Continuous realized variance) *Suppose the set of payoff functions (indexed by T) $g_T : \mathbb{R}_+ \to \mathbb{R}$, $T \geq 0$, are continuous, uniformly bounded and satisfy $\|g_T - g_0\|_\infty \to 0$ as $T \to 0^+$. Asymptotically, we have*

$$\lim_{T\to 0^+} E\left[g_T\left(\frac{1}{T}[X,X]_T\right)\right] = g_0(\sigma^2). \tag{5.58}$$

Proof *First, we observe*

$$\begin{aligned} |E[g_T([X,X]_T/T)] - g_0(\sigma^2)| \quad &\leq \quad E[|g_T([X,X]_T/T) - g_0([X,X]_T/T)|] \\ &\quad + E[|g_0([X,X]_T/T) - g_0(\sigma^2)|]. \end{aligned}$$

Since $\|g_T - g_0\| \to 0$ as $T \to 0^+$, by the dominated convergence theorem, we have the first term converge to zero. On the other hand, since

$$[X,X]_T/T \to \sigma^2 + \lim_{T\to 0^+} [J,J]_T/T$$

and it can be shown that the latter limit is indeed zero. The dominated convergence theorem and continuity of g_0 imply that the second term converges to zero as well.

Theorem 7 (Discrete realized variance) *Suppose that the set of payoff functions (indexed by T) $g_{N,T} : \mathbb{R}_+ \to \mathbb{R}$, $T \geq 0$, $N \in \mathbb{N}$, are continuous, uniformly bounded and satisfy $\|g_{N,T} - g_{N,0}\|_\infty \to 0$ as $T \to 0^+$ for each N. We observe the following asymptotic small time limit property:*

$$\lim_{T\to 0^+} E\left[g_{N,T}\left(I(0,T;N)\right)\right] = E\left[g_{N,0}(Z_N)\right], \tag{5.59}$$

where Z_N has a gamma distribution with shape parameter $N/2$ and scale parameter $2\sigma^2/N$.

The proof of Theorem 5.3 can be found in Keller-Ressel and Muhle-Karbe (2013). Interestingly, the limiting distribution of $I(0,T;N)$ depends only on the number of monitoring intervals N and variance σ^2 of the continuous part of X_t. The jump part, however, may affect the value indirectly through the payoff functions $g_{N,T}$. For example, the jump part can affect the at-the-money strike of an option on discrete realized variance.

Suppose there exists a function $g : \mathbb{R}_+ \to \mathbb{R}$ such that $\|g_T - g\|_\infty \to 0$ and $\|g_{N,T} - g\|_\infty \to 0$ as $T \to 0^+$ for all $N \in \mathbb{N}$, then we have

$$\begin{aligned} &\lim_{T\to 0^+} E\left[g_{N,T}(I(0,T;N)) - g_T\left(\frac{1}{T}[X,X]_T\right)\right] \\ = \quad & E[g(Z_N) - g(\sigma^2)] = \Delta_g(N), \end{aligned} \tag{5.60}$$

where $\Delta_g(N)$ is called the discretization gap. It can well capture the impact of discrete sampling.

Proposition 5.1 *Assume that the function g is convex, then $\Delta_g(N)$ has the following properties:*

1. *$\Delta_g(N) \geq 0$ for all $N \in \mathbb{N}$.*

2. *$\Delta_g(N) = 0$ if and only if $\sigma^2 = 0$ or g is affine-linear.*

3. *The mapping $N \mapsto \Delta_g(N)$ is decreasing in N and converges to 0 as $N \to \infty$.*

Proof *Since g is convex and $E[Z_N] = \sigma^2$, Jensen's inequality gives*

$$E[g(Z_N)] \geq g(E[Z_N]) = g(\sigma^2).$$

This proves the first property. The proof of the if part of (2) is straightforward. To show the only if part, suppose $\sigma^2 = 0$ and g is not affine-linear, then g is strictly convex on some interval (a,b). Since $\sigma^2 > 0$, the interval (a,b) has strictly positive measure under the law of Z_N and the strict Jensen inequality implies that $\Delta_g(N) > 0$, which is a contradiction. Part (3) is linked to the property of the gamma distribution and can be verified easily.

Intuitively, the first property implies that under the Lévy model, a call or put option on discrete realized variance is at least as expensive as its counterpart on quadratic variation for small maturities. The second property says that the discretization gap disappears under a pure jump Lévy model or the payoff function takes a special linear form in small time limit. The latter observation implies that the continuous variance swap is indeed quite close to its discrete counterpart. The third one is in effect a natural consequence of the convergence of the realized variance toward the quadratic variation. Using $\Delta_g(n)$, we have the following small time asymptotic approximation:

$$E[g_{N,T}(I(0,T;N))] \approx E\left[g_T\left(\frac{1}{T}[X,X]_T\right)\right] + \Delta_g(N). \tag{5.61}$$

5.4.2 Small time asymptotics under the semimartingale models

The small time asymptotic approximation can be extended to the general semimartingale framework. Suppose the log asset price under a risk neutral measure Q is driven by the following semimartingale process:

$$dX_t = b_t \, dt + \sigma_t \, dW_t + k_t(x) * (N(dt,dx) - F(dx)dt), \tag{5.62}$$

where W_t is a standard Brownian motion, $N(dt,dx)$ is a Poisson random measure with absolutely continuous compensator $F(dx)dt$ while b_t, σ_t, and k_t are all predictable. We further assume that X_t is square-integrable so that

$$\int_0^T \left(E^Q[b_t^2] + E^Q[\sigma_t^2] + E^Q\left[\int k_t^2(x)F(dx)\right]\right) < \infty.$$

Since we take the small time limit $T \to 0^+$, it is natural to approximate X_t by the square-integrable Lévy process as governed by

$$d\tilde{X}_t = b_0 dt + \sigma_0 dW_t + k_0(x) * (N(dt,dx) - F(dx)dt),$$

which is obtained by freezing all the coefficients of X_t at time 0. Under some additional convergence conditions for the coefficients, we can deduce analogous results on the small time asymptotics as those under the Lévy models.

Theorem 8 *Let X_t be a semimartingale of the form (5.62).*

1. *Suppose the set of payoff functions (indexed by T) $g_T : \mathbb{R}_+ \to \mathbb{R}$, $T \geq 0$, are continuous, uniformly bounded and satisfy $\|g_T - g_0\|_\infty \to 0$ as $T \to 0^+$. Moreover, suppose that g_0 is Lipschitz continuous. We have the following small time asymptotic approximation result:*

$$\lim_{T\to 0^+} E^Q\left[g_T\left(\frac{1}{T}[X,X]_T\right)\right] = g_0(\sigma_0^2). \tag{5.63}$$

2. *Suppose that the set of payoff functions (indexed by T) $g_{N,T} : \mathbb{R}_+ \to \mathbb{R}$, $T \geq 0$, $N \in \mathbb{N}$, are continuous, uniformly bounded and satisfy $\|g_{N,T} - g_{N,0}\|_\infty \to 0$ as $T \to 0^+$ for each N. Moreover, suppose that $g_{N,0}$ is Lipschitz continuous. We observe the following small time asymptotic limit property:*

$$\lim_{T\to 0^+} E^Q[g_{N,T}(I(0,T;N))] = E^Q[g_{N,0}(Z_N)], \tag{5.64}$$

where Z_N has the gamma distribution with shape parameter $N/2$ and scale parameter $2\sigma_0^2/N$.

The proof of Theorem 5.4 is quite similar to that of Theorems 6 and 7. In fact, most of the conclusions made in the previous Lévy setting can be transferred to the semimartingale setting after "freezing" the coefficients, and most of the stochastic volatility models. The Heston model, 3/2 model and their extensions with jumps are covered under this semimartingale framework.

5.4.3 Option pricing using small time asymptotic approximation

We illustrate how to use small time asymptotics to derive an approximation to the price of a put option on the discrete realized variance under the Lévy model as specified by (5.57). Consider the two puts on continuous realized variance and discrete realized variance, respectively. Their payoff functions are specified as follows:

$$\begin{aligned} g_T(x) &= (mK(0,T;\infty) - x)^+, \\ g_{N,T}(x) &= (mK(0,T;N) - x)^+. \end{aligned}$$

Here, m is the moneyness ($m = 1$ for at-the-money options), and

$$\begin{aligned} K(0,T;\infty) &= E^Q\left[\frac{1}{T}[X,X]_T\right] = \sigma^2 + \nu^2, \\ K(0,T;N) &= E^Q[I(0,T;N)] = \sigma^2 + \nu^2 + \tilde{b}^2 T/N, \end{aligned}$$

where

$$\nu^2 = \int_{-\infty}^{\infty} x^2 \, F(\mathrm{d}x)$$

and

$$\tilde{b} = b + r - d - \psi_L(1).$$

Apparently, g_T and $g_{n,T}$ share the same asymptotic limit

$$g(x) = (m(\sigma^2 + \nu^2) - x)^+.$$

The small time asymptotic limit of the price of the put on continuous realized variance is given by

$$\lim_{T\to 0^+} E^Q\left[\left(mK(0,T;\infty) - \frac{1}{T}[X,X]_T\right)^+\right] = [\sigma^2(m-1) + \nu^2 m]^+. \tag{5.65}$$

Next, we compute the small time asymptotic limit of the put option on discrete realized variance. By Theorem 7, we have

$$\begin{aligned}
&\lim_{T\to 0^+} E^Q[(mK(0,T;N) - I(0,T;N))^+] \\
&= E^Q[(m(\sigma^2 + \nu^2) - Z_N)^+] \\
&= \sigma^2 P_{m,N}\left(\frac{\nu^2}{\sigma^2}\right) + [\sigma^2(m-1) + \nu^2 m] Q_{m,N}\left(\frac{\nu^2}{\sigma^2}\right).
\end{aligned} \tag{5.66}$$

Here, the functions $P_{m,N}(x)$ and $Q_{m,N}(x)$ are strictly decreasing and increasing on $[0,\infty)$, respectively. They are given by

$$P_{m,N} = \frac{2/N}{\Gamma(N/2)}\left[\frac{N}{2}\frac{m(1+x)}{\exp(m(1+x))}\right]^{N/2} \quad \text{and} \quad Q_{m,N}(x) = \frac{\gamma(m(1+x)N/2, N/2)}{\Gamma(N/2)},$$

where $\Gamma(x) = \int_0^\infty z^{x-1} e^{-z} \, \mathrm{d}z$ is the complete gamma function and $\gamma(x,u) = \int_0^u z^{x-1} e^{-z} \, \mathrm{d}z$ is the lower incomplete gamma function. Next, we calculate the discretization gap for the put option price as follows:

$$\begin{aligned}
\Delta_{m,N} = {} & \sigma^2 P_{m,N}\left(\frac{\nu^2}{\sigma^2}\right) + [\sigma^2(m-1) + \nu^2 m] Q_{m,N}\left(\frac{\nu^2}{\sigma^2}\right) \\
& - [\sigma^2(m-1) + \nu^2 m]^+.
\end{aligned} \tag{5.67}$$

The small time asymptotic approximation of the undiscounted price of the put option on discrete realized variance is given by

$$\begin{aligned}
& E^Q[(mK(0,T;N) - I(0,T;N))^+] \\
= \; & E^Q[(mK(0,T;\infty) - I(0,T;\infty))^+] + \Delta_{m,N},
\end{aligned} \tag{5.68}$$

where $\Delta_{m,N}$ is defined in (5.67).

Note that the payoff function of the call option on discrete or continuous realized variance is unbounded. Consequently, we cannot directly perform the small time asymptotic approximation. However, by the put-call parity relation, the price of the call option is equal to the price of the put option plus a forward contract. Since the forward contract has a linear payoff, then by part (2) in Proposition 5.1, we conclude that the discretization gap for the forward contract is zero. As a result, the discretization gap for the call option is equal to that of the put option. We then obtain similar small time asymptotic property of the price of the call option counterpart as follows:

$$E[(I(0,T;N) - mK(0,T;N))^+] = E[(I(0,T;\infty) - mK(0,T;\infty))^+] + \Delta_{m,N}, \quad (5.69)$$

where $\Delta_{m,N}$ is defined in (5.67).

Remark

The merit of the small time asymptotic approximation approach is that it works for any payoff function that satisfies the technical conditions in Theorems 6 and 7. Hence, it can be used as the fundamental approximation technique in dealing with discrete realized variance. Moreover, the small time asymptotic approximation method works under the general semimartingale framework that covers most of the popular stochastic volatility models. Though the small time asymptotic approximation is only guaranteed to work well theoretically for short-maturity options, numerical tests show that it is still able to deliver good approximation for options with longer maturities.

Chapter 6

Timer Options

Timer options were first introduced into the financial market in April 2007, though Neuberger (1990) has initiated the theoretical study of derivative products with features that are similar to the timer options. The basic structural features and uses of the timer options have been presented in Sec. 1.3.3. In essence, a timer option resembles its European vanilla counterpart, except that earlier expiration of the timer option may occur if the accumulated discrete realized variance exceeds a given variance budget on a monitoring date. That is, the expiration date may come earlier than the mandated expiration date. In essence, the earlier knock-out feature floats with the realized variance.

A timer option provides the buyer the combination of directional bet and volatility bet in single product. The product feature of a timer option can be related to the quadratic variation-based strategies. Bick (1995) discusses a family of dynamic trading strategies that rely on the current levels of stock price and cumulative quadratic variation. He proposes potential applications of these volatility free trading strategies in quadratic variation-based portfolio insurance. Replication of portfolio payoffs are done at the stopping time at which the realized cumulative variance hits a predetermined level. DeMarzo *et al.* (2016) propose the gradient strategies that minimize asymptotic regret in their robust option pricing procedures. The resulting pricing bounds on options depend only on the realized quadratic variation of the price process. Their pricing procedures can be applied to bound the price of a timer option without any assumption on the underlying price process.

The earlier timer option pricing models focus on the perpetual timer options that have no mandatory expiry date and assume continuous monitoring of potential premature knock-out. To reflect the actual contractual specifications, later research works manage to price finite maturity timer options under discrete monitoring of premature knock-out. This chapter discusses the different approaches of pricing timer options, which include analytic closed-form formulas, approximation methods, numerical algorithms, and analytic-simulation methods.

This chapter is summarized as follows: In Sec. 6.1, we present the model formulation of timer options under perpetuity within the framework of stochastic volatility models. Nice closed-form analytic formulas can be derived for the perpetual timer options under two special cases: (i) uncorrelated asset price and stochastic variance processes, (ii) zero interest rate and dividend yield. In Sec 6.2, we show how the perpetual timer option prices can be represented as conditional expectation on the Black-Scholes type formulas. Based on these results, we derive the analytic-simulation approach for effective pricing of the perpetual timer options. Alternatively, we also

DOI: 10.1201/9781003263524-6

show how the perpetual timer call and put option prices under the Heston model can be expressed in terms of integral transform formulas. In addition, we illustrate the use of the effective perturbation methods to obtain highly accurate analytic approximations to the perpetual timer option prices under the Heston model and 3/2-model. In the last section, we show how to derive the Fourier integral representation of the price functions of the finite maturity timer options under the 3/2-model. Also, we discuss the numerical approach of using the Fourier space time stepping algorithm to price the finite maturity discrete timer options under the Heston model and 3/2-model.

6.1 Model formulation

In the actual contracts of timer options, mandatory expiry date is specified and accumulated realized variance is monitored discretely (typically daily) for premature knock-out. To achieve better analytical tractability, most of the earlier works on pricing timer options assume perpetuity and continuous monitoring of premature knock-out. Under the assumption of continuously sampled integrated variance, the random maturity time of a timer option τ_B is defined to be the first hitting time that the integrated variance reaches the variance budget B, where

$$\tau_B = \inf\left\{t > 0: \int_0^t V_s \, ds = B\right\}. \tag{6.1}$$

Here, V_t is the instantaneous variance process and B is the variance budget set in the timer option contract. Under a risk neutral measure Q, the time-t price of a finite maturity timer call option is given by

$$c_t = E_t^Q[e^{-r(\tau_B \wedge T - t)} \max(S_{\tau_B \wedge T} - K,\, 0)], \quad 0 \leq t < T. \tag{6.2}$$

where r is the constant riskless rate, K is the strike price, T is the mandatory expiry date, and $\tau_B \wedge T$ denotes $\min(\tau_B, T)$. The challenge in finding the timer call price c_0 is the determination of the joint dynamics of $\tau_B \wedge T$ and $S_{\tau_B \wedge T}$.

The pricing of timer options under general stochastic volatility model is a three-dimensional problem. By employing the time change technique, Bernard and Cui (2011) manage to reduce the pricing of perpetual timer option to an one-dimensional problem. This section presents some of their key results, including the derivation of the Black-Scholes type formulas under the following two special cases: (i) zero interest rate and dividend yield, (ii) zero correlation between the asset price process and variance process.

6.1.1 Governing partial differential equation

Under a risk neutral measure Q, we consider a general stochastic volatility framework for the underlying asset process S_t and instantaneous variance process V_t that are specified by

$$\frac{dS_t}{S_t} = (r-q)dt + \sqrt{V_t}(\rho \, dW_t^1 + \sqrt{1-\rho^2} \, dW_t^2) \tag{6.3a}$$

$$dV_t = \alpha(V_t)dt + \beta(V_t)dW_t^1, \tag{6.3b}$$

where W_t^1 and W_t^2 are uncorrelated Brownian motions. Here, ρ is the correlation coefficient between the stochastic processes of S_t and V_t, r is the interest rate and q is the dividend yield. The parameters r, q, and ρ are assumed to be constant. The drift function $\alpha(V_t)$ and volatility function $\beta(V_t)$ are measurable functions with respect to the natural filtration generated by the two correlated Brownian motions.

Given the known integrated variance $I_t = \int_0^t V_s \, ds$ from time 0 to time t, τ_B is modified from (6.1) to become

$$\tau_B = \inf\left\{u > t : \int_t^u V_s \, ds = B - I_t\right\}. \tag{6.4}$$

The call price function $c_t = c(S,V,I,t)$ in (6.2) satisfies the following three-dimensional partial differential equation:

$$\begin{aligned} &\frac{\partial c}{\partial t} + (r-q)S\frac{\partial c}{\partial S} + \frac{S^2 V}{2}\frac{\partial^2 c}{\partial S^2} + V\frac{\partial c}{\partial I} + \alpha(V)\frac{\partial c}{\partial V} \\ &+ \frac{\beta^2(V)}{2}\frac{\partial^2 c}{\partial V^2} + \rho S\sqrt{V}\beta(V)\frac{\partial^2 c}{\partial S \partial V} - rc = 0, \end{aligned} \tag{6.5}$$

where $(S,V,I,t) \in [0,\infty) \times (0,\infty) \times [0,B] \times [0,\infty)$. Under the assumption of perpetuity, we drop the dependence on t. The domain of definition becomes $(S,V,I) \in [0,\infty) \times (0,\infty) \times [0,B]$. The perpetual timer option price $c(S,V,I)$ is completely determined by the state variables (S,V,I). The pricing problem involves finding the first hitting time that the three-dimensional Markov process (S_t,V_t,I_t) hits the plane:

$$\Gamma = \{(x_1,x_2,B) : x_1 \in \mathbb{R}_+, \; x_2 \in \mathbb{R}_+\}.$$

When $I = B$, the perpetual timer option is terminated and the payoff at knock-out is given by

$$c(S,V,B) = \max(S-K,0). \tag{6.6}$$

Under perpetuity, we take $T \to \infty$ and set $t = 0$ in (6.2). The time-0 price of a perpetual timer call option admits the following expectation representation under a risk neutral measure Q

$$c(S,V,I) = E_0^Q[e^{-r\tau_B}\max(S_{\tau_B} - K,0)], \tag{6.7}$$

where S, V, and I are the time-0 asset price, instantaneous variance and integrated variance, respectively. Accordingly, τ_B is given by setting $t = 0$ and I_t becomes I in (6.4).

Theorem 9 *Under the general stochastic volatility model as specified by (6.3a,b), the time-0 price of the perpetual timer call option is given by*

$$c_0 = E_0^Q \left[\max\left(S_0 e^{\widetilde{W}_B - B/2 - q\tau_B} - Ke^{-r\tau_B}, 0\right)\right], \tag{6.8}$$

where $\widetilde{W}$ is a Q-standard Brownian motion and τ_B is defined in (6.1).

Proof *The underlying asset price can be written as*

$$S_t = S_0 \exp\left((r-q)t - \frac{1}{2}\int_0^t V_s \, ds + M_t\right),$$

where

$$M_t = \int_0^t \sqrt{V_s}\left(\sqrt{1-\rho^2}\, dW_s^2 + \rho \, dW_s^1\right),$$

Note that M_t is a martingale, and it can be decomposed into a linear combination of the two martingale processes:

$$\int_0^t \sqrt{V_s}\, dW_s^1 \quad \textit{and} \quad \int_0^t \sqrt{V_s}\, dW_s^2.$$

Both of the above martingale processes share the same variance process V_s. By the Dubins-Schwarz theorem (Karatzas and Shreve, 1991), there exists a Q-standard Brownian motion $\widetilde{W}$ such that

$$M_t = \widetilde{W}_{[M,M]_t},$$

where $[M,M]_t$ is the quadratic variation process associated with M_t. In the current context, we deduce that

$$[M,M]_t = \int_0^t V_s \, ds.$$

At τ_B, we observe

$$[M,M]_{\tau_B} = \int_0^{\tau_B} V_s \, ds = B$$

and

$$\int_0^{\tau_B} \sqrt{V_s}\, dW_s^1 = \int_0^{\tau_B} \sqrt{V_s}\, dW_s^2 = \widetilde{W}_B.$$

As a result, we obtain

$$S_{\tau_B} = S_0 \exp\left((r-q)\tau_B - \frac{B}{2} + \widetilde{W}_B\right). \tag{6.9}$$

The expectation representation of the perpetual timer call option price in (6.8) *then follows by substituting S_{τ_B} in* (6.9) *into formula* (6.8).

From (6.8), we observe that evaluation of the perpetual timer call price c_0 remains to involve the two-dimensional expectation with correlated $\widetilde{W}_B$ and τ_B. Under the following two special cases, the expectation calculations can be simplified and analytic price formulas can readily be derived (Bernard and Cui, 2011).

Zero interest rate and dividend yield

Suppose $r = q = 0$, then dependence on τ_B vanishes since an investor is indifferent at the instant at which she receives the future cashflows. Hence, the random maturity no longer plays a role in the price of the timer option. Since the integrated variance up to τ_B is simply $[M,M]_{\tau_B} = B$, we can compute the price of the timer call option explicitly in the form of the Black-Scholes formula with the familiar integrated variance term $\sigma^2 T$ being replaced by B. Therefore, (6.8) degenerates to become

$$c_0 = E_0^Q[\max(S_0 e^{\widetilde{W}_B - B/2} - K,\ 0)] = S_0 N(d_1) - KN(d_2), \tag{6.10}$$

where

$$d_1 = \frac{\ln \frac{S_0}{K} + \frac{B}{2}}{\sqrt{B}}, \quad d_2 = d_1 - \sqrt{B}.$$

Apparently, when the target volatility budget B observes $B < \sigma_{imp}^2 T$, where σ_{imp} is the implied volatility, the timer call option is cheaper than its vanilla counterpart. Due to the presence of the volatility risk premium, it is common to set $B < \sigma_{imp}^2 T$, so a timer option is cheaper than its vanilla counterpart.

Zero correlation

When $\rho = 0$, $\widetilde{W}_B$ and τ_B are independent so that the expectation in (6.8) can be calculated using iterated expectation as follows:

$$\begin{aligned} c_0 &= E_0^Q\left[E^Q\left[\max\left(S_0 e^{\widetilde{W}_B - B/2 - q\tau_B} - Ke^{-r\tau_B}, 0\right)\right] \middle| \tau_B\right] \\ &= S_0 E_0^Q[e^{-q\tau_B} N(d_1(\tau_B))] - KE_0^Q[e^{-r\tau_B} N(d_2(\tau_B))], \end{aligned} \tag{6.11}$$

where

$$d_1(\tau_B) = \frac{\ln \frac{S_0}{K} + \frac{B}{2} + (r-q)\tau_B}{\sqrt{B}}, \quad d_2(\tau_B) = d_1(\tau_B) - \sqrt{B}.$$

Now, the timer call option price can be computed once the distribution of the random maturity τ_B is known. One may derive the characteristic function of τ_B in closed form so that the expectation in (6.11) can be expressed as a Fourier integral, the valuation of which can be effected by numerical integration.

6.2 Pricing perpetual timer options

In this section, we consider two general approaches for pricing perpetual timer options. In the conditional expectation approach, we derive the Black-Scholes type formula with random drift term μ_{τ_B}. The randomness of μ_{τ_B} arises from the random quantities: τ_B, V_{τ_B} and some functional on V_t expressed as $\int_0^{\tau_B} g(V_t)\, dt$, where $g(V)$

is defined later [see (6.12)]. The joint law of these three random quantities is derived under various stochastic volatility models. The perpetual timer call option price can be computed by taking expectation of the Black-Scholes type formula under the joint law. Under the Heston model, by observing that the distribution of (V_{τ_B}, τ_B) can be characterized by a Bessel process with constant drift, the price formulas of the perpetual timer option prices can be expressed in terms of integral transform formulas. Besides the conditional expectation approach, we also show how to use the perturbation approach to derive analytic approximation to the prices of perpetual timer options under the Heston model and $3/2$-model by considering the perturbation of the maturity and variance variables in the Black-Scholes type formula.

6.2.1 Conditional expectation based on Black-Scholes type formula

We show how to compute the perpetual timer call option price c_0 by taking expectation with respect to the joint distribution of τ_B and V_{τ_B}, and some functional on V_t under the general stochastic volatility models. By virtue of (6.9), given the realization of the instantaneous variance process up to the first hitting time τ_B, we would like to show that S_{τ_B} has a lognormal distribution of the form:

$$\ln \frac{S_{\tau_B}}{S_0} \sim N(\mu_{\tau_B}, (1-\rho^2)B).$$

Here, $N(\mu, \sigma^2)$ denotes the normal distribution with mean μ and variance σ^2. Given the dynamic of V_t as depicted in (6.3), we show how to relate μ_{τ_B} to $\alpha(V_t)$ and $\beta(V_t)$. Recall the following definition of $f(V)$ and $g(V)$ in terms of $\alpha(V)$ and $\beta(V)$, where

$$f(V) = \int_0^V \frac{\sqrt{u}}{\beta(u)}\, du \quad \text{and} \quad g(V) = \alpha(V) f'(V) + \frac{\beta^2(V)}{2} f''(V). \tag{6.12}$$

Recall from (4.8), that S_T admits the following integral representation

$$\begin{aligned} S_T = S_t \exp((r-q)(T-t) - \frac{1}{2}\int_t^T V_s \,\mathrm{d}s + \sqrt{1-\rho^2}\int_t^T \sqrt{V_s}\,\mathrm{d}W_s) \\ + \rho \left[f(V_T) - f(V_t) - \int_t^T g(V_s)\,\mathrm{d}s \right]. \end{aligned} \tag{6.13}$$

By setting $t = 0$ and $T = \tau_B$, we deduce that $\ln \frac{S_{\tau_B}}{S_0}$ is lognormally distributed with mean

$$\mu_{\tau_B} \quad = \quad (r-q)\tau_B - \frac{B}{2} + \rho \left[f(V_{\tau_B}) - f(V_0) - \int_0^{\tau_B} g(V_s)\,\mathrm{d}s \right], \tag{6.14}$$

and variance $(1-\rho^2)B$. Consequently, conditional on the realization of the instantaneous variance process, the perpetual timer call option can be visualized as a vanilla call option with maturity τ_B on a lognormal process with drift μ_{τ_B} and unannualized volatility $\sqrt{(1-\rho^2)B}$. As a result, the perpetual timer call price formula can be expressed in the form of the Black-Scholes type formula.

Theorem 10 *Under the general stochastic volatility model as specified in (6.3a,b), the time-0 price of a perpetual timer call option can be expressed as (Bernard and Cui, 2011)*

$$c_0 = E_0^Q\left[S_0 e^{\mu_{\tau_B} - r\tau_B + (1-\rho^2)B/2} N(d_1) - K e^{-r\tau_B} N(d_2)\right], \tag{6.15}$$

where μ_{τ_B} is given by (6.14) and

$$d_1 = \frac{\ln\frac{S_0}{K} + \mu_{\tau_B} + (1-\rho^2)B}{\sqrt{(1-\rho^2)B}}, \quad d_2 = d_1 - \sqrt{(1-\rho^2)B}.$$

Proof *The proof of* (6.15) *follows by first computing the expectation of the discounted exercise payoff conditioning on the triple $\left(\tau_B, V_{\tau_B}, \int_0^{\tau_B} g(V_s)\, \mathrm{d}s\right)$, then expressing in the form of the Black-Scholes formula. We then have*

$$c_0 = E_0^Q\left[E^Q\left[e^{-r\tau_B}\max(S_{\tau_B} - K, 0)\middle|\, \tau_B, V_{\tau_B}, \int_0^{\tau_B} g(V_s)\, \mathrm{d}s\right]\right]. \tag{6.16}$$

The result follows when the inner conditional expectation is calculated explicitly by applying the Black-Scholes formula.

To evaluate (6.16), it is necessary to find the joint law of $\left(\tau_B, V_{\tau_B}, \int_0^{\tau_B} g(V_s)\, \mathrm{d}s\right)$. The joint law is related to another process Y_t, the details are presented in Lemma 6.1.

Lemma 6.1 *Assuming that the stochastic variance process is driven by a general stochastic volatility model specified in (6.3b), which observes*

$$\lim_{t\to\infty}\int_0^t V_s\, \mathrm{d}s = \infty, \quad a.s.$$

so that τ_B defined in (6.1) is finite. The joint law of $\left(\tau_B, V_{\tau_B}, \int_0^{\tau_B} g(V_s)\, \mathrm{d}s\right)$ is given by (Bernard and Cui, 2011)

$$\left(\tau_B, V_{\tau_B}, \int_0^{\tau_B} g(V_s)\, \mathrm{d}s\right) \overset{d}{\sim} \left(\int_0^B Y_s^{-1}\, \mathrm{d}s, Y_B, \int_0^B \frac{g(Y_s)}{Y_s}\, \mathrm{d}s\right), \tag{6.17a}$$

where Y_t is governed by the following dynamic equation:

$$\mathrm{d}f(Y_t) = \frac{g(Y_t)}{Y_t}\, \mathrm{d}t + \mathrm{d}\hat{W}_t, \quad Y_0 = V_0. \tag{6.17b}$$

Here, $\hat{W}_t$ is a standard Brownian motion, and f and g are defined in (6.12).

The proof of Lemma 6.1 is given in the Appendix. To illustrate the implementation procedure under some popular stochastic volatility models, we derive the joint law of the triple under the Heston model and Hull-White model.

Heston model

Under the Heston model, V_t follows a square-root process, where $\alpha(x) = \kappa(\theta - x)$ and $\beta(x) = \varepsilon\sqrt{x}$. We observe $f(V) = \frac{V}{\varepsilon}$, $g(V) = \frac{\kappa}{\varepsilon}(\theta - V)$ and $\int_0^{\tau_B} V_t \, \mathrm{d}t = B$. Furthermore, we have

$$\int_0^{\tau_B} g(V_t) \, \mathrm{d}t = \int_0^{\tau_B} \frac{\kappa}{\varepsilon}(\theta - V_t) \, \mathrm{d}t = \frac{\kappa}{\varepsilon}(\theta\tau_B - B),$$

so it is completely determined by τ_B. Hence, the joint law of the triple degenerates to the joint law of (τ_B, V_{τ_B}). The corresponding μ_{τ_B} [see (6.14)] becomes

$$\mu_{\tau_B} = (r-q)\tau_B - \frac{B}{2} + \frac{\rho}{\varepsilon}(V_{\tau_B} - V_0 - \kappa\theta\tau_B + \kappa B).$$

By virtue of Lemma 6.1, the corresponding joint law is summarized in Proposition 6.1.

Proposition 6.1 *The joint law of (τ_B, V_{τ_B}) in the Heston model is given by*

$$(\tau_B, V_{\tau_B}) \overset{d}{\sim} \left(\int_0^B Y_s^{-1} \mathrm{d}s, \, Y_B \right), \tag{6.18}$$

where Y_t is a Bessel process governed by

$$\mathrm{d}Y_t = \left(\frac{\kappa\theta}{Y_t} - \kappa \right) \mathrm{d}t + \varepsilon \, \mathrm{d}\hat{W}_t, \quad Y_0 = V_0,$$

and $\hat{W}_t$ is a standard Brownian motion.

Hull-White model

Under the Hull-White model, $\alpha(x) = ax$ and $\beta(x) = \nu x$, where a and ν are constants. We obtain $f(V) = \frac{2\sqrt{V}}{\nu}$ and $g(V) = \left(\frac{a}{\nu} - \frac{\nu}{4}\right)\sqrt{V}$. The corresponding μ_{τ_B} becomes

$$\mu_{\tau_B} = (r-q)\tau_B - \frac{B}{2} + \frac{2\rho}{\nu}(\sqrt{V_{\tau_B}} - \sqrt{V_0}) - \rho\left(\frac{a}{\nu} - \frac{\nu}{4}\right)\int_0^{\tau_B} \sqrt{V_t} \, \mathrm{d}t.$$

After performing some calculus procedures, we obtain the joint law under the Hull-White model as summarized in Proposition 6.2.

Proposition 6.2 *The joint law of $\left(\tau_B, V_{\tau_B}, \int_0^{\tau_B} g(V_t) \, \mathrm{d}t\right)$ in the Hull-White model is given by*

$$\left(\tau_B, V_{\tau_B}, \int_0^{\tau_B} g(V_t) \, \mathrm{d}t\right) \overset{d}{\sim} \left(\frac{4}{\nu^2} \int_0^B Y_s^{-2} \, \mathrm{d}s, \frac{\nu^2}{4} Y_B^2, \left(\frac{2a}{\nu^2} - \frac{1}{2}\right) \int_0^B Y_s^{-1} \, \mathrm{d}s \right), \tag{6.19}$$

where Y_t is a standard Bessel process governed by

$$\mathrm{d}Y_t = \left(\frac{2a}{\nu^2} - \frac{1}{2}\right)\frac{1}{Y_t} \, \mathrm{d}t + \mathrm{d}\hat{W}_t, \quad Y_0 = \frac{2}{\nu}\sqrt{V_0},$$

and $\hat{W}_t$ is a standard Brownian motion.

Analytic-simulation procedure

Lemma 6.1 and the results in Propositions 6.1 and 6.2 provide the theoretical basis for the following analytic-simulation approach for pricing the perpetual timer call option under the Heston model and Hull-White model.

1. Based on (6.17b), we generate a random path of $\{Y_t, 0 \le t \le B\}$ and calculate $\left(\int_0^B Y_s^{-1} \mathrm{d}s, Y_B, \int_0^B \frac{g(Y_s)}{Y_s} \mathrm{d}s\right)$.
2. By virtue of (6.17a), given the simulated values $\left(\tau_B, V_{\tau_B}, \int_0^{\tau_B} g(V_t)\, \mathrm{d}t\right)$ from step 1, we compute the price of the perpetual timer call using (6.16).
3. Repeat steps 1 and 2 sufficiently in many simulation runs and take the sampled average.

As a remark, one can improve convergence of the Monte Carlo simulation using the control variate technique. Since there exists closed-form price formula of the timer call option when $r = q = 0$, this would serve as a good candidate for the control variate.

6.2.2 Integral price formulas under the Heston model

Let R_t^μ denote the Bessel process with order $\nu \ge 0$ and drift μ. The dynamic of R_t^μ is governed by

$$\mathrm{d}R_t^\mu = \left(\frac{2\nu+1}{2}\frac{1}{R_t^\mu} + \mu\right) \mathrm{d}t + \mathrm{d}W_t, \quad R_0^\mu > 0, \tag{6.20}$$

where W_t is a standard Brownian motion. Under the Heston model, the dynamics of V_t is governed by

$$\mathrm{d}V_t = \kappa(\theta - V_t)\, \mathrm{d}t + \varepsilon\sqrt{V_t}\, \mathrm{d}W_t^V. \tag{6.21}$$

Imposing the usual Feller condition: $2\kappa\theta \ge \varepsilon^2$, Proposition 6.1 reveals that the joint distribution of (V_{τ_B}, τ_B) can be characterized by a Bessel process with constant drift, where

$$(V_{\tau_B}, \tau_B) \overset{d}{\sim} \left(\varepsilon X_B, \int_0^B \frac{1}{\varepsilon X_u}\, \mathrm{d}u\right).$$

Here, X_t is a Bessel process with index $\nu = \frac{\kappa\theta}{\varepsilon^2} - \frac{1}{2}$ and drift $\mu = -\frac{\kappa}{\varepsilon}$, and whose dynamics is governed by

$$\mathrm{d}X_t = \left(\frac{\kappa\theta}{\varepsilon^2}\frac{1}{X_t} - \frac{\kappa}{\varepsilon}\right) \mathrm{d}t + \mathrm{d}W_t, \quad X_0 = V_0/\varepsilon.$$

For any arbitrary scalar $s > 0$, the joint density function of the zero-drift Bessel process is defined by

$$p(x,t;s)\, \mathrm{d}x\, \mathrm{d}t = P\left(R_s \in \mathrm{d}x, \int_0^s \frac{1}{R_u}\, \mathrm{d}u \in \mathrm{d}t\right). \tag{6.22}$$

It is desirable to derive the analytic representation of the joint law of the Bessel process R_t with index $\nu \geq 0$ and starting at $R_0 > 0$. Let T be an independent exponential random variable with intensity $\lambda > 0$. According to Borodin and Salminen (2002), the joint distribution of the Bessel process and the integral functional of its reciprocal stopped at an independent exponential time admits the following representation

$$P\left(R_T \in \mathrm{d}x, \int_0^T \frac{1}{R_u}\,\mathrm{d}u \in \mathrm{d}t\right) = q(x,t;\lambda)\,\mathrm{d}x\,\mathrm{d}t, \tag{6.23}$$

where

$$q(x,t;\lambda) = \frac{\lambda\sqrt{2\lambda}x^{\nu+1}}{R_0^{\nu}\sinh\left(t\sqrt{\frac{\lambda}{2}}\right)}\exp\left(-\frac{(R_0+x)\sqrt{2\lambda}\cosh\left(t\sqrt{\frac{\lambda}{2}}\right)}{\sinh\left(t\sqrt{\frac{\lambda}{2}}\right)}\right)I_{2\nu}\left(\frac{2\sqrt{2\lambda R_0 x}}{\sinh\left(t\sqrt{\frac{\lambda}{2}}\right)}\right).$$

Here, I_ν is the modified Bessel function for the first kind as defined by

$$I_\nu(z) = \sum_{k=0}^{\infty}\frac{1}{k!\Gamma(\nu+k+1)}\left(\frac{z}{2}\right)^{\nu+k}.$$

Importantly, it can be shown that $p(x,t;s)$ and $q(x,t;\lambda)$ are related by

$$\mathcal{H}(\lambda) = \int_0^\infty e^{-\lambda s}p(x,t;s)\,\mathrm{d}s = \frac{q(x,t;\lambda)}{\lambda}. \tag{6.24}$$

By taking the inversion of the Laplace transform, we obtain the following Bromwich integral representation of $p(x,t;B)$:

$$p(x,t;B) = \lim_{z\to\infty}\frac{1}{2\pi i}\int_{\gamma-iz}^{\gamma+iz} e^{B\lambda}\mathcal{H}(\lambda)\,\mathrm{d}\lambda, \tag{6.25}$$

where $\gamma > 0$ is the damping factor. Once $p(x,t;B)$ is available, Li (2016) shows that the perpetual timer call and put options admit the analytic Black-Scholes type price formulas as presented in Theorem 6.3.

Theorem 11 *Assume that the variance process of the underlying asset price follows the Heston model [see (6.21)], the price formulas of the perpetual timer call and put options with strike K and variance budget B are given by*

$$\begin{aligned} c_0 &= S_0\Pi_1^c - K\Pi_2^c, \\ p_0 &= K\Pi_2^p - S_0\Pi_1^p, \end{aligned} \tag{6.26}$$

where

$$\begin{aligned} \Pi_i^c &= \int_0^\infty\int_0^\infty \Omega_i^c(\varepsilon x, \frac{t}{\varepsilon})p(x,t;B)\,\mathrm{d}x\,\mathrm{d}t, \quad i=1,2, \\ \Pi_i^p &= \int_0^\infty\int_0^\infty \Omega_i^p(\varepsilon x, \frac{t}{\varepsilon})p(x,t;B)\,\mathrm{d}x\,\mathrm{d}t, \quad i=1,2. \end{aligned}$$

Here, the integrand functions are given by

$$\begin{aligned}
\Omega_1^c(\nu,\xi) &= N(d_1(\nu,\xi))\exp(d_0(\nu,\xi)+c(\nu,\xi)),\\
\Omega_2^c(\nu,\xi) &= N(d_2(\nu,\xi))\exp(-r\xi+c(\nu,\xi)),\\
\Omega_1^p(\nu,\xi) &= N(-d_1(\nu,\xi))\exp(d_0(\nu,\xi)+c(\nu,\xi)),\\
\Omega_2^p(\nu,\xi) &= N(-d_2(\nu,\xi))\exp(-r\xi+c(\nu,\xi)),
\end{aligned}$$

where $\nu = \frac{\kappa\theta}{\varepsilon^2} - \frac{1}{2}$, $R_0 = V_0/\varepsilon$ *and*

$$\begin{aligned}
c(\nu,\xi) &= \frac{\kappa}{\varepsilon^2}(V_0-\nu)+\frac{\kappa^2\theta}{\varepsilon^2}\xi-\frac{\kappa^2}{2\varepsilon^2}B,\\
d_0(\nu,\xi) &= \frac{\rho}{\varepsilon}(\nu-V_0-\kappa\theta\xi+\kappa B)-\frac{\rho^2}{2}B,\\
d_1(\nu,\xi) &= \frac{1}{\sqrt{(1-\rho^2)B}}\left[\ln\frac{S_0}{K}+r\xi+\frac{B(1-\rho^2)}{2}+d_0(\nu,\xi)\right],\\
d_2(\nu,\xi) &= d_1(\nu,\xi)-\sqrt{(1-\rho^2)B}.
\end{aligned}$$

The details of the proof of Theorem 6.3 can be found in Appendix A in Li (2016). The ideas of the proof stem from computing the conditional expectation in (6.16) using the known joint density $p(x,t;s)$.

Stochastic time change formulation

As a final remark, by following a similar conditional expectation approach and using the stochastic time change technique, Cui *et al.* (2017b) show that analytic pricing of the perpetual timer options under various stochastic volatility models can be reduced to taking conditional expectation on some Black-Scholes type formula [see (6.16)] on the joint probability law of functionals related to the new time change process. Based on their time change approach of deriving the joint law required in (6.17a,b), they manage to obtain analytic price formulas of the perpetual timer call option in terms of double integrals under the Heston model, 3/2-model and α-Hypergeometric stochastic volatility model [see Propositions 4.3-4.5 in Cui *et al.* (2017b)]. Besides, Zhang *et al.* (2017) also manage to derive Black-Scholes type formula for the price of perpetual timer call option under the Hull-White model.

6.2.3 Perturbation approximation

Though the integral price formulas of the perpetual timer options are available, the numerical valuation of these price formulas can be quite daunting. Li and Mercurio (2014, 2015) derive analytic perturbation approximation price formulas of the perpetual and finite maturity timer options based on the perturbation expansion of several carefully chosen parameter functions in powers of the volatility of variance parameter. Numerical accuracy of these approximate price formulas is seen to be promising, typically with percentage error being well within a small fraction of one percent even with finite (not so small) value of the volatility of variance parameter.

In this subsection, we focus on the derivation of the perturbation approximation price formulas for the perpetual timer call option under the Heston model and 3/2-model. Recall that the maturity T appears in three different scenarios in the Black-Scholes price formula for the vanilla call option, each has its unique interpretation. Li and Mercurio (2014) propose the following Black-Scholes type representation of the perpetual timer call option price function:

$$c(S,V,I) = Se^{-q\tilde{T}}N(d_+) - Ke^{-rT}N(d_-), \tag{6.27}$$

where

$$d_\pm(S,\tilde{T},T,\Sigma) = \frac{\ln\frac{S}{K} + rT - q\tilde{T}}{\Sigma} \pm \frac{\Sigma}{2}.$$

When $r = q = 0$, the price formula (6.27) agrees with the closed-form price formula (6.10), where $\Sigma = \sqrt{B-I}$. The functional dependence of $\widetilde{T}$, T and Σ on the state variables V, I, and S will be revealed once we obtain the partial differential equations of these three parameter functions. The proposed form of the price formula retains homogeneity of degree one in S and K, a desirable property to be observed.

Due to uncertain maturity in the timer call option, it is natural to assume $\widetilde{T}$, T and Σ to be dependent on V and I. Note that the correlation between S_{τ_B} and $e^{-r\tau_B}$ plays an important role. More precisely, $\tilde{T}$ is related to the calculation of the expectation $E[S_{\tau_B}e^{-r\tau_B}]$, which exhibits an obvious dependency on the correlation. On the other hand, T is related to the calculation of the expectation $E[e^{-r\tau_B}]$, which shows no direct dependency on the correlation. This explains why we have to assume $\widetilde{T}$ and T to be different parameter functions. The remaining parameter function Σ, which takes the form $\sigma\sqrt{T}$ in the classical Black-Scholes formula with constant volatility σ, is another parameter function that shows dependence on V, I, and S.

The solution form proposed in (6.27) is an elegant choice for several good reasons. It observes homogeneity of degree one in S and K, which is consistent with usual option price function. As long as $\Sigma > 0$, we always have $C(S,V,I) > 0$ since $C(0^+,V,I) = 0$ and $\frac{\partial C}{\partial S} > 0$ for all $S > 0$. The price formula has sensible limits when S or K goes to 0 or ∞. Note that $\Sigma \to 0$ when $I \to B$ and $\Sigma \to \infty$ when $B \to \infty$, so the price formula also has consistent limiting behavior for B. In addition, the price formula preserves another desirable property of the Black-Scholes formula, namely,

$$Se^{-q\tilde{T}}n(d_+) - Ke^{-rT}n(d_-) = 0, \tag{6.28}$$

where $n(\cdot)$ is the standard normal density function. By virtue of this property, the greeks of the price function $c(S,V,I)$ admit simple forms similar to the classical Black-Scholes formula. For instance, we obtain

$$\Delta = \frac{\partial c}{\partial S} = e^{-q\tilde{T}}N(d_+), \tag{6.29a}$$

$$\Gamma = \frac{\partial^2 c}{\partial S^2} = \frac{e^{-q\tilde{T}}}{S\Sigma}n(d_+). \tag{6.29b}$$

Finally, the put-call parity for the perpetual timer options resembles closely to that of the European vanilla options, where

$$c(S,V,I) - P(S,V,I) = Se^{-q\tilde{T}} - Ke^{-rT}. \tag{6.30}$$

We introduce the volatility of variance ε explicitly in the dynamic equation of V_t by modifying (6.3b) as follows:

$$\mathrm{d}V_t = \alpha(V_t)\mathrm{d}t + \varepsilon\beta(V_t)\mathrm{d}W_t^1. \tag{6.31}$$

With this new assumed dynamic equation of V_t, $c(S,V,I)$ satisfies the following partial differential equation:

$$\begin{aligned}(r-q)S\frac{\partial c}{\partial S} + \frac{S^2V}{2}\frac{\partial^2 c}{\partial S^2} + V\frac{\partial c}{\partial I} + \alpha(V)\frac{\partial c}{\partial V}& \\ + \frac{\varepsilon^2}{2}\beta^2(V)\frac{\partial^2 c}{\partial V^2} + \rho\varepsilon S\sqrt{V}\beta(V)\frac{\partial^2 c}{\partial S\partial V} - rc = 0.&\end{aligned} \tag{6.32}$$

In the current context of perpetual timer call option, it is not appropriate to consider the perturbation expansion of the option price function in powers of the volatility of variance parameter ε as follows:

$$c(S,V,I) = c_0(S,V,I) + \varepsilon c_1(S,V,I) + \varepsilon^2 c_2(S,V,I) + \cdots.$$

To facilitate the solution procedure, it is more appropriate to consider the perturbation expansion of $\tilde{T}$, T, and Σ in powers of ε.

Governing equations of the parameter functions

Recall that $\tilde{T} = \tilde{T}(V,I)$, $T = T(V,I)$ and $\Sigma = \Sigma(S,V,I)$. We would like to derive the governing partial differential equations for $\widetilde{T}$, T and Σ. Based on the assumed price formula (6.27), we obtain the following first-order and second-order partial derivatives of $c(S,V,I)$ as follows:

$$\begin{aligned}
\frac{\partial c}{\partial I} &= -q\frac{\partial \tilde{T}}{\partial I}Se^{-q\tilde{T}}N(d_+) + r\frac{\partial T}{\partial I}Ke^{-rT}N(d_-) + \frac{\partial \Sigma}{\partial I}Se^{-q\tilde{T}}n(d_+), \\
\frac{\partial c}{\partial V} &= -q\frac{\partial \tilde{T}}{\partial V}Se^{-q\tilde{T}}N(d_+) + r\frac{\partial T}{\partial V}Ke^{-rT}N(d_-) + \frac{\partial \Sigma}{\partial V}Se^{-q\tilde{T}}n(d_+), \\
\frac{\partial c}{\partial S} &= e^{-q\tilde{T}}N(d_+), \quad \frac{\partial^2 c}{\partial S^2} = \frac{e^{-q\tilde{T}}}{S\Sigma}n(d_+), \\
\frac{\partial^2 c}{\partial S\partial V} &= -q\frac{\partial \tilde{T}}{\partial V} + \frac{\partial d_+}{\partial V}e^{-q\tilde{T}}n(d_+), \\
\frac{\partial^2 c}{\partial V^2} &= q\left[q\left(\frac{\partial \tilde{T}}{\partial V}\right)^2 - \frac{\partial^2 \tilde{T}}{\partial V^2}\right]Se^{-q\tilde{T}}N(d_+) - r\left[r\left(\frac{\partial T}{\partial V}\right)^2 - \frac{\partial^2 T}{\partial V^2}\right]Ke^{-rT}N(d_-) \\
&\quad + R(S,V,I)Se^{-q\tilde{T}}n(d_+),
\end{aligned} \tag{6.33}$$

where

$$\begin{aligned}
\frac{\partial d_+}{\partial V} &= \frac{r\frac{\partial T}{\partial V} - q\frac{\partial \tilde{T}}{\partial V} - d_- \frac{\partial \Sigma}{\partial V}}{\Sigma}, \quad \frac{\partial d_-}{\partial V} = \frac{r\frac{\partial T}{\partial V} - q\frac{\partial \tilde{T}}{\partial V} - d_+ \frac{\partial \Sigma}{\partial V}}{\Sigma}, \\
R(S,V,I) &= r\frac{\partial T}{\partial V}\frac{\partial d_-}{\partial V} - q\frac{\partial \tilde{T}}{\partial V}\frac{\partial d_+}{\partial V} + \frac{\partial^2 \Sigma}{\partial V^2} - q\frac{\partial \tilde{T}}{\partial V}\frac{\partial \Sigma}{\partial V} - \frac{\partial \Sigma}{\partial V}\frac{\partial d_+}{\partial V} d_+.
\end{aligned}$$

We substitute these partial derivatives of c into (6.32). By collecting the like terms involving $N(d_+)$, $N(d_-)$, and $n(d_+)$ and setting them to be zero in order that (6.32) holds for all possible values of K, we obtain the corresponding governing partial differential equations for the parameter functions. Collecting all like terms involving $N(d_+)$, we obtain the following partial differential equation for $\tilde{T}$:

$$V\frac{\partial \tilde{T}}{\partial I} + \tilde{\alpha}(V)\frac{\partial \tilde{T}}{\partial V} + \frac{\varepsilon^2}{2}\beta^2(V)\left[\frac{\partial^2 \tilde{T}}{\partial V^2} - q\left(\frac{\partial \tilde{T}}{\partial V}\right)^2\right] + 1 = 0, \tag{6.34}$$

with boundary condition: $\tilde{T}(V,B) = 0$. Note that we introduce the modified drift $\tilde{\alpha}(V)$, where

$$\tilde{\alpha}(V) = \alpha(V) + \varepsilon\rho\sqrt{V}\beta(V). \tag{6.35}$$

Similarly, collecting all like terms involving $N(d_-)$ gives the partial differential equation for T as follows:

$$V\frac{\partial T}{\partial I} + \alpha(V)\frac{\partial T}{\partial V} + \frac{\varepsilon^2}{2}\beta^2(V)\left[\frac{\partial^2 T}{\partial V^2} - r\left(\frac{\partial T}{\partial V}\right)^2\right] + 1 = 0, \tag{6.36}$$

with boundary condition: $T(V,B) = 0$. Lastly, collecting all like terms involving $n(d_+)$ gives the partial differential equation for Σ as follows:

$$V\frac{\partial \Sigma}{\partial I} + \alpha(V)\frac{\partial \Sigma}{\partial V} + \frac{V}{2\Sigma} + \varepsilon\rho\sqrt{V}\beta(V)\frac{\partial d_+}{\partial V} + \frac{\varepsilon^2}{2}\beta^2(V)R(S,V,I) = 0, \tag{6.37}$$

with boundary condition: $\Sigma(S,V,B) = 0$. Note that the imposed boundary conditions are determined such that the timer call option price becomes $(S-K)^+$ for any $K \geq 0$ when $I = B$.

It is seen that the governing equations for $\widetilde{T}$ and T are very similar except that $\widetilde{T}$ is dependent on ρ via the modified drift $\widetilde{\alpha}(V)$ while T has no dependence on ρ. Both $\widetilde{T}$ and T have functional dependence on V and I but not S, while Σ has additional dependence on S due to $R(S,V,I)$ that appears in the quadratic power term in ε. If we approximate solution to Σ only up to the first order in ε, we may assume $\Sigma = \Sigma(V,I)$ by dropping dependence on S.

There are several interesting degenerate cases. When $r = 0$, there is no term involving T in the first and second-order derivatives of $c(S,V,I)$, and this leaves us with only two equations for $\widetilde{T}$ and Σ. Similarly, when $q = 0$, there is no term involving $\tilde{T}$ in the first and second-order derivatives of $c(S,V,I)$. Also, since the equation for Σ involves ρ, so the effect of ρ can be captured by Σ. This means that the crude approximation of taking $\Sigma^2 = B - I$ would not be sufficient for nonzero ρ.

Perturbation expansion

Since the governing partial differential equations for $\tilde{T}$, T, and Σ are nonlinear, we solve these parameter functions using perturbation approximation in powers of ε. We consider the following perturbation expansions for $\tilde{T}$ and T in powers of ε, where

$$\tilde{T}(V,I) = \tilde{T}_0(V,I)+\varepsilon\tilde{T}_1(V,I)+\varepsilon^2\tilde{T}_2(V,I)+o(\varepsilon^2), \tag{6.38a}$$
$$T(V,I) = T_0(V,I)+\varepsilon T_1(V,I)+\varepsilon^2 T_2(V,I)+o(\varepsilon^2). \tag{6.38b}$$

Putting the above expressions into (6.34) and (6.36), respectively, and setting the coefficients of the zeroth, first and second-order terms in ε to zero, we obtain the partial differential equations for the successive order functions.

For the zeroth order terms $\tilde{T}_0$ and T_0, the governing partial differential equations are

$$V\frac{\partial\tilde{T}_0}{\partial I}+\tilde{\alpha}(V)\frac{\partial\tilde{T}_0}{\partial V}+1=0, \tag{6.39a}$$
$$V\frac{\partial T_0}{\partial I}+\alpha(V)\frac{\partial T_0}{\partial V}+1=0, \tag{6.39b}$$

subject to the boundary conditions: $\tilde{T}_0(V,B)=T_0(V,B)=0$. Later, we illustrate how to solve the equations with specified $\tilde{\alpha}(V)$ and $\alpha(V)$ under specific stochastic volatility models. For the first-order terms $\tilde{T}_1$ and T_1, the governing partial differential equations are

$$V\frac{\partial\tilde{T}_1}{\partial I}+\tilde{\alpha}(V)\frac{\partial\tilde{T}_1}{\partial V}=0, \tag{6.40a}$$
$$V\frac{\partial T_1}{\partial I}+\alpha(V)\frac{\partial T_1}{\partial V}=0, \tag{6.40b}$$

subject to the boundary conditions: $\tilde{T}_1(V,B)=T_1(V,B)=0$. We observe that the solutions to the above equations are the trivial solutions:

$$\tilde{T}_1(V,I)=T_1(V,I)=0. \tag{6.41}$$

Finally, for the second-order terms $\tilde{T}_2$ and T_2, the governing partial differential equations have nonhomogeneous terms involving the zeroth order solutions, where

$$V\frac{\partial\tilde{T}_2}{\partial I}+\tilde{\alpha}(V)\frac{\partial\tilde{T}_2}{\partial V}+\frac{\beta^2(V)}{2}\left[\frac{\partial^2\tilde{T}_0}{\partial V^2}-q\left(\frac{\partial\tilde{T}_0}{\partial V}\right)^2\right]=0, \tag{6.42a}$$
$$V\frac{\partial T_2}{\partial I}+\alpha(V)\frac{\partial T_2}{\partial V}+\frac{\beta^2(V)}{2}\left[\frac{\partial^2 T_0}{\partial V^2}-r\left(\frac{\partial T_0}{\partial V}\right)^2\right]=0, \tag{6.42b}$$

subject to the boundary conditions: $\tilde{T}_2(V,B)=T_2(V,B)=0$.

In summary, since the first-order terms $\tilde{T}_1$ and T_1 are zero, the approximate solutions to $\widetilde{T}$ and T up to the second order in ε is given by

$$\tilde{T}(V,I) = \tilde{T}_0(V,I)+\varepsilon^2\tilde{T}_2(V,I), \quad (6.43a)$$
$$T(V,I) = T_0(V,I)+\varepsilon^2 T_2(V,I). \quad (6.43b)$$

It can be easily shown that they satisfy the following partial differential equations, respectively, where

$$V\frac{\partial \tilde{T}}{\partial I}+\tilde{\alpha}(V)\frac{\partial \tilde{T}}{\partial V}+\frac{\varepsilon^2}{2}\beta^2(V)\left[\frac{\partial^2 \tilde{T}_0}{\partial V^2}-q\left(\frac{\partial \tilde{T}_0}{\partial V}\right)^2\right]+1=0, \quad (6.44a)$$
$$V\frac{\partial T}{\partial I}+\alpha(V)\frac{\partial T}{\partial V}+\frac{\varepsilon^2}{2}\beta^2(V)\left[\frac{\partial^2 T_0}{\partial V^2}-r\left(\frac{\partial T_0}{\partial V}\right)^2\right]+1=0, \quad (6.44b)$$

with boundary conditions: $\tilde{T}(V,B)=T(V,B)=0$.

It is easier to solve for Σ^2 instead of Σ. We reformulate (6.37) as another partial differential equation for Σ^2, where

$$V\frac{\partial \Sigma^2}{\partial I}+\alpha(V)\frac{\partial \Sigma^2}{\partial V}+V+2\varepsilon\rho\sqrt{V}\beta(V)\frac{\partial d_+}{\partial V}\Sigma+\varepsilon^2\beta^2(V)R(S,V,I)\Sigma=0. \quad (6.45)$$

Again, we consider the following perturbation expansion for Σ^2 in powers of ε, where

$$\Sigma^2(V,I,S)=\Sigma_0^2(S,V,I)+\varepsilon\Sigma_1^2(S,V,I)+\varepsilon^2\Sigma_2^2(S,V,I)+o(\varepsilon^2).$$

The zeroth order approximation Σ_0^2 satisfies the following partial differential equation:

$$V\frac{\partial \Sigma_0^2}{\partial I}+\alpha(V)\frac{\partial \Sigma_0^2}{\partial V}+V=0, \quad (6.46)$$

with boundary condition: $\Sigma_0(S,V,B)=0$. It is seen that the solution to the above equation is given by

$$\Sigma_0^2=B-I, \quad (6.47)$$

which is the unannualized accumulated variance under the Black-Scholes model.

To proceed with the solution of the first-order approximation, we first make an approximation for $\frac{\partial d_+}{\partial V}$. Note that $\frac{\partial}{\partial V}(T-T_0)=O(\varepsilon^2)$ and $\frac{\partial}{\partial V}(\tilde{T}-T)=O(\varepsilon)$. Moreover, we should have $\frac{\partial \Sigma}{\partial V}=O(\varepsilon)$ by virtue of (6.47). As a result, we obtain

$$\frac{\partial d_+}{\partial V}=\frac{r\frac{\partial T}{\partial V}-q\frac{\partial \tilde{T}}{\partial V}-d_-\frac{\partial \Sigma}{\partial V}}{\Sigma}=(r-q)\frac{\frac{\partial T_0}{\partial V}}{\Sigma}+O(\varepsilon). \quad (6.48)$$

Consequently, by virtue of the approximation in (6.48), (6.45) reduces to

$$V\frac{\partial \Sigma^2}{\partial I}+\alpha(V)\frac{\partial \Sigma^2}{\partial V}+V+2\varepsilon\rho(r-q)\sqrt{V}\beta(V)\frac{\partial T_0}{\partial V}=O(\varepsilon^2). \quad (6.49)$$

To take advantage of the specific structure of the above equation, it would be more convenient to consider the first-order approximation to the solution (with no dependence on S) that takes the form:

$$\Sigma^2(V,I) = B - I + 2\varepsilon\rho(r-q)G(V,I). \tag{6.50}$$

By direct substitution of (6.50) into (6.49), the governing partial differential equation for G is given by

$$V\frac{\partial G}{\partial I} + \alpha(V)\frac{\partial G}{\partial V} + \sqrt{V}\beta(V)\frac{\partial T_0}{\partial V} = 0, \tag{6.51}$$

with boundary condition: $G(V,B) = 0$.

The approximation of Σ^2 to the second order in ε requires much tedious efforts. First, we are forced to introduce dependency on S due to the term involving $R(S,V,I)$. Second, the governing partial differential equation for Σ^2 to higher order in ε is more complicated. Though the approximation of Σ^2 only up to the first order introduces some inconsistency in order approximation with $\tilde{T}$ and T, the numerical tests performed by Li and Mercurio (2014) show that the first-order approximation of Σ^2 is indeed acceptable.

The solutions to $\widetilde{T}$, T, and Σ^2 at various orders involve partial differential equations that all share the following common form:

$$V\frac{\partial f}{\partial I} + A(V)\frac{\partial f}{\partial V} = g(V,I), \tag{6.52}$$

for given functions $A(V)$ and $g(V,I)$, together with boundary condition: $f(V,B) = 0$. We illustrate the general solution technique for solving (6.52) using the method of characteristic in the Appendix. We then show how to solve for $\widetilde{T}$, T, and Σ^2 under the Heston model and 3/2-model.

Heston stochastic volatility model

We assume that the usual Feller condition $2\kappa\theta > \varepsilon^2$ is satisfied and $\kappa - \rho\varepsilon > 0$. Under the Heston model, the partial differential equation for T_0 becomes

$$V\frac{\partial T_0}{\partial I} + \kappa(\theta - V)\frac{\partial T_0}{\partial V} + 1 = 0. \tag{6.53}$$

To solve (6.53), we choose the following pair of transformation of variables:

$$z = W\left(\frac{V-\theta}{\theta}\exp\left(-\frac{\kappa}{\theta}(B-I) + \frac{V-\theta}{\theta}\right)\right), \quad z_0 = \frac{V-\theta}{\theta}, \tag{6.54}$$

Here, $W(x)$ is Lambert's product log function that solves

$$W(xe^x) = x. \tag{6.55}$$

When $I = B$, it is seen that $z = z_0$ by virtue of the property of Lambert's product log function.

Note that $g(V,I) = -1$ with regard to (6.53), by applying the formula shown in the Appendix, we can obtain the solution:

$$T_0(V,I) = -\int_z^{z_0} \frac{\theta}{\kappa[\theta - \theta(1+x)]}\, dx = \frac{1}{\kappa} \ln \frac{z_0}{z}. \tag{6.56}$$

By observing an alternative representation of Lambert's product log function: $W(x)e^{W(x)} = x$, we obtain an alternative representation for T_0:

$$T_0(V,I) = \frac{z - z_0}{\kappa} + \frac{B - I}{\theta}. \tag{6.57}$$

Once we have obtained T_0, we can proceed to compute the second-order term T_2, which satisfies

$$V\frac{\partial T_2}{\partial I} + \kappa(\theta - V)\frac{\partial T_2}{\partial V} + \frac{V}{2}\left[\frac{\partial^2 T_0}{\partial V^2} - r\left(\frac{\partial T_0}{\partial V}\right)^2\right] = 0, \tag{6.58}$$

with boundary condition: $T_2(V,B) = 0$. By following similar techniques, T_2 is found to be (Li and Mercurio, 2014)

$$\begin{aligned} T_2(V,I) = {} & \frac{Y-1}{4\kappa^3 Y^2 (1+z)^3 \theta} \{\kappa[2Y^2 z^2 + Y(2 - 5z - 2z^2) - 2 - z] \\ & - r(1+z)[1 + 2Y^2 z + Y(2z - 3)]\} + \frac{[3\kappa z + r(2z^2 + z - 1)] \ln Y}{2\kappa^3 (1+z)^3 \theta}, \end{aligned} \tag{6.59}$$

where

$$Y = \frac{z_0}{z} = e^{z - z_0 + \kappa \frac{B-I}{\theta}}. \tag{6.60}$$

The second-order approximation solution is given by

$$T(V,I) = T_0(V,I) + \varepsilon^2 T_2(V,I), \tag{6.61}$$

where T_0 and T_2 are given by (6.57) and (6.59), respectively.

We let $\tilde{\kappa} = \kappa - \rho\varepsilon$ and $\tilde{\theta} = \frac{\kappa\theta}{\tilde{\kappa}}$. By replacing κ, θ, and r by $\tilde{\kappa}$, $\tilde{\theta}$, and q, respectively, in the above expressions, we can obtain the second-order approximation of $\tilde{T}(V,I)$.

Next, we calculate the first-order approximation of Σ^2. Under the Heston model, (6.51) becomes

$$V\frac{\partial G}{\partial I} + \kappa(V - \theta)\frac{\partial G}{\partial V} + V\frac{\partial T_0}{\partial V} = 0, \tag{6.62}$$

with boundary condition: $G(V,B) = 0$. It can be shown that the solution to the above equation is given by

$$G(V,I) = \frac{Y(z-1)\ln Y - (Y-1)(Yz-1)}{\kappa^2 Y(1+z)}, \tag{6.63}$$

where Y is defined in (6.60). Finally, the first-order approximation of Σ^2 is given by

$$\Sigma^2(V,I) = B - I + 2\varepsilon\rho(r-q)G(V,I). \tag{6.64}$$

Since $G(V,I)$ does not depend on ρ, so the first-order approximate solution has a linear dependency on ρ.

3/2 stochastic volatility model

Under the 3/2 stochastic volatility model, where the instantaneous variance process is governed by

$$dV_t = \kappa V_t(\theta - V_t)dt + \varepsilon V_t^{3/2}\, dW_t^1, \tag{6.65}$$

then (6.39b) becomes

$$V\frac{\partial T_0}{\partial I} + \kappa V(\theta - V)\frac{\partial T_0}{\partial V} + 1 = 0. \tag{6.66}$$

We choose the new coordinates as follows:

$$z = \frac{V-\theta}{\theta}e^{-\kappa(B-I)}, \quad z_0 = \frac{V-\theta}{\theta}.$$

The solution to $T_0(V,I)$ is given by

$$T_0(V,I) = \int_z^{z_0} \frac{1}{\kappa\theta x(1+x)}\, dx = \frac{1}{\kappa\theta}\ln\frac{z_0(1+z)}{z(1+z_0)} = \frac{1}{\kappa\theta}\ln\frac{V+\theta(e^{B-I}-1)}{V}. \tag{6.67}$$

The second-order term can be obtained by solving the following partial differential equation:

$$V\frac{\partial T_2}{\partial I} + \kappa V(\theta - V)\frac{\partial T_2}{\partial V} + \frac{1}{2}V^3\left[\frac{\partial^2 T_0}{\partial V^2} - r\left(\frac{\partial T_0}{\partial V}\right)^2\right] = 0, \tag{6.68}$$

with boundary condition: $T_2(V,B) = 0$. Again, the partial differential equation can be solved explicitly to give

$$\begin{aligned} T_2(V,I) = {} & \frac{1-4R+(3-2\ln R)R^2}{4\kappa^3[V+\theta(R-1)]^2}r + \frac{V[1+(\ln R-1)]R}{\kappa^2[V+\theta(R-1)]^2} \\ & + \frac{\theta[4(1-\ln R)R+(2\ln R-1)R^2-3]}{4\kappa^2[V+\theta(R-1)]^2}, \end{aligned} \tag{6.69}$$

where $R = e^{\kappa(B-I)}$. The second-order approximation to T is given by

$$T(V,I) = T_0(V,I) + \varepsilon^2 T_2(V,I), \tag{6.70}$$

where T_0 and T_2 are given by (6.67) and (6.69), respectively.

Recall that $\widetilde{\kappa} = \kappa - \rho\varepsilon$ and $\widetilde{\theta} = \frac{\kappa\theta}{\widetilde{\kappa}}$. Similarly, by replacing κ, θ, and r by $\tilde{\kappa}$, $\tilde{\theta}$, and q, respectively, in the above expression, one can obtain the second-order approximation to $\tilde{T}(V,I)$.

Next, we calculate the first-order approximation of Σ^2. Under the 3/2 stochastic volatility model, we have

$$V\frac{\partial G}{\partial I} + \kappa V(V-\theta)\frac{\partial G}{\partial V} + V^2\frac{\partial T_0}{\partial V} = 0, \tag{6.71}$$

with boundary condition: $G(V,B) = 0$. It can be shown that the solution to the above partial differential equation is given by

$$G(V,I) = -\frac{1+R[\ln R - 1]}{\kappa^2\{V+\theta[R-1]\}}. \tag{6.72}$$

Finally, the first-order approximation of Σ^2 is given by

$$\Sigma^2(V,I) = B - I + 2\varepsilon\rho(r-q)G(V,I), \tag{6.73}$$

where $G(V,I)$ is given by (6.72). The detailed calculations for the above results can be found in Li and Mercurio (2014).

As a final remark, there are two other research papers that use the perturbation method to price timer options. Ma *et al.* (2015) consider pricing perpetual timer options under the Heston model and Hull-White model together with stochastic interest rates. They consider the perturbation approximation of the price function using the volatility of variance as the perturbation parameter and derive analytic approximation formulas up to the first-order approximation. When compared with the benchmark Monte Carlo simulation results, high level of numerical accuracy is achieved using their approximation formulas. Li and Mercurio (2015) extend their perturbation method to price finite maturity timer options under the Heston model and 3/2-model. They consider the perturbation approximation of the density functions of τ_B and I_T using the volatility of variance as the perturbation parameter. The computing times required to implement their formulas are typically very short since the analytic expressions in the formulas are simple functions (no Bessel functions or hypergeometric functions are involved). Numerical accuracy is typically within 0.5%, and the error becomes vanishingly small when the volatility of variance is small or maturity is long.

6.3 Finite maturity discrete timer options

The analytic results on pricing timer options presented in the earlier sections are based on the assumption of perpetuity and continuous monitoring of the asset price process. Pricing the discrete timer option with finite maturity poses more challenges mathematically. Liang *et al.* (2011) apply the path integral framework in quantum mechanics to derive the relevant transition probability function (propagator in the context of quantum mechanics) of the underlying stochastic volatility process using the Duru-Kleinert time transformation. They manage to derive integral price formulas for both continuously monitored perpetual and finite maturity timer options under the Heston model (related to the Kratzer potential) and 3/2-model (related to the Morse potential). In this section, we discuss two methods for pricing discrete timer options with finite maturity under the 3/2 model using the Fourier transform techniques. In one method, we derive the two-dimensional Fourier inversion integral for the price function and computation of the price function amounts to numerical valuation of the Fourier inversion integral. In the other method, we use the direct Fourier space time stepping algorithm via the knowledge of the joint moment generating function of log asset price and integrated variance.

6.3.1 Fourier inversion integral price formula

Instead of setting a deterministic expiry date as in a vanilla option, a discrete timer option with finite maturity expires on either the first time when the pre-specified variance budget is fully consumed by the realized variance of the underlying asset price or the pre-specified mandated expiry date, depending on which one comes earlier. The accumulation of discrete realized variance over successive monitoring dates and potential premature knock out of the timer option prior to the mandatory expiry date are illustrated in Fig. 6.1.

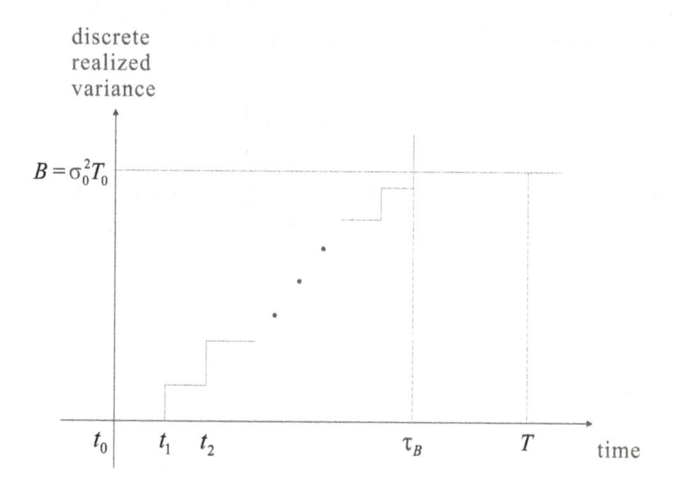

FIGURE 6.1: The timer option is knocked out at τ_B prior to the mandatory expiry date T once the pre-specified variance budget has been exceeded.

Let T be the mandatory expiry of the timer option. We denote the tenor of the monitoring dates for the discrete realized variance by $\{t_0, t_1, \cdots, t_N\}$. For brevity, we take $t_0 = 0$ and $t_N = T$ and assume equally spaced monitoring interval so that $t_j = j\Delta = j\frac{T}{N}$, $j = 1, 2, \cdots, N$. At initiation of the timer option, the investor specifies the variance budget

$$B = \sigma_0^2 T, \tag{6.74}$$

based on an expected investment horizon T and a target volatility σ_0. Let τ_B be the first time in the tenor of monitoring dates at which the discrete realized variance exceeds the variance budget B, namely,

$$\tau_B = \min\left\{ j \left| \sum_{i=1}^{j} \left(\ln \frac{S_{t_i}}{S_{t_{i-1}}} \right)^2 \geq B \right. \right\} \Delta. \tag{6.75}$$

The time-0 price of a discretely monitored finite maturity timer call option can be decomposed into two components, depending on whether the variance budget is exceeded by the mandatory expiry date T or otherwise, where

$$\begin{aligned} c_0 &= E_0^Q[e^{-r(T\wedge\tau_B)}\max(S_{T\wedge\tau_B}-K,0)] \\ &= E_0^Q[e^{-rT}\max(S_T-K,0)\mathbf{1}_{\{\tau_B>T\}}+e^{-r\tau_B}\max(S_{\tau_B}-K,0)\mathbf{1}_{\{\tau_B\leq T\}}], \end{aligned} \tag{6.76}$$

where K is the strike price and r is the constant interest rate.

To enhance tractability in calculating the discretely monitored timer option price, we use the continuous integrated variance $I_t = \int_0^t V_s \, ds$ as a proxy of the discrete realized variance for monitoring of the knock-out time. Accordingly, we redefine τ_B as follows:

$$\tau_B = \min\left\{ j \left| I_{t_j} \geq B \right. \right\} \Delta. \tag{6.77}$$

With this simplification, the price of the discretely monitored finite maturity timer call option can be conveniently computed by further decomposing into a sequence of timerlets as follows:

$$\begin{aligned} c_0 = {} & E_0^Q[e^{-rT}\max(S_T-K,0)\mathbf{1}_{\{I_T<B\}}] \\ & + E_0^Q\left[\sum_{j=0}^{N-1} e^{-rt_{j+1}}\left[\max(S_{t_{j+1}}-K,0)\mathbf{1}_{\{I_{t_j}<B\}} - \max(S_{t_{j+1}}-K,0)\mathbf{1}_{\{I_{t_{j+1}}<B\}}\right]\right]. \end{aligned} \tag{6.78}$$

The key observation is that the event $\{\tau_B > t\}$ is equivalent to $\{I_t < B\}$. The first term in (6.78) corresponds to no knock-out prior to T; that is, $\{\tau_B > T\} \Leftrightarrow \{I_T < B\}$. Note that knock-out at t_{j+1}, $j = 0,1,\ldots,N-1$ can be formulated as $\{\tau_B = t_{j+1}\} \Leftrightarrow \{I_{t_j} < B\}\backslash\{I_{t_{j+1}} < B\}$. Therefore, we have

$$\{\tau_B \leq T\} = \bigcup_{j=0}^{N-1} \{\tau_B = t_{i+1}\} = \bigcup_{j=0}^{N-1} \{I_{t_j} < B\}\backslash\{I_{t_{j+1}} < B\}.$$

This leads to the formulation in (6.78) that is expressed in terms of $\mathbf{1}_{\{I_{t_j}<B\}}$, $j = 0,1,\ldots,N-1$.

Analytic valuation in the Fourier domain

We write $X_t = \ln S_t$. According to (6.78), pricing of the timerlets involves the joint process of S_t and I_t (may or may not at the same time point). Let x represent $\ln S_{t_{j+1}}$ and y represent either I_{t_j} or $I_{t_{j+1}}$. The generalized Fourier transform of $(S_{t_{j+1}} - K)^+ \mathbf{1}_{\{I_{t_j} < B\}}$ and $(S_{t_{j+1}} - K)^+ \mathbf{1}_{\{I_{t_{j+1}} < B\}}$ admit the same analytic representation

$$\hat{F}(\omega,\eta) = \int_{-\infty}^{\infty}\int_{-\infty}^{\infty} e^{-\mathrm{i}\omega x - \mathrm{i}\eta y}(e^x - K)^+ \mathbf{1}_{\{y<B\}}\, \mathrm{d}x\mathrm{d}y = \frac{K^{1-\mathrm{i}\omega} e^{-\mathrm{i}\eta B}}{(\mathrm{i}\omega + \omega^2)\mathrm{i}\eta}, \tag{6.79}$$

where the Fourier transform variables ω and η are complex. Note that $\mathrm{Im}\{\omega\} = \omega_I < -1$ and $\mathrm{Im}\{\eta\} = \eta_I > 0$ are imposed to ensure existence of the two-dimensional generalized Fourier transform. By Parseval's theorem, the discretely monitored finite maturity timer option price can be derived as follows

$$\begin{aligned} c_0 &= \frac{1}{4\pi^2}\int_{-\infty}^{\infty}\int_{-\infty}^{\infty} e^{-rT}\hat{F}(\omega,\eta) E_0^Q[e^{\mathrm{i}\omega X_{t_N} + \mathrm{i}\eta I_{t_N}}]\, \mathrm{d}\omega_R \eta_R \\ &\quad + \sum_{j=0}^{N-1} \frac{1}{4\pi^2}\int_{-\infty}^{\infty}\int_{-\infty}^{\infty} e^{-rt_{j+1}}\Big\{\hat{F}(\omega,\eta) E_0^Q[e^{\mathrm{i}\omega X_{t_{j+1}} + \mathrm{i}\eta I_{t_j}}] \\ &\qquad\qquad - \hat{F}(\omega,\eta) E_0^Q[e^{\mathrm{i}\omega X_{t_{j+1}} + \mathrm{i}\eta I_{t_{j+1}}}]\Big\}\, \mathrm{d}\omega_R \eta_R, \end{aligned} \tag{6.80}$$

where $\mathrm{Re}\{\omega\} = \omega_R$ and $\mathrm{Re}\{\eta\} = \eta_R$. To evaluate the above sum of expectations, we derive the explicit representation for the characteristic functions of (X_{t_j}, I_{t_j}) and $(X_{t_{j+1}}, I_{t_j})$ under the relevant stochastic volatility model.

Let $G(X_0, I_0, V_0, t_0; X_{t_j}, I_{t_j}, V_{t_j}, t_j)$ denote the transition density function of the triple (X, I, V) from state (X_0, I_0, V_0) at time t_0 to state $(X_{t_j}, I_{t_j}, V_{t_j}, t_j)$ at time t_j, $t_j > t_0$. Recall from Theorem 2.1 that the partial Fourier transform of G can be expressed as

$$\begin{aligned} &\int_{-\infty}^{\infty}\int_0^{\infty} e^{i\omega x + i\eta y} G(X_0, I_0, V_0, t_0; x, y, V_{t_j}, t_j)\, \mathrm{d}y\, \mathrm{d}x \\ &= e^{i\omega X_0 + i\eta I_0} g(V_0, t_0; \omega, \eta, V_{t_j}, t_j). \end{aligned} \tag{6.81}$$

Also, we recall from Theorem 2.3 that

$$E_0^Q[e^{\mathrm{i}\omega X_{t_j} + \mathrm{i}\eta I_{t_j}}] = e^{\mathrm{i}\omega X_0 + \mathrm{i}\eta I_0} h(V_0, t_0; \omega, \eta, t_j), \tag{6.82a}$$

where

$$h(V_0, t_0; \omega, \eta, t_j) = \int_0^{\infty} g(V_0, t_0; \omega, \eta, v, t_j)\, \mathrm{d}v.$$

By setting $\eta = 0$ in (6.82a), we deduce the following expectation formula that is useful for later calculation:

$$E_{t_j}^Q\left[e^{i\omega X_{t_{j+1}}}\right] = e^{i\omega X_{t_j}} h(V_j, t_j; \omega, 0, t_{j+1}). \tag{6.82b}$$

Lastly, we would like to establish that

$$E_0^Q[e^{\mathrm{i}\omega X_{t_{j+1}} + \mathrm{i}\eta I_{t_j}}] = e^{\mathrm{i}\omega X_0 + \mathrm{i}\eta I_0}\int_0^{\infty} g(V_0, t_0; \omega, \eta, v, t_j) h(v, t_j; \omega, 0, t_{j+1})\, \mathrm{d}v. \tag{6.83}$$

To show (6.83), we use the iterated expectation procedure. Performing expectation calculation backward in time from t_{j+1} to t_j, we compute $E_{t_j}^Q[e^{\mathrm{i}\omega X_{t_{j+1}}}]$; then from t_j

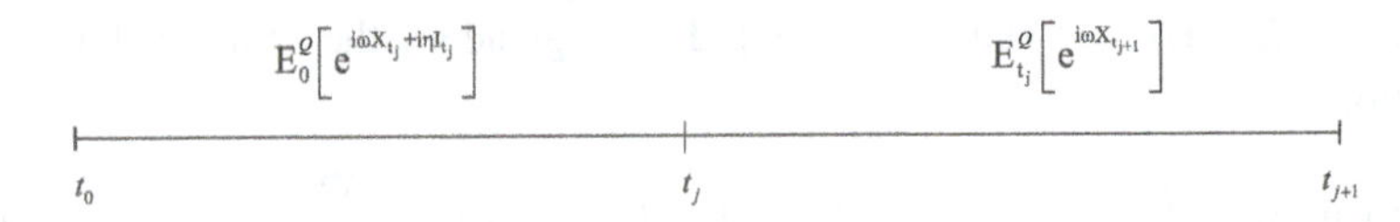

FIGURE 6.2: Iterated expectation procedure to compute $E_0^Q[e^{i\omega X_{t_{j+1}}+i\eta I_{t_j}}]$.

to t_0, we compute $E_0^Q[e^{i\omega X_{t_j}+i\eta I_{t_j}}]$. This is done by setting $\eta = 0$ in $h(v,t_j;\omega,\eta,t_{j+1})$ [see (6.82b)] and integrating $g(V_0,t_0;\omega,\eta,v,t_j)\ h(v,t_j;\omega,0,t_{j+1})$ with respect to v from 0 to ∞ (see Fig. 6.2).

Putting the above results together, we obtain the time-0 price of the discretely monitored finite maturity timer call option as follows:

$$c_0 = \frac{1}{4\pi^2}\int_{-\infty}^{\infty}\int_{-\infty}^{\infty}\hat{F}(\omega,\eta)H(\omega,\eta)\,\mathrm{d}\omega_R\eta_R,$$

where

$$\begin{aligned} H(\omega,\eta) = {} & e^{-rT}e^{i\omega X_0+i\eta I_0}h(V_0,t_0;\omega,\eta,t_N) \\ & + e^{i\omega X_0+i\eta I_0}\sum_{j=0}^{N-1} e^{-rt_{j+1}} \\ & \left(\int_0^{\infty} g(V_0,t_0;\omega,\eta,v,t_j)h(v,t_j;\omega,0,t_{j+1})\,\mathrm{d}v - h(V_0,t_0;\omega,\eta,t_{j+1})\right). \end{aligned} \tag{6.84}$$

For the 3/2-model, the analytic formulas for $g(V_0,t_0;\omega,\eta,v,t_j)$ and $h(v,t_j;\omega,\eta,t_{j+1})$ are given in (2.149) and (2.158), respectively. The pricing calculation of the discretely monitored timer call option amounts to numerical evaluation of the three-dimensional Fourier inversion integrals in (6.84).

6.3.2 Fourier space time stepping numerical algorithm

We illustrate the use of the standard Fourier space time stepping numerical algorithm for pricing discretely monitored finite maturity timer options. The fast Fourier transform algorithm is used in the space domain since the joint moment generating functions of the log asset price and integrated variance are known under the Heston model and 3/2 model. A variant of the Fourier transform algorithm for pricing barrier style options is the Hilbert transform algorithm. Since the timer options are essentially barrier style options in the volatility space, where the knock-out condition is dependent on the discrete realized variance, Zeng *et al.* (2015) apply the fast Hilbert transform algorithms for pricing discretely monitored timer options under the Heston model and 3/2 model. The major advantage of the Hilbert transform algorithm is that it computes a sequence of Hilbert transforms on all monitoring dates and only one final step of the Fourier inversion transform is required to recover the option price and state variables in the real domain. This avoids the nuisance of recovering the option prices and the underlying state variables in the real domain at each monitoring instant in order to check for the premature expiration condition of the discretely monitored timer option.

Fast Fourier transform algorithm

We follow the usual Fourier space time stepping algorithms for pricing options under stochastic volatility and Lévy models (Fang and Oosterlee, 2011). The underlying asset price process S_t and its instantaneous variance V_t under a risk neutral measure Q are assumed to follow the following stochastic volatility model

$$\frac{\mathrm{d}S_t}{S_t} = (r-q)\mathrm{d}t + \sqrt{V_t}\,\mathrm{d}W_t^S, \tag{6.85a}$$

$$\mathrm{d}V_t = \alpha(V_t)\mathrm{d}t + \beta(V_t)\mathrm{d}W_t^V, \tag{6.85b}$$

where $\mathrm{d}W_t^S\mathrm{d}W_t^V = \rho\ \mathrm{d}t$. The model parameters and parameter functions have the same interpretation as those in (6.3a,b). We define

$$\gamma_t = \ln V_t \quad \text{and} \quad x_t = \ln\frac{S_t}{K}, \tag{6.86}$$

where K is the strike price of the timer option. We choose the log-variance instead of the usual variance as the state variable since the corresponding conditional density of log-variance exhibits two advantages. First, the left tail of the conditional density of log-variance decays to zero more rapidly. Second, the conditional density of the log-variance process for varying parameter values is more symmetric than that of the variance process.

Let $V_{t_k}(x_{t_k}, \gamma_{t_k}, I_{t_k})$ be the option value of the discretely monitored finite maturity timer call option at the monitoring time t_k, $k = 1,2,\cdots,N$, where x_{t_k}, γ_{t_k}, and I_{t_k} denote the time-t_k normalized log-asset price, log-variance, and realized continuous integrated variance, respectively. Let Δt denote the uniform time interval between successive monitoring times. Note that $V_{t_k}(x_{t_k}, \gamma_{t_k}, I_{t_k})$ is the sum of the following two terms:

$$\begin{aligned} &V_{t_k}(x_k,\gamma_k,I_k) \\ &= e^{-r\Delta t}U_{t_k}(x_k,\gamma_k,I_k)\mathbf{1}_{\{I_k<B\}} + K(e^{x_k}-1)^+\mathbf{1}_{\{I_k\geq B\}},\ \ k=1,2,\cdots,N-1. \end{aligned} \tag{6.87}$$

The first term is the continuation value conditional on $\{I_k < B\}$, where

$$e^{-r\Delta t}U_{t_k}(x_k,\gamma_k,I_k) = e^{-r\Delta t}E_{t_k}^Q[V_{t_{k+1}}(x_{k+1},\gamma_{k+1},I_{k+1})].$$

The second term is the timer option payoff upon knock out conditional on $\{I_k \geq B\}$.

By the tower property and conditional on the log-variance process $\gamma_{t_{k+1}}$ at time t_{k+1}, it follows that

$$U_{t_k}(x_k,\gamma_k,I_k) = E^Q\left[E^Q[V_{t_{k+1}}(x_{k+1},\gamma_{k+1},I_{k+1})|\mathcal{F}_{t_k},\gamma_{k+1}]|\mathcal{F}_{t_k}\right], \tag{6.88}$$

where $\mathcal{F}_{t_k}$ denotes the filtration at time t_k. The outer expectation in (6.88) involves integration over the density function $p_\gamma(\gamma_{t_{k+1}}|\gamma_{t_k})$, which has analytic closed form under the Heston model [see (2.115)] and 3/2 model [see (2.156)]. By performing discretization along the dimension of log-variance $\gamma_{t_{k+1}}$ at the discrete nodes ζ_j, $j = 1,2,\cdots,J$, we obtain the quadrature formula:

$$U_{t_k}(x_k,\gamma_k,I_k) \approx \sum_{j=1}^{J} W_j p_\gamma(\zeta_j|\gamma_k) E^Q\left[V_{t_{k+1}}(x_{k+1},\gamma_{k+1},I_{k+1})|\mathcal{F}_{t_k},\gamma_{k+1}=\zeta_j\right], \quad (6.89)$$

where W_j is the weight at the quadrature node ζ_j, $j = 1,2,\cdots,J$.

Given the known joint conditional characteristic function of x_{k+1} and I_{k+1}, we can apply the fast Fourier transform method to perform the inner two-dimensional expectation calculations in (6.88). In order to guarantee the existence of the fast Fourier transform, we need to introduce the proper exponential damping factor in the Fourier transform variables w and u. Let $w = \alpha_1 + i\beta_1$ and $u = \alpha_2 + i\beta_2$, where α_1 and α_2 are the damping parameters. At $\gamma_{t_{k+1}} = \zeta_j$, $x_{t_{k+1}} = x$ and $I_{t_{k+1}} = y$ for given α_1 and α_2, we define

$$V_{t_{k+1}}^{\alpha_1,\alpha_2}(x,\zeta_j,y) = e^{\alpha_1 x+\alpha_2 y} V_{t_{k+1}}(x,\zeta_j,y).$$

The generalized two-dimensional Fourier transform of $V_{t_{k+1}}(x,\zeta_j,y)$ is defined by

$$\hat{V}_{t_{k+1}}^{\alpha_1,\alpha_2}(\zeta_j;\beta_1,\beta_2) = \int_{-\infty}^{\infty}\int_{-\infty}^{\infty} e^{i\beta_1 x+i\beta_2 y} V_{t_{k+1}}^{\alpha_1,\alpha_2}(x,\zeta_j,y)\,\mathrm{d}\beta_1\mathrm{d}\beta_2. \quad (6.90)$$

By Parseval's theorem, the inner expectation is found to be

$$\begin{aligned} & E^Q[V_{t_{k+1}}(x_{k+1},\gamma_{k+1},I_{k+1})|\mathcal{F}_{t_k},\gamma_{k+1}=\zeta_j] \\ &= \int_{-\infty}^{\infty}\int_{-\infty}^{\infty} V_{t_{k+1}}(x,\zeta_j,y)\, p(x,y|\mathcal{F}_{t_k},\gamma_{k+1}=\zeta_j)\,\mathrm{d}x\mathrm{d}y \\ &= \frac{1}{4\pi^2}\int_{-\infty}^{\infty}\int_{-\infty}^{\infty} \hat{V}_{t_{k+1}}^{\alpha_1,\alpha_2}(\zeta_j;\beta_1,\beta_2)\check{p}(w,u|\mathcal{F}_{t_k},\gamma_{k+1}=\zeta_j)\,\mathrm{d}\beta_1\mathrm{d}\beta_2, \end{aligned} \quad (6.91)$$

where

$$\begin{aligned} \check{p}(w,u|\mathcal{F}_{t_k},\gamma_{k+1}=\zeta_j) &= E^Q\left[e^{-wx_{k+1}-uI_{k+1}}|\mathcal{F}_{t_k},\gamma_{k+1}=\zeta_j\right] \\ &= e^{-wx_k-uI_k}E[e^{-w(x_{k+1}-x_k)-u(I_{k+1}-I_k)}|\mathcal{F}_{t_k},\gamma_{k+1}=\zeta_j] \end{aligned}$$

is the generalized inverse Fourier transform of the joint conditional density function $p(x_{k+1},I_{k+1}|\mathcal{F}_{t_k},\gamma_{k+1}=\zeta_j)$.

For notational convenience, we express $\check{p}$ in terms of the conditional moment generating function as defined by

$$\Psi(w,u;\gamma_t,\gamma_s) = E^Q[e^{w(x_t-x_s)+u(I_t-I_s)}|\mathcal{F}_s,\gamma_t]. \quad (6.92)$$

Here, we have suppressed the dependency of Ψ on $t-s$ for notational convenience. We may express the inner expectation integral at $\gamma_{t_{k+1}} = \zeta_j$ by the following two-dimensional inverse Fourier transform

$$\begin{aligned} & E^Q[V_{t_{k+1}}(x_{k+1},\gamma_{k+1},I_{k+1})|\mathcal{F}_{t_k},\gamma_{k+1}=\zeta_j] \\ &= \frac{1}{4\pi^2}\int_{-\infty}^{\infty}\int_{-\infty}^{\infty} e^{-wx_k-uI_k}\hat{V}_{t_{k+1}}^{\alpha_1,\alpha_2}(\zeta_j;\beta_1,\beta_2)\Psi(-w,-u;\zeta_j,\gamma_k)\,\mathrm{d}\beta_1\mathrm{d}\beta_2. \end{aligned} \quad (6.93)$$

Here, we have set $\gamma_{t_{k+1}} = \zeta_j$ in $\Psi(w,u;\gamma_{k+1},\gamma_k)$. By the tower property, for $s < t$, we have

$$\Psi(w,u;\gamma_t,\gamma_s) = E^Q\left[E^Q[e^{w(x_t-x_s)+u(I_t-I_s)}|\mathcal{F}_s,\gamma_t,I_t-I_s]|\mathcal{F}_s,\gamma_t\right]$$
$$= E^Q\left[E^Q[e^{w(x_t-x_s)}|\mathcal{F}_s,\gamma_t,I_t-I_s]e^{u(I_t-I_s)}|\mathcal{F}_s,\gamma_t\right]. \tag{6.94}$$

Next, we show that the conditional moment generating function $\Psi(w,u;\gamma_t,\gamma_s)$ admits closed-form analytic representation under the Heston model and 3/2 model.

Heston model

For the Heston model, the dynamics of its instantaneous variance is governed by

$$\mathrm{d}V_t = \kappa(\theta - V_t)\,\mathrm{d}t + \varepsilon\sqrt{V_t}\,\mathrm{d}W_t^V.$$

This corresponds to $f(V_t) = \frac{V_t}{\varepsilon}$ and $h(V_t) = \frac{\kappa(\theta - V_t)}{\varepsilon}$ in (6.85b). Recall that one can express the normalized log-asset price process as follows [see (4.8)]:

$$\ln\frac{S_t}{K} = \ln\frac{S_0}{K} + (r-q)t + \frac{\rho}{\varepsilon}(e^{\gamma_t} - e^{\gamma_0} - \kappa\theta t) + \left(\frac{\rho\kappa}{\varepsilon} - \frac{1}{2}\right)I_t + \sqrt{(1-\rho^2)I_t}\,W,$$

where W is a standard normal random variable. We obtain

$$\Psi(w,u;\gamma_t,\gamma_s) = E^Q\left[E^Q[e^{w(x_t-x_s)}|\mathcal{F}_s,\gamma_t,I_t-I_s]e^{u(I_t-I_s)}|\mathcal{F}_s,\gamma_t\right]$$
$$= e^{w\{(r-q)(t-s)+\frac{\rho}{\varepsilon}[e^{\gamma_t}-e^{\gamma_s}-\kappa\theta(t-s)]\}}$$
$$\Phi\left(-iw\left(\frac{\rho\kappa}{\varepsilon}-\frac{1}{2}\right)-\frac{iw^2}{2}(1-\rho^2)-iu;e^{\gamma_t},e^{\gamma_s}\right), \tag{6.95}$$

where

$$\Phi(\xi;\gamma_t,\gamma_s) = E^Q[e^{i\xi\int_s^t V_u\,\mathrm{d}u}|\gamma_t,\gamma_s] \tag{6.96}$$

is the conditional characteristic function of the time integrated log-variance process $\int_s^t V_u\,\mathrm{d}u$. The analytic closed-form expression of $\Phi(\xi;\gamma_t,\gamma_s)$ is given by (2.114). As a result, we manage to express the two-dimensional conditional moment generating function Ψ in terms of the one-dimensional conditional characteristic function Φ.

3/2 model

The instantaneous variance process under the $3/2$ model evolves according to the following dynamic equation:

$$\mathrm{d}V_t = \kappa V_t(\theta - V_t)\,\mathrm{d}t + \varepsilon V_t^{3/2}\,\mathrm{d}W_t^V.$$

The use of Itô's formula gives the corresponding dynamics for $\frac{1}{V_t}$:

$$\mathrm{d}\left(\frac{1}{V_t}\right) = \kappa\theta\left(\frac{\kappa+\varepsilon^2}{\kappa\theta} - \frac{1}{V_t}\right)\mathrm{d}t - \frac{\varepsilon}{\sqrt{V_t}}\,\mathrm{d}W_t^V.$$

The reciprocal of the variance process of the 3/2 model follows a mean-reversion square root process with parameters $(\kappa\theta, \frac{\kappa+\varepsilon^2}{\kappa\theta}, -\varepsilon)$.

Similarly, the normalized log-asset price process under the 3/2 model can be expressed as

$$\ln\frac{S_t}{K} = \ln\frac{S_0}{K} + (r-q)t + \frac{\rho}{\varepsilon}[\gamma_t - \gamma_0 - \kappa\theta t] + \left[\frac{\rho\kappa}{\varepsilon}\left(1+\frac{\varepsilon^2}{2\kappa}\right) - \frac{1}{2}\right] I_t + \sqrt{(1-\rho^2)I_t}\, W,$$

where W is a standard normal random variable. Again, we manage to express the two-dimensional moment generating function in terms of the one-dimensional conditional characteristic function Φ as follows:

$$\begin{aligned}&\Psi(w,u;\gamma_t,\gamma_s)\\ &= e^{w\{(r-q)(t-s)+\frac{\rho}{\varepsilon}[\gamma_t-\gamma_s-\kappa\theta(t-s)]\}}\\ &\quad \Phi\left(-iw\left[\frac{\rho\kappa}{\varepsilon}\left(1+\frac{\varepsilon^2}{2\kappa}\right)-\frac{1}{2}\right] - \frac{iw^2}{2}(1-\rho^2) - iu; e^{\gamma_t}, e^{\gamma_s}\right),\end{aligned} \tag{6.97}$$

where $\Phi(\xi;\gamma_t,\gamma_s)$ is defined in (6.96).

Now, all the expectation formulas in the form of inverse Fourier transform integrals required in the Fourier space time stepping procedures are available. Given N discrete monitoring times t_k, $k = 1,2,\ldots,N$, the backward induction procedures for pricing the discretely monitored finite maturity timer call option involve numerical evaluation of the above expectation integrals and they are summarized as follows:

1. The backward induction is initiated by the terminal payoff condition:

$$V_{t_N}(x_N,\gamma_N,I_N) = K(e^{x_N}-1)^+. \tag{6.98}$$

2. The time stepping calculations between two consecutive monitoring times are performed sequentially. For $k = N-1, N-2, \ldots, 1$, the numerical approximation of $V_{t_k}(x_k,\gamma_k,I_k)$ is recursively calculated by computing a sequence of Fourier transforms and inverse Fourier transforms:

$$V_{t_k}(x_k,\gamma_k,I_k) = e^{-r\Delta t}U_{t_k}(x_k,\gamma_k,I_k)\mathbf{1}_{\{I_k<B\}} + K(e^{x_k}-1)^+\mathbf{1}_{\{I_k\ge B\}}, \tag{6.99}$$

where

$$\begin{aligned}U_{t_k}(x_k,\gamma_k,I_k) \approx \sum_{j=1}^{J}\frac{W_j}{4\pi^2}\int_{-\infty}^{\infty}\int_{-\infty}^{\infty} & e^{-wx_k-uI_k}\hat{V}_{t_{k+1}}^{\alpha_1,\alpha_2}(\zeta_j;\beta_1,\beta_2)\\ &\Psi(-w,-u;\zeta_j,\gamma_k)p_\gamma(\zeta_j|\gamma_k)\,\mathrm{d}\beta_1\mathrm{d}\beta_2.\end{aligned} \tag{6.100}$$

Recall that $\hat{V}_{t_{k+1}}^{\alpha_1,\alpha_2}(\zeta_j;\beta_1,\beta_2)$ is computed by the two-dimensional Fourier integral in (6.90).

3. The discretely monitored finite maturity timer call option value is approximated by

$$V_{t_0}(x_0,\gamma_0,I_0) \approx \sum_{j=1}^{J} \frac{e^{-r\Delta t}W_j}{4\pi^2} \int_{-\infty}^{\infty}\int_{-\infty}^{\infty} e^{-wx_0}e^{-uI_0}\hat{V}_{t_1}^{\alpha_1,\alpha_2}(\zeta_j;\beta_1,\beta_2)$$
$$\Psi(-w,-u;\zeta_j,\gamma_0)p_\gamma(\zeta_j|\gamma_0)\,\mathrm{d}\beta_1\mathrm{d}\beta_2. \tag{6.101}$$

The evaluation of the above Fourier transform integrals can be performed using the FFT method. Since $V_{t_k}(x_k,\gamma_k,I_k)$ involves the barrier condition: $\mathbf{1}_{\{I_k<B\}}$, we may adopt an extended version of the Fourier cosine transform algorithm to deal effectively with the embedded barrier condition (Fang and Oosterlee, 2011). Alternatively, one may use the effective Hilbert transform method (enhanced Fourier transform algorithm), the details of which can be found in Zeng *et al.* (2015).

Appendix

Proof of Lemma 6.1

We define $\theta(t) = \int_0^t V_s\,\mathrm{d}s$ for any $t<\infty$. Since $\theta(t)$ is a continuous and increasing function on $\mathbb{R}_+$, it is invertible. Let $\tau(y)$ denote the inverse function of $\theta(t)$ so that

$$\tau(\theta(t)) = t \quad \text{and} \quad \theta(\tau(y)) = y.$$

The first hitting time of $\theta(u)$ for a fixed level y is given by

$$\tau(y) = \inf\{u>0 : \theta(u) = y\}.$$

By virtue of the assumption

$$\lim_{t\to\infty}\int_0^t V_s\,\mathrm{d}s = \infty, \text{ a.s.}$$

the stopping time $\tau(y)<\infty$ a.s. for any $y\in(0,\infty)$. Differentiation of both sides of $\theta(\tau(y)) = y$ with respect to y yields

$$\theta'(\tau(y))\tau'(y) = 1.$$

On the other hand, we have $\theta'(t) = V_t$. Consequently, we obtain

$$\theta'(\tau(y)) \;=\; \frac{1}{\tau'(y)} = V_{\tau(y)}$$

so that

$$\tau(y) \;=\; \int_0^y \tau'(x)\,\mathrm{d}x = \int_0^y \frac{1}{\theta'(\tau(x))}\,\mathrm{d}x = \int_0^y \frac{1}{V_{\tau(x)}}\,\mathrm{d}x.$$

Recall the two functions f and g as defined by (6.12), where

$$f(x)=\int_0^x \frac{\sqrt{u}}{\beta(u)}\,du \quad \text{and} \quad g(x)=\alpha(x)f'(x)+\frac{\beta^2(x)}{2}f''(x).$$

By virtue of Itô's lemma, we have

$$df(V_t)=g(V_t)dt+f'(V_t)\beta(V_t)dW_t^1=g(V_t)dt+\sqrt{V_t}\,dW_t^1.$$

Integrating both sides of the above equation from 0 to $\tau(y)$ yields

$$\int_0^{\tau(y)}\sqrt{V_s}\,dW_s^1=f(V_{\tau(y)})-f(V_0)-\int_0^{\tau(y)}g(V_s)\,ds.$$

On the other hand, by the Dubins-Schwartz theorem [see Karatzas and Shreve (1991, p.174)], we have

$$\int_0^{\tau(y)}\sqrt{V_s}\,dW_s^1=\hat{W}_y,$$

where $\hat{W}$ is a standard Brownian motion. Differentiating both sides with respect to y of the following equation:

$$\hat{W}_y=f(V_{\tau(y)})-f(V_0)-\int_0^{\tau(y)}g(V_s)\,ds,$$

we obtain

$$df(V_{\tau(y)})=g(V_{\tau(y)})\tau'(y)\,dy+d\hat{W}_y=\frac{g(V_{\tau(y)})}{V_{\tau(y)}}\,dy+d\hat{W}_y.$$

Let $Y_t=V_{\tau(t)}$, then Y_t satisfies the dynamic equation [see (6.17b)]

$$df(Y_t)=\frac{g(Y_t)}{Y_t}\,dt+d\hat{W}_t,$$

with $Y_0=V_0$. We write $\tau_B=\tau(B)$. Apparently, $V_{\tau_B}=Y_B$ and $\tau_B=\int_0^B\frac{1}{Y_s}\,ds$. Lastly, by applying the change of variable $s=\tau(t)$ in the integral $\int_0^{\tau(B)}g(V_s)\,ds$, we observe $ds=\tau'(t)\,dt$ and $s=\tau(B)$ when $t=B$. Finally, we obtain

$$\int_0^{\tau_B}g(V_s)\,ds=\int_0^B g(V_{\tau(t)})\tau'(t)\,dt=\int_0^B\frac{g(Y_t)}{Y_t}\,dt.$$

Putting the results together, we have [see (6.17a)]

$$\left(\tau_B,V_{\tau_B},\int_0^{\tau_B}g(V_s)\,ds\right)\overset{d}{\sim}\left(\int_0^B Y_s^{-1}ds,Y_B,\int_0^B\frac{g(Y_s)}{Y_s}\,ds\right),$$

which proves Lemma 6.1.

Solution of (6.52) via method of characteristics

Let $V^* > 0$ be a constant and Φ be a smooth function. We define

$$z = \Phi\left(B - I + \int_{V^*}^{V} \frac{u}{A(u)}\,\mathrm{d}u\right), \quad z_0 = \frac{V - V^*}{V^*},$$

where Φ is determined such that $z = z_0$ when $I = B$. We change the coordinate from (V, I) to (z, z_0) and write $\tilde{f}(z, z_0) = f(V, I)$. By differentiating z with respect to I and V, we obtain

$$\frac{\partial z}{\partial I} = -\Phi'\left(B - I + \int_{V^*}^{V} \frac{u}{A(u)}\,\mathrm{d}u\right), \quad \frac{\partial z}{\partial V} = \Phi'\left(B - I + \int_{V^*}^{V} \frac{u}{A(u)}\,\mathrm{d}u\right)\frac{V}{A(V)}.$$

It then follows that

$$V\frac{\partial z}{\partial I} + A(V)\frac{\partial z}{\partial V} = 0.$$

We write $\tilde{f}(z, z_0) = f(V, I)$ and $\tilde{g}(z, z_0) = g(V, I)$. Now, (6.52) can be rewritten as

$$\begin{aligned}
\tilde{g}(z, z_0) &= V\frac{\partial f}{\partial I} + A(V)\frac{\partial f}{\partial V} \\
&= V\frac{\partial \tilde{f}}{\partial z}\frac{\partial z}{\partial I} + A(V)\left[\frac{\partial \tilde{f}}{\partial z}\frac{\partial \tilde{z}}{\partial V} + \frac{\partial \tilde{f}}{\partial z_0}\frac{1}{V^*}\right] \\
&= \frac{\partial \tilde{f}}{\partial z_0}\frac{A(V^*(1 + z_0))}{V^*}.
\end{aligned}$$

Treating $\tilde{f}$ as a function of single variable z_0, the above equation is identified as an ordinary differential equation with boundary condition $\tilde{f}(z_0, z_0) = f(V, B) = 0$. The solution is found to be

$$f(V, I) = \tilde{f}(z, z_0) = \int_{z}^{z_0} \frac{V^*}{A(V^*(1 + x))}\tilde{g}(z, x)\,\mathrm{d}x.$$

In the solution procedure, the determination of Φ is vital and challenging. Fortunately, Φ can be written explicitly in terms of the Lambert product log function under the Heston model while it has explicit expression under the 3/2 model.

Bibliography

Aguilar, J.P. (2020). Some pricing tools for the Variance Gamma model. *International Journal of Theoretical and Applied Finance*, **23(4)**, 2050025 (35 pages).

Ahallal, A., Torné, O. (2019). Knocking-out corridor variance. *Working paper of Barclays.*

Aït-Sahalia, Y. (2004). Telling from discrete data whether the underlying continuous time model is a diffusion. *Journal of Finance*, **42**, 2075–2112.

Ait-Sahalia, Y., Karaman, M., Mancini, L. (2020). The term structure of equity and variance risk premia. *Journal of Econometrics*, **219**, 204–230.

Albrecher, H., Mayer, P., Schoutens, W., Tistaert, J. (2007). The little Heston trap. *Wilmott Magazine*, January issue, 83–92.

Alexander, C., Korovilas, D. (2013). Volatility exchange-traded notes: Curse or cure? *Journal of Alternative Investments*, **16(2)**, 52–70.

Almendral, A., Oosterlee, C.W. (2007). On American options under the Variance Gamma process. *Applied Mathematical Finance*, **14(2)**, 131–152.

Alòs, E., Chatterjee, R., Tudor, S., Wang, T.H. (2019). Target volatility option pricing in lognormal fractional SABR model. *Quantitative Finance*, **19**, 1–18.

Andersen, L.B.G., Brotherton-Ratcliffe, R. (1998). The equity option volatility smile: An implicit finite-difference approach. *Journal of Computational Finance*, **1**, 5–38.

Andersen, L.B.G., Piterbarg, V.V. (2007). Moment explosions in stochastic volatility models. *Finance and Stochastics*, **11**, 29–50.

Applebaum, D. (2009). *Lévy Processes and Stochastic Calculus*. Cambridge University Press, New York.

Arai, T. (2019). Pricing and hedging of VIX options for Barndorff-Nielsen and Shephard models. *International Journal of Theoretical and Applied Finance*, **22(8)**, 1950043 (26 pages).

Avramidis, A.N., L'Ecuyer, P. (2006). Efficient Monte Carlo and quasi-Monte Carlo option pricing under the Variance Gamma model. *Management Science*, **52(12)**, 1930–1944.

Badescu, A., Chen, Y., Couch, M., Cui, Z.Y. (2019). Variance swaps valuation under non-affine GARCH models and their diffusion limits. *Quantitative Finance*, **19**, 227–246.

Badescu, A., Cui, Z.Y., Ortega, J.P. (2020). Closed-form variance swap prices under general affine GARCH models and their continuous-time limits. *Annals of Operations Research.*

Bakshi, G., Cao, C., Chen, Z.W. (1997). Empirical performance of alternative option pricing models. *Journal of Finance*, **52(5)**, 2003–2049.

Bakshi, G., Ju, N., Ou-Yang, H. (2006). Estimation of continuous-time models with an application to equity volatility dynamics. *Journal of Financial Economics*, **82(1)**, 227–249.

Baldeaux, J., Badran, A. (2014). Consistent modeling of VIX and equity derivatives using a 3/2 plus jumps model. *Applied Mathematical Finance*, **21(4)**, 299–312.

Bardgett, C., Gourier, E., Leippold, M. (2019). Inferring volatility dynamics and risk premia from the S&P 500 and VIX markets. *Journal of Financial Economics*, **131**, 593–618.

Barletta, A., Nicolato, E. (2018). Orthogonal expansions for VIX options under affine jump diffusion. *Quantitative Finance*, **18(6)**, 951–967.

Barndorff-Nielsen, O.E. (1997). Normal inverse Gaussian distributions and stochastic volatility modelling. *Scandinavian Journal of Statistics*, **24**, 1–13.

Barndorff-Nielsen, O.E. (1998). Processes of normal inverse Gaussian type. *Finance and Stochastics*, **2**, 41–68.

Barndorff-Nielsen, O.E., Shephard, N. (2001). Non-Gaussian Ornstein-Uhlenbeck-based models and some of their uses in financial economics. *Journal of Royal Statistical Society*, **63**, 167–241.

Barndorff-Nielsen, O.E., Shephard, N. (2003). Realized power variation and stochastic volatility models. *Bernuolli*, **9(2)**, 243–265.

Benth, F.E., Groth, M., Kufakunesu, R. (2007). Valuing volatility and variance swaps for a non-Gaussian Ornstein-Uhlenbeck stochastic volatility model. *Applied Mathematical Finance*, **14(4)**, 347–363.

Bernard, C., Cui, Z.Y. (2011). Pricing timer options. *Journal of Computational Finance*, **12(1)**, 69–104.

Bernard, C., Cui, Z.Y. (2014). Prices and asymptotics for discrete variance swaps. *Applied Mathematical Finance*, **21**, 140–173.

Bick, A. (1995). Quadratic-variation based dynamic strategies. *Management Science*, **41**, 722–732.

Black, F., Scholes, M. (1973). The pricing of options and corporate liabilities. *Journal of Political Economy*, **81**, 637–654.

Borodin, A.N., Salminen, P. (2002). Probability and its applications, in *Handbook of Brownian motion–facts and formulae*, second edition, Birkhäuser Verlag, Basel.

Bossu, S. (2006). Introduction to variance swap. *Wilmott Magazine*, March issue, 50–55.

Bouzoubaa, M., Osseiran, A. (2010). *Exotic Options and Hybrids: A Guide to Structuring, Pricing and Trading*. Wiley, New York.

Branger, N., Kraftschik, A., Volkert, C. (2016). The fine structure of variance: Pricing VIX derivatives in consistent and log-VIX models. *Working paper of University of Muenster.*

Brenner, M., Galai, D. (1989). New financial instruments for hedging changes in volatility. *Financial Analyst Journal*, July/August issue, 61–65.

Britten-Jones, M., Neuberger, A. (2000). Option prices, implied price processes, and stochastic volatility. *Journal of Finance*, **55(2)**, 839–866.

Broadie, M., Kaya Ö. (2006). Exact simulation of stochastic volatility and other affine jump diffusion processes. *Operations Research*, **54**, 217–231.

Broadie, M., Jain A. (2008). The effect of jumps and discrete sampling on volatility and variance swaps. *International Journal of Theoretical and Applied Finance*, **11(8)**, 761–797.

Brockhaus, O., Long, D. (2000). Volatility swaps made simple. *Risk*, January issue, 92–95.

Burgard, C., Torné, O. (2018). Efficient pricing and super replication of corridor variance swaps and related products. *Journal of Computational Finance*, **21(4)**, 79–96.

Bühler, H. (2006). *Volatility Markets: Consistent Modeling, Hedging and Implementation*. PhD thesis, TU Berlin.

Cao, H.K., Badescu, A., Cui, Z.Y. (2020). Valuation and calibration of VIX options and target volatility options with affine GARCH models. *Journal of Futures Markets*, in press.

Cao, J.L., Lian, G.H., Roslan, T.R.N. (2016). Pricing variance swaps under stochastic volatility and stochastic interest rate. *Applied Mathematics and Computation*, **277**, 72–81.

Cao, J.L., Ruan, X.F., Su, S., Zhang, W.J. (2020). Pricing VIX derivatives with infinite-activity jumps. *Journal of Futures Markets*, **40**, 329–354.

Cao, J.P., Fang, Y.B. (2017). An analytical approach for variance swaps with an Ornstein-Uhlenbeck process. *ANZIAM Journal*, **59**, 83–102.

Carr, P., Geman, H., Madan, D., Yor, M. (2002). The fine structure of asset returns: An empirical investigation. *Journal of Business*, **75(2)**, 305–332.

Carr, P., Geman, H., Madan, D., Yor, M. (2003). Stochastic volatility for Lévy processes. *Mathematical Finance*, **13(3)**, 345–382.

Carr, P., Geman, H., Madan, D., Yor, M. (2005). Pricing options on realized variance. *Finance and Stochastics*, **9**, 453–475.

Carr, P., Lee, R. (2007). Realized volatility and variance: Options via swaps. *Risk*, May issue, 76–83.

Carr, P., Lee, R. (2008). Robust replication of volatility derivatives. *Working paper of New York University*.

Carr, P., Lee, R. (2009). Volatility derivatives. *Annual Review of Finance and Economics*, **1**, 1–21.

Carr, P., Lee, R., Wu, L.R. (2012). Variance swaps on time-changed Lévy processes. *Finance and Stochastics*, **16**, 335–355.

Carr, P., Lee, R., Lorig, M. (2021). Pricing variance swaps on time-changed Markov processes. *SIAM Journal on Financial Mathematics*, **12(2)**, 672–689.

Carr, P., Lewis, K. (2004). Corridor variance swaps. *Risk*, February issue, 67–72.

Carr, P., Madan, D. (1998). Towards a theory of volatility trading. In R. Jarrow (ed.), *Volatility* (London: Risk Books), 417–427.

Carr, P., Madan, D. (1999). Option valuation using the fast Fourier transform. *Journal of Computational Finance*, **2(4)**, 61–73.

Carr, P., Sun, J. (2007). A new approach for option pricing under stochastic volatility. *Review of Derivatives Research*, **10(2)**, 87–150.

Carr, P., Wu, L.R. (2004). Time-changed Lévy processes and option pricing. *Journal of Financial Economics*, **71**, 113–141.

Carr, P., Wu, L.R. (2006). A tale of two indices. *Journal of Derivatives*, Spring issue, **13(3)**, 13–29.

Carr, P., Wu, L.R. (2009). Variance risk premium. *Review of Financial Studies*, **22(3)**, 1311–1341.

Cboe (2019). Cboe volatility index. *White Paper, Cboe Exchange*.

Chacko, G., Das, S. (2002). Pricing interest rate derivatives: A general approach. *Review of Financial Studies*, **15(1)**, 195–241.

Chacko, G., Viceira, L.M. (2003). Spectral GMM estimation of continuous-time processes. *Journal of Econometrics*, **116(1–2)**, 259–292.

Chan, L.L., Platen, E. (2015). Pricing and hedging of long dated variance swaps under a $3/2$ volatility model. *Journal of Computational and Applied Mathematics*, **278**, 181–196.

Cheang, G., Chiarella, C., Ziogas, A. (2013). The representation of American option prices under stochastic volatility and jump-diffusion dynamics. *Quantitative Finance*, **13(2)**, 241–253.

Cherubini, U., Lunga, G.D., Mulinacci, S., Rossi, P. (2010). *Fourier Transform Methods in Finance*. Wiley, Chichester.

Chourdakis, K. (2005). Lévy processes driven by stochastic volatility. *Asia-Pacific Financial Markets*, **12**, 333–352.

Christoffersen, P., Heston, S., Jacobs, K. (2009). The shape and term structure of the index option smirk: Why multifactor stochastic volatility models work so well. *Management Science*, **55(12)**, 1914–1932.

Connolly, K.B. (1997). *Buying and Selling Volatility*, Wiley, New York.

Cont, R., Tankov, P. (2003). *Financial Modelling with Jump Processes*, Chapman and Hall/CRC, London.

Cox, J.C., Ingersoll, E., Ross, S.A. (1985). A theory of the term structure of interest rates. *Econometrica*, **53(2)**, 385–407.

Crosby, J., Davis, M.H. (2011). Variance derivatives: Pricing and convergence. *Working paper of Imperial College*.

Cui, Z.Y., Kirby, J.L., Nguyen, D. (2017a). A general framework for discretely sampled realized variance derivatives in stochastic volatility models with jumps. *European Journal of Operational Research*, **262**, 381–400.

Cui, Z.Y., Kirkby, J.L., Lian, G.H., Nguyen, D. (2017b). Integral representation of probability density of stochastic volatility models and timer options. *International Journal of Theoretical and Applied Finance*, **20(8)**, 1750055 (32 pages).

Cui, Z.Y., Lee, C.H., Liu, M.Z. (2021). Valuation of VIX derivatives through combined Itô-Taylor expansion and Markov chain approximation. *Working paper of Stevens Institute of Technology*.

Da Fonseca, J., Grasselli, M., Tebaldi, C. (2008). A multifactor volatility Heston model. *Quantitative Finance*, **8(6)**, 591–604.

Da Fonseca, J., Gnoatto, A., Grasselli, M. (2015). Analytic pricing of volatility-equity options within Wishart-based stochastic volatility models: An efficient conditioning technique. *Operations Research Letters*, **43**, 601–607.

Da Fonseca, J., Martini, C. (2016). The α-Hypergeometric stochastic volatility model. *Stochastic Processes and their Applications*, **126(5)**, 1472–1502.

Dash, S., Moran, M.T. (2005). VIX as companion for hedge fund portfolios. *Journal of Alternative Investments*, Winter issue, 75–80.

Davis, M., ObLój, J., Raval, V. (2014). Arbitrage bounds for prices of weighted variance swap. *Mathematical Finance*, **24**, 821–854.

Demeterfi, K., Derman, E., Kamal, M., Zou, J. (1999). A guide to volatility and variance swaps. *Journal of Derivatives*, summer issue, 9–32.

Detemple, J., Osakwe, C. (2000). The valuation of volatility options. *European Finance Review*, **4**, 21–50.

Detemple, J., Kitapbayev, Y. (2018). On American VIX options under the generalized $3/2$ and $1/2$ models. *Mathematical Finance*, **28(2)**, 550–581.

DeMarzo, P.M., Kremer, I., Mansour, Y. (2016). Robust option pricing: Hannan and Blackwell meet Black and Scholes. *Journal of Economic Theory*, **163**, 410–434.

Di Graziano, G., Torricelli, L. (2012). Target volatility option pricing. *International Journal of Theoretical and Applied Finance*, **15**, 1250005 (17 pages).

Dong, F.Y., Wong, H.Y. (2017). Variance swaps under the threshold Ornstein-Uhlenbeck model. *Applied Stochastic Models in Business and Industry*, **33**, 507–521.

Drimus, G.G. (2012). Options on realized variance by transform methods: A non-affine stochastic volatility model. *Quantitative Finance*, **12(11)**, 1679–1694.

Drimus, G.G., Farkas, W., Gourier, E. (2016). Valuation of options on discretely sampled variance: A general analytic approximation. *Journal of Computational Finance*, **20(2)**, 39–66.

Duan, J.C. (1995). The GARCH option pricing model. *Mathematical Finance*, **5**, 13–32.

Duan, J.C., Yeh, C.Y. (2010). Jump and volatility risk premiums implied by VIX. *Journal of Economic Dynamics and Control*, **34**, 2232–2244.

Duffie, D., J. Pan, K. Singleton (2000). Transform analysis and option pricing for affine jump-diffusion. *Econometrica*, **68**, 1343–1376.

Dupire, B. (1994). Pricing with a smile. *Risk*, January issue, 18–20.

Eberlein, E., Keller, U., Prause, K. (1998). New insights into smile, mispricing and Value at Risk: The Hyperbolic model. *Journal of Business*, **71**, 371–405.

Elliott, R.J., Siu, T.K., Chan, L.L. (2007). Pricing volatility swaps under Heston's stochastic volatility model with regime switching. *Applied Mathematical Finance*, **14(1)**, 41–62.

Elliott, R.J., Lian, G.H. (2013). Pricing variance and volatility swaps in a stochastic volatility model with regime switching: Discrete observations case. *Quantitative Finance*, **13(5)**, 687–698.

Eraker, B., Johannes, M., Polson, N. (2003). The impact of jumps in volatility and return. *Journal of Finance*, **58(3)**, 1269–1300.

Fahrner, I. (2007). Modern logarithms for the Heston model. *International Journal of Theoretical and Applied Finance*, **10(1)**, 23–30.

Fang, F., Oosterlee, C.W. (2011). A Fourier-based valuation method for Bermudan and barrier options under Heston's model. *SIAM Journal of Financial Mathematics*, **2**, 439–463.

Fitzgerald, M.D. (1996). Trading volatility. In: Alexander C. (editor), *Handbook of Risk Management and Analysis*, 329–359, Wiley, New York.

Fouque, J.P., Saporito, Y.F. (2018). Heston stochastic vol-of-vol model for joint calibration of VIX and S&P 500 options. *Quantitative Finance*, **18(6)**, 1003–1016.

Gehricke, S.A., Zhang, J. (2018). Modeling VXX. *Journal of Futures Markets*, **38**, 958–976.

Geman, H., Madan, D., Yor, M. (2001). Time changes for Lévy processes. *Mathematical Finance*, **11**, 79–96.

Goard, J. (2011). Pricing of volatility derivatives using 3/2-stochastic models. *Proceedings of the World Congress on Engineering, July 6–8, 2011, London.*

Goard, J., Mazur, M. (2013). Stochastic volatility models and the pricing of VIX options. *Mathematical Finance*, **23(3)**, 439–458.

Gonzalez-Perez, M.T. (2015). Model-free volatility indexes in the financial literature: A review. *International Review of Economics and Finance*, **40**, 141–159.

Grasselli, M. (2017). The 4/2 stochastic volatility model: A unified approach for the Heston and the 3/2 model. *Mathematical Finance*, **27(4)**, 1013–1034.

Grasselli, M., Romo, J.B. (2015). Stochastic skew and target volatility options. *Journal of Futures Markets*, **36(2)**, 174–193.

Grasselli, M., Wagalath, L. (2020). VIX vs VXX: A joint analytic framework. *International Journal of Theoretical and Applied Finance*, **23(5)**, 2050033, 39 pages.

Grünbichler, A., Longstaff, F.A. (1996). Valuing futures and options on volatility. *Journal of Banking and Finance*, **20**, 985–1001.

Guo, J.H., Hung, M.W. (2007). A note on the discontinuity problem in Heston's stochastic volatility model. *Applied Mathematical Finance*, **14(4)**, 339–345.

Habtemicael, S., SenGupta, I. (2016). Pricing variance and volatility swaps for Barndorff-Nielsen and Shephard process driven financial markets. *International Journal of Financial Engineering*, **3(4)**, 1650027(35 pages).

Hao, J.J., Zhang, J. (2013). GARCH option pricing models, the CBOE VIX, and variance risk premium. *Journal of Financial Econometrics*, **11(3)**, 556–580.

He, X.J., Zhu, S.P. (2018). A series-form solution for pricing variance and volatility swaps with stochastic volatility and stochastic interest rate. *Computers and Mathematics with Applications*, **76**, 2223–2234.

He, X.J., Zhu, S.P. (2019). Variance and volatility swaps under two-factor stochastic volatility model with regime switching. *International Journal of Theoretical and Applied Finance*, **22(1)**, 1950009 (19 pages).

Heston, S. (1993). Closed-form solution for options with stochastic volatility, with application to bond and currency options. *Review of Financial Studies*, **6**, 327–343.

Heston, S., Nandi, S. (2000). A closed form GARCH option valuation model. *Review of Financial Studies*, **13(3)**, 585–625.

Hirsa, A., Madan, D.B. (2004). Pricing American options under variance gamma. *Journal of Computational Finance*, **7**, 63–80.

Hitaj, A., Mercuri, L., Rroji, E. (2017). VIX computation based on affine stochastic volatility models in discrete time. In: Consigli, G., Stefani, S., Zambruno, G. (editors), *Handbook of Recent Advances in Commodity and Financial Modeling*, 141–164, Springer, Berlin.

Hobson, D., Klimmek, M. (2012). Model-independent hedging strategies for variance swaps. *Finance and Stochastics*, **16**, 611–649.

Hong, G. (2017). Efficient switch corridor variance swap pricing. *Working paper of Credit Suisse*.

Howison, S., Rafailidis, A., Rasmussen, H. (2004). On the pricing and hedging of volatility derivatives. *Applied Mathematical Finance*, **11**, 317–346.

Huang, J.Z., Wu, L.R. (2004). Specification analysis of option pricing models based on time changed Lévy processes. *Journal of Finance*, **59(3)**, 1405–1440.

Huang, Z., Tong, C., Wang, T.Y. (2019). VIX term structure and VIX futures pricing with realized volatility. *Journal of Futures Markets*, **39**, 72–93.

Hull, J., White, A. (1987). The pricing of options on assets with stochastic volatilities. *Journal of Finance*, **42(2)**, 281–300.

Itkin, A., Carr, P. (2010). Pricing swaps and options on quadratic variation under stochastic time change models – discrete observations case. *Review of Derivatives Research*, **13**, 141–176.

Jarrow, R., Kchia, Y., Larsson, M., Protter, P. (2013). Discretely sampled variance and volatility swaps versus their continuous approximation. *Finance and Stochastics*, **17**, 305–324.

Javaheri, A., Wilmott, P., Haug, E.G. (2004). GARCH and volatility swaps. *Quantitative Finance*, **4**, 589–595.

Jeanblanc, M., Yor, M., Chesney, M. (2009). *Mathematical Methods for Financial Markets*, Springer-Verlag, London.

Jiang, G., Tian, Y.S. (2005). The model-free implied volatility and its information content. *Review of Financial Studies*, **18**, 1305–1342.

Jiang, G., Tian, Y.S. (2007). Extracting model-free volatility from option prices: An examination of the VIX index. *Journal of Derivatives*, **14**, 35–60.

Jing, B., Li, S.H., Ma, Y. (2021). Consistent pricing of VIX options with the Hawkes jump-diffusion model. *North American Journal of Economics and Finance*, **36**, 101326.

Jones, C.S. (2003). The dynamics of stochastic volatility: Evidence from underlying and options markets. *Journal of Econometrics*, **116(1–2)**, 181–224.

Kaeck, A., Alexander, C. (2012). Volatility dynamics for the S&P 500: Further evidence from non-affine, multi-factor jump diffusions. *Journal of Banking and Finance*, **36**, 3110–3121.

Kaeck, A., Alexander, C. (2013). Continuous-time VIX dynamics: On the role of stochastic volatility of volatility. *International Review of Financial Analysis*, **28**, 46–56.

Kaeck, A., Seeger, N.J. (2020). VIX derivatives, hedging and vol-of-vol risk. *European Journal of Operational Research*, **283**, 767–782.

Kahalé, N. (2016). Model-independent lower bound on variance swaps. *Mathematical Finance*, **26(4)**, 939–961.

Kahl, C., Jäckel, P. (2005). Not-so-complex logarithms in the Heston model. *Wilmott Magazine*, September issue, 94–103.

Kallsen, J., Muhle-Karbe, J., Voß, M. (2011). Pricing options on variance in affine stochastic volatility models. *Mathematical Finance*, **21(4)**, 627–641.

Kanniainen, J., Lin, B., Yang, H. (2014). Estimating and using GARCH models with VIX data for option valuation. *Journal of Banking and Finance*, **43**, 200–211.

Karatzas, I., Shreve, S.E. (1991). *Brownian Motion and Stochastic Calculus*, Springer.

Keller-Ressel, M., Muhle-Karbe, J. (2013). Asymptotic and exact pricing of options on variance. *Finance and Stochastics*, **17**, 107–133.

Kim, S.W., Kim, J.H. (2018). Analytic solutions for variance swaps with double-mean-reverting volatility. *Chaos, Solitons and Fractals*, **114**, 130–144.

Kim, S.W., Kim, J.H. (2019). Variance swaps with double exponential Ornstein-Uhlenbeck stochastic volatility. *North American Journal of Economics and Finance*, **48**, 149–169.

Kokholm, T., Stisen, M. (2015). Joint pricing of VIX and SPX options with stochastic volatility and jump models. *Journal of Risk Finance*, **16(1)**, 27–48.

Kou, S. (2002). A jump-diffusion model for option pricing. *Management Science*, **48**, 1086–1101.

Kou, S. (2008). Jump-diffusion models for asset pricing in financial engineering. In: Birge, J.R., Linetsky, V. (editors), *Handbooks in OR & MS*, **15**, 73–116, Elsevier, New York.

Kwok, Y.K., Zheng, W.D. (2018). *Saddlepoint Approximation Methods in Financial Engineering*. Springer, Berlin.

Kyprianou, A.E. (2006). *Introductory Lectures on Fluctuations of Lévy Processes with Applications*. Springer.

Langrené, N., Lee, G., Zhu, Z.L. (2016). Switching to nonaffine stochastic volatility: A closed-form expansion for the inverse gamma function. *International Journal of Theoretical and Applied Finance*, **19(5)**, 1650031 (37 pages).

Le Floc'h, F. (2018). Variance swap replication: Discrete or continuous. *Journal of Risk and Financial Management*, **6**, 11–15.

Lee, R. (2001). Implied and local volatilities under stochastic volatility. *International Journal of Theoretical and Applied Finance*, **4(1)**, 45–89.

Lee, R. (2010). Gamma swap and corridor variance swap. *Encyclopedia of Quantitative Finance*, Wiley, New York.

Leippold, M., Cheng, J., Ibraimi, M., Zhang, J. (2012). A remark on Lin's and Chang's paper 'consistent modeling of S&P 500 and VIX derivatives'. *Journal of Economic Dynamics and Control*, **36(5)**, 716–718.

Leung, T., Lorig, M. (2016). Optimal static quadratic hedging. *Quantitative Finance*, **16**, 1341–1355.

Lewis, A. (2016). *Option Valuation under Stochastic Volatility II: With Mathematica Code*. Finance Press.

Li, C.X., Li, C.X. (2015). A closed-form expansion approach for pricing discretely monitored variance swaps. *Operations Research Letters*, **43**, 450–455.

Li, C.X. (2016a). Bessel processes, stochastic volatility, and timer options. *Mathematical Finance*, **26(1)**, 122–148.

Li, J. (2016b). Trading VIX futures under mean reversion with regime switching. *International Journal of Financial Engineering*, **3(3)**, 1650021 (20 pages).

Li, J., Li, L.F., Zhang, G.Q. (2017). Pure jump models for pricing and hedging VIX derivatives. *Journal of Economic Dynamics and Control*, **74**, 28–55.

Li, M.Q., Mercurio, F. (2014). Closed-form approximation of perpetual timer option prices. *International Journal of Theoretical and Applied Finance*, **17(4)**, 1450026.

Li, M.Q., Mercurio, F. (2015). Analytic approximation of finite-maturity timer option prices. *Journal of Futures markets*, **35(3)**, 245–273.

Lian, G.H., Zhu, S.P. (2013). Pricing VIX options with stochastic volatility and random jumps. *Decisions in Finance and Economics*, **36(1)**, 71–88.

Lian, G.H., Chiarella, C., Kalev, P.S. (2014). Volatility swaps and volatility options on discretely sampled realized variance. *Journal of Economic Dynamics and Control*, **47**, 239–262.

Liang, L.Z.J., Lemmens, D., Tempere, J. (2011). Path integral approach to the pricing of timer options with the Duru-Kleinert time transformation. *Physical Review E*, **83**, 056112.

Lin, W., Li, S.H., Luo, X.G., Chern, S. (2017). Consistent pricing of VIX and equity derivatives with the 4/2 stochastic volatility plus jumps model. *Journal of Mathematical Analysis and Applications*, **447**, 778–797.

Lin, W., Li, S.H., Chern, S., Zhang, J. (2019). Pricing VIX derivatives with free stochastic volatility model. *Review of Derivatives Research*, **22**, 41–75.

Little, T., Pant, V. (2001). A finite-difference method for the valuation of variance swaps. *Journal of Computational Finance*, **5(1)**, 81–103.

Liu, W.Y., Zhu, S.P. (2019). Pricing variance swaps under the Hawkes jump-diffusion process. *Journal of Futures Markets*, **39(6)**, 635–655.

Lo, C.L., Shih, P.T., Wang, Y.H., Yu, M.T. (2019). VIX derivatives: Valuation models and empirical evidence. *Pacific-Basin Finance Journal*, **53**, 1–21.

Lord, R., Kahl, C. (2010). Complex logarithms in Heston-like models. *Mathematical Finance*, **4**, 671–694.

Luo, X.G., Zhang, J. (2012). The term structure of VIX. *Journal of Futures Markets*, **32(12)**, 1092–1123.

Luo, X.G., Zhang, J., Zhang, W.J. (2019). Instantaneous squared VIX and VIX derivatives. *Journal of Futures Markets*, **39**, 1193–1213.

Ma, J.T., Deng, D.Y., Lai, Y.Z. (2015). Explicit approximate analytic formulas for timer option pricing with stochastic interest rates. *North American Journal of Economics and Finance*, **34**, 1–21.

Ma, C.F., Xu, W., Kwok, Y.K. (2020). Willow tree algorithms for pricing VIX derivatives under stochastic volatility models. *International Journal of Financial Engineering*, **7(1)**, 2050003, 28 pages.

Madan, D.B., Carr, P., Chang, E.C. (1998). The Variance Gamma process and option pricing. *European Finance Review*, **2**, 79–105.

Madan, D.B., Milne, F. (1991). Option pricing with V.G. martingale components. *Mathematical Finance*, **1**, 39–55.

Madan, D.B., Seneta, E. (1990). The Variance-Gamma (V.G.) model for share market returns. *Journal of Business*, **63**, 511–524.

Madan, D.B., Yor, M. (2008). Respecting the CGMY and Meixner Lévy processes as time changed Brownian motions. *Journal of Computational Finance*, **12(1)**, 27–47.

Mencía, J., Sentana, E. (2013). Valuation of VIX derivatives. *Journal of Financial Economics*, **108**, 367–391.

Merton, R.C. (1973). Theory of rational option pricing. *Bell Journal of Economics and Management Science*, **4**, 141–183.

Merton, R.C. (1976). Option pricing when the underlying stock returns are discontinuous. *Journal of Financial Economics*, **3**, 125–144.

Mougeot, N. (2005). Variance swaps and beyond. *Volatility Investing Handbook, BNP Paribus*.

Neuberger, A. (1990). Voltaility trading. Working paper of London Business School.

Neuberger, A. (1994). The log contract. *Journal of Portfolio Management*, **20(2)**, 74–80.

Papanicolaou, A., Sircar, R. (2014). A regime-switching Heston model for VIX and S&P 500 implied volatilities. *Quantitative Finance*, **14(10)**, 1811–1827.

Park, Y.H (2016). The effects of asymmetric volatility and jumps on the pricing of VIX derivatives. *Journal of Econometrics*, **192**, 313–328.

Protter, P. (2004). *Stochastic Integration and Differential Equations*, second edition. Springer, Berlin.

Psychoyios, D., Dotsis, G., Markellos, R. (2010). A jump diffusion model for VIX volatility options and futures. *Review of Quantitative Finance and Accounting*, **35(3)**, 245–269.

Pun, C.S., Chung, S.F., Wong, H.Y. (2015). Variance swap with mean reversion, multifactor stochastic volatility and jumps. *European Journal of Operational Research*, **245**, 571–580.

Rujivan, S., Zhu, S.P. (2012). A simplified analytical approach for pricing discretely-sampled variance swaps with stochastic volatility. *Applied Mathematics Letters*, **25**, 1644–1650.

Rujivan, S., Zhu, S.P. (2014). A simple closed-form formula for pricing discretely-sampled variance swaps under the Heston model. *ANZIAM Journal*, **56**, 1–27.

Rujivan, S. (2016). A novel analytic approach for pricing discretely sampled gamma swaps in the Heston model. *ANZIAM Journal*, **57**, 244–268.

Sato, K.I. (1999). *Lévy Processes and Infinitely Divisible Distributions*. Cambridge University Press.

Sawyer, N. (2007). SG CIB launches timer options. *Risk*, July issue, 6.

Schöbel, R., Zhu, J.W. (1999). Stochastic volatility with an Ornstein-Uhlenbeck process: An extension. *European Finance Review*, **3**, 23–46.

Schoutens, W. (2003). *Lévy Processes in Finance: Pricing Financial Derivatives*. John Wiley and Sons, Chichester.

Seneta, E. (2007). The early years of Variance-Gamma process. In: Fu, M.C., Jarrow, R.A., Yen, J.Y., Elliott, R.J. (editors), *Advances in Mathematical Finance*, Springer.

Sepp, A. (2007). Variance swaps under no conditions. *Risk*, January issue, 82–87.

Sepp, A. (2008). Pricing options on realized variance in the Heston model with jumps in returns and volatility. *Journal of Computational Finance*, **11(4)**, 33–70.

Sepp, A. (2012). Pricing options on realized variance in the Heston model with jumps in returns and volatility II: An approximation of the discrete variance. *Journal of Computational Finance*, **16(2)**, 3–32.

Shen, Y., Siu, T.K. (2013). Pricing variance swaps under a stochastic interest rate and volatility model with regime-switching. *Operations Research Letters*, **41**, 180–187.

Tan, X.Y., Wang, C.X., Huang, W.L., Li, S.H. (2018). Pricing VIX options in a 3/2 plus jumps model. *Applied Mathematics, Journal of Chinese Universities*, **33(3)**, 323–334.

Tong, C., Huang, Z. (2021). Pricing VIX options with realized volatility. *Journal of Futures Markets*, **41**, 1180–1200.

Tong, Z.G., Liu, A. (2017). Analytic pricing formulas for discretely sampled generalized variance swaps under stochastic time change. *International Journal of Financial Engineering*, **4**, 1750028 (24 pages).

Torricelli, L. (2013). Pricing joint claims on an asset and its realized variance under stochastic volatility models. *International Journal of Theoretical and Applied Finance*, **16(1)**, 1350005 (18 pages).

Torricelli, L. (2016). Valuation of asset and volatility derivatives using decoupled time-changed Lévy processes. *Review of Derivatives Research*, **19**, 1–39.

Wang, C.X., Huang, W.L., Li, S.H., Bao, Q.F. (2016). Pricing VIX options in a stochastic vol-of-vol model. *Applied Stochastic Models in Business and Industry*, **32**, 168–183.

Wang, Q., Wang, Z.R. (2020). VIX valuation and its futures pricing through a generalized affine realized volatility model with hidden components and jump. *Journal of Banking and Finance*, **116**, 105845.

Wang, T.Y., Shen, Y.W., Jiang, Y.T., Huang, Z. (2017). Pricing the CBOE VIX futures with the Heston-Nandi GARCH model. *Journal of Futures Markets*, **37(7)**, 641–659.

Wang, X.C., Wang, Y.J. (2014). Variance-optimal hedging for target volatility options. *Journal of Industrial and Management Optimization*, **10(1)**, 207–218.

Whaley, R.E. (1993). Derivatives on market volatility: Hedging tools long overdue. *Journal of Derivatives*, **1**, 71–84.

Whaley, R.E. (2000). The investor fear gauge. *Journal of Portfolio Management*, Spring issue, 75–80.

Windcliff, H., Forsyth, P.A., Vetzal, K.R. (2006). Pricing methods and hedging strategies for volatility derivatives. *Journal of Banking and Finance*, **30**, 409–431.

Wu, D., Liu, T.X. (2018). New approach to estimating VIX truncation errors using corridor variance swaps. *Journal of Derivatives*, summer issue, 54–70.

Wu, L.R. (2008). Modeling financial security returns using Lévy processes. In: Birge, J.R., Linetsky, V. (editors), *Handbooks in OR & MS*, **15**, 117–162, Elsevier, New York.

Xie, H.B., Zhou, M., Ruan, T.H. (2020). Pricing VIX futures under the GJR-GARCH process: An analytical approximation method. *Journal of Derivatives*, **27(4)**, 77–88.

Yan, C., Zhao, B. (2019). A general jump-diffusion process to price volatility derivatives. *Journal of Futures Markets*, **39**, 15–37.

Yang, B.Z., Yue, J., Huang, N.J. (2019a). Equilibrium price of variance swaps under stochastic volatility with Lévy jumps and stochastic interest rate. *International Journal of Theoretical and Applied Finance*, **22(4)**, 1950016 (33 pages).

Yang, B.Z., Yue, J., Wang, M.H., Huang, N.J. (2019b). Volatility swaps valuation under stochastic volatility with jumps and stochastic intensity. *Applied Mathematics and Computation*, **355**, 73–84.

Yang, X.L., Wang, P. (2018). VIX futures pricing with conditional skewness. *Journal of Futures Markets*, **38(9)**, 1126–1151.

Yang, X.L., Wang, P., Chen, J. (2019c). VIX futures pricing with affine jump-GARCH dynamics and variance-dependent pricing kernels. *Journal of Derivatives*, **27(1)**, 110–126.

Yuen, C.H., Zheng, W.D., Kwok, Y.K. (2015). Pricing exotic variance swaps under 3/2-stochastic volatility models. *Applied Mathematical Finance*, **22(5)**, 421–449.

Zang, X., Ni, J., Huang, J.Z., Wu, L. (2017). Double-jump stochastic volatility model for VIX: Evidence from VVIX. *Quantitative Finance*, **17(2)**, 227–240.

Zeng, P.P., Kwok, Y.K., Zheng, W.D. (2015). Fast Hilbert transform algorithms for pricing discrete timer options under stochastic volatility models. *International Journal of Theoretical and Applied Finance*, **18**, 1550046 (26 pages).

Zhang, J.C., Lu, X.P., Han, Y.C. (2017). Pricing perpetual timer option under the stochastic volatility model of Hull-White. *ANZIAM Journal*, **58**, 406–416.

Zhang, J., Shu, J.H., Brenner, M. (2010). The new market for volatility trading. *Journal of Futures Markets*, **30(9)**, 809–833.

Zhang, J., Zhu, Y.Z. (2006). VIX futures. *Journal of Futures Markets*, **26(6)**, 521–531.

Zhang, L.W. (2014). A closed-form pricing formula for variance swaps with mean-reverting Gaussian volatility. *ANZIAM Journal*, **55**, 362–382.

Zhao, Z., Cui, Z.Y., Florescu, I. (2018). VIX derivatives valuation and estimation based on closed-form series expansions. *International Journal of Financial Engineering*, **5(2)**, 1850020 (18 pages).

Zheng, W.D., Kwok, Y.K. (2014a). Closed form pricing formulas for discretely sampled generalized variance swaps. *Mathematical Finance*, **24(4)**, 855–881.

Zheng, W.D., Kwok, Y.K. (2014b). Fourier transform algorithms for pricing and hedging discretely sampled exotic variance products and volatility derivatives under additive processes. *Journal of Computational Finance*, **18**, 3–30.

Zheng, W.D., Kwok, Y.K. (2014c). Saddlepoint approximation methods for pricing derivatives on discretely sampled realized variance. *Applied Mathematical Finance*, **21(1)**, 1–31.

Zheng, W.D., Kwok, Y.K. (2015). Pricing options on discrete realized variance with partially exact and bounded approximation. *Quantitative Finance*, **15(12)**, 2011–2019.

Zheng, W.D., Yuen, C.H., Kwok, Y.K. (2016). Recursive algorithms for pricing discrete variance options and volatility swaps under time-changed Lévy processes. *International Journal of Theoretical and Applied Finance*, **19**, 1650011 (29 pages).

Zheng, W.D., Zeng, P.P. (2016). Pricing timer options and variance derivatives with closed-form partial transform under the 3/2-model. *Applied Mathematical Finance*, **23**, 344–373.

Zhu, J.W. (2010). *Applications of Fourier Transform to Smile Modeling: Theory and Implementation*. Springer, Berlin.

Zhu, S.P., Lian, G.H. (2011). A closed-form exact solution for pricing variance swaps with stochastic volatility. *Mathematical Finance*, **21(2)**, 233–256.

Zhu, S.P., Lian, G.H. (2012a). On the valuation of variance swaps with stochastic volatility. *Applied Mathematics and Computation*, **219**, 1654–1669.

Zhu, S.P., Lian, G.H. (2012b). An analytical formula for VIX futures and its applications. *Journal of Futures Markets*, **32(2)**, 166–190.

Zhu, S.P., Lian, G.H. (2015a). Pricing forward-start variance swaps with stochastic volatility. *Applied Mathematics and Computation*, **250**, 920–933.

Zhu, S.P., Lian, G.H. (2015b). Analytically pricing volatility swaps under stochastic volatility. *Journal of Computational and Applied Mathematics*, **288**, 332–340.

Zhu, S.P., Lian, G.H. (2018). On the convexity correction approximation in pricing volatility swaps and VIX futures. *New Mathematics and Natural Computation*, **14(3)**, 383–401.

Zhu, W.L., Ruan, X.F. (2019). Pricing swaps on discrete realized higher moments under the Lévy process. *Computational Economics*, **53**, 507–532.

Zhu, Y.Z., Zhang, J. (2007). Variance term structure and VIX futures pricing. *International Journal of Theoretical and Applied Finance*, **10(1)**, 111–127.

Index